W9-CXD-986

The Standard Handbook
of Textiles

The Standard Handbook of Textiles

A. J. HALL, B.SC., F.R.I.C., F.T.I., F.S.D.C.
Consulting Chemist to the Textile and Allied Industries,
Gold Medallist for Research (The Worshipful Company of Dyers),
Sometime Examiner to the City and Guilds of London Institute in the
Dyeing and Finishing of Textiles

A HALSTED PRESS BOOK

JOHN WILEY & SONS
New York–Toronto

Published in the U.S.A. and Canada by Halsted Press,
a division of John Wiley & Sons Inc., New York

Library of Congress Cataloging in Publication Data

Hall, Archibald John.
    The standard handbook of textiles.

    "A Halsted Press book."
    Bibliography: p.
    Includes index.
      1. Textile industry and fabrics. 2. Textile fibers.
I. Title.
TS1445.H27 1975   677'.028   75–1320
ISBN 0–470–34297–8

First published by National Trade Press, 1946.
Eighth edition published by Newnes-Butterworths,
an imprint of the Butterworth Group, 1975.

Printed in Great Britain

PREFACE TO THE EIGHTH EDITION

The need for preparing this revised and enlarged edition is evidence of the rapid changes in textile manufacture which are now taking place. They concern developments in all stages of fibres, yarn, fabric and garment production and the related processes of dyeing and finishing these materials as they stem from world-wide intensive effort and research.

While rapid progress is being made in the manufacture and utilisation of synthetic fibres adequate notice must be taken of current important changes in the long established, historically interesting machinery conventionally employed for yarn spinning, weaving and knitting.

Much success is being obtained in attempts to spin yarn with simplified machines quite different from the usual types; to weave fabric with shuttleless looms, and even without looms; and to modify knitting machines so as to produce fabric having a firmer structure to take the place of woven fabric for garment making.

The automatic control of manufacturing, dyeing and finishing operations is advancing steadily and leading to the more economical use of labour and the production of higher quality textile goods.

Cotton remains the most largely used textile fibre and is using research to discover means of improving its usefulness in order to keep ahead of the synthetic fibres. Wool, aided by its unique properties and recently devised improvement treatments, is strongly resisting displacement by other fibres; its rate of production is almost static.

The author hopes that this book will be both interesting and useful to the general reader and also to anyone concerned with the distribution and marketing of fabrics and garments. It should also be a useful introduction to those contemplating entering into textile manufacture in so far as it describes the essential technical features of this very important industry which offers interest and advancement to those who welcome problems to solve and opportunities to grasp.

Minehead   A. J. HALL

Contents

CHAPTER ONE

The natural and man-made textile fibres

Had this book been written somewhat less than 100 years ago there would have been no point in distinguishing between natural and artificial textile fibres, but it is definitely otherwise today. The number of different types of fibres mainly used for domestic clothing and industrial purposes has recently been very much extended by man-made fibres (Figure 1.1).

THE NATURAL FIBRES

The chief natural fibres now in use are cotton, linen, wool and silk. There are several others such as kapok, hemp, jute and ramie, but these have specified uses which are not on the whole connected with clothes or fabrics used in the home. They vary considerably as regards their properties and their production. Each type of fibre has its particular interest in spite of the fact that they are all characterised in being extremely fine and tenuous.

Cotton

In many parts of the world, especially in the tropical regions of North America, South America, India, Egypt, the West Indies and Africa, the climate is suitable for the growth of a bushy plant 3 to 5 ft (0·9 to 1·5 m) high. This is the cotton plant. It is cultivated under appropriate agricultural conditions. Thus, seed is sown on ploughed land during the spring, and later the plants are thinned out into rows. In due course each plant bears numerous pinkish-white flowers and when these die they leave capsules containing up to eight cotton seeds.

As the season of growth continues the seeds become covered with cotton fibres, the whole being packed within a pod or 'boll' as it is more commonly termed. Usually each seed is enclosed completely within the 4 000 or so fibres which grow outwardly from it. Later the growing mass of cotton fibres causes the boll to burst to expose a mass of fluffy white cotton.

At this period of ripening of the bolls workers go round and pick the cotton. It is necessary to go over the cotton fields several times since the bolls do not all ripen at the same time. In this way huge masses of cotton, all the fibres being attached to seeds, are collected. The bolls of cotton are also picked mechanically by means of

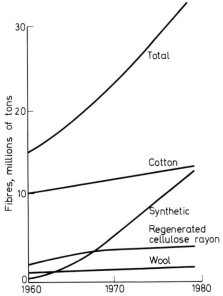

Figure 1.1 Progress in production of textile fibres (in millions of tons). Production for 1980 is anticipated. (Courtesy Textile Business Press Ltd, Manchester)

machines which have been specially designed for this somewhat complicated operation. Unfortunately, the mechanical picking-machines cannot be so selective as hand pickers so that cotton bolls of different degrees of maturity become mixed and a larger amount of so-called 'trash' (leaf, twigs, etc.) is drawn in with the cotton. The mechanical picking of cotton is now very largely practised.

As picking proceeds the cotton obtained is led through a machine known as a 'gin', the purpose of which is to remove the fibres from the seeds. These cotton seeds are useful as cattle food and also as a source of cotton-seed oil which can be used for a variety of purposes

Figure 1.2 Automatic picking of cotton in the USA

including the manufacture of soaps and edible fats. However, it is the separated cotton fibre which is more important, and in this state, containing various impurities such as dried leaf, twigs, earth and dust, it constitutes the raw cotton of commerce. In order to facilitate transport to the various places where such cotton can be manufactured into yarn and fabric, this loose raw cotton is highly compressed into bales each weighing about 500 lb (230 kg).

The cotton fibres which grow on the seeds vary somewhat in length up to a maximum of about 2 in (50 mm), but in addition there is a downy growth of very short fibres. The ginning operation is conducted so that these short fibres are separated from the longer ones, since they are much too short to be of use in cotton manufacture. These short fibres, known as cotton lint, however, form an important raw material to manufacturers of rayon since they are a convenient and cheap source of the cellulose which is the basis of many rayons.

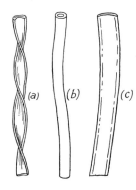

Figure 1.3 Cotton fibres with their cross-sections (a) normal, showing their collapsed, twisted, ribbon-like form, (b) as growing on cotton seeds before drying and collapse and (c) abnormal with thin walls

As we shall see later, there are different qualities of cotton, the differences between them arising mainly from the conditions of growth. Indian cotton fibres are generally short and coarse compared with the Egyptian, which are more valuable because they are long and fine. American cotton fibres have intermediate characteristics.

The scientific aspects of cotton growing are subject to continuous supervision and research. Not only is there the necessity for improving cotton types, but there is an urgent need for methods of controlling the various diseases and pests to which the cotton plant is exposed. From time to time new types of cotton are discovered but, as may be anticipated, this is a line of progress which is necessarily somewhat slow.

The nature of cotton fibres   Each cotton fibre is a very interesting mass of cellulose formed by the interplay of natural forces so diverse that the physical form of each fibre differs subtly from that of its neighbours in the same cotton boll. Yet all cotton fibres have a family relationship. Thus, each fibre closely resembles a length of twisted ribbon having thickened edges. Every two or three twists the direction of twist reverses so that they occur in groups alternately right- and left-handed.

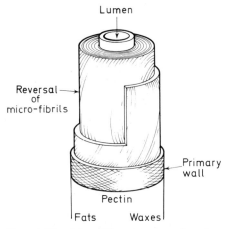

Figure 1.4 Special multi-layer structure of a cotton fibre before its collapse to a ribbon form by drying in the cotton boll. A particular feature is that the micro-fibrils in each secondary layer are spirally aligned at an angle of about 23°, but this angle increases further within the fibre (it is very high at those points where the twist of the fibre reverses)

During growth in the cotton boll the cotton fibres are rod-like and cylindrical. Growth takes place each night so that the fibre is built up progressively of about forty concentric layers of cellulose. Right through the centre of each growing fibre is a canal or space through which nutriment can pass. When the cotton boll ripens and bursts the fibres are exposed to the hot sun and they dry and collapse to lose their cylindrical form, acquiring the twisted ribbon-like appearance. By application of special swelling treatments such as mercerisation with caustic soda or by liquid ammonia, or by steaming under pressure, these collapsed fibres can again be made to become cylindrical and attain their form as in the cotton boll just before ripening and drying, but in ordinary cotton materials the fibres are always present in their ribbon-like form.

The latest evidence obtained by research on the fine structure of a cotton fibre indicates that it is not solid throughout but that it is built up of *fibrils* twisted about each other in much the same manner as the individual cotton fibres are twisted to form cotton thread. These fibrils are exceptionally fine and tenuous so that some thousands of them go to make up a single cotton fibre. It is this complex stress and strain accommodating structure which gives

Figure 1.5 Bundles, each composed of about 400 micro-fibrils spirally aligned in the secondary wall of a cotton fibre. (Courtesy Shirley Institute)

each cotton fibre great tensile strength and resistance to repeated flexing and wear.

The cellulose of which a cotton fibre is made is very similar to the cellulose which is found in many other natural products such as wood, linen and the harder parts of most plant stalks and leaves. In the cotton fibre it is associated with a number of other substances, notably waxes, pectic products and mineral substances. These are quite small in amount, say not more than 4 per cent altogether. They are referred to as impurities by the manufacturer of cotton goods, but actually they fulfil a useful function in the cotton plant. These impurities give the raw cotton a yellowish colour and make it somewhat harsh in handle. The waxy impurities give water-repellency. Generally these are objectionable effects and would make it difficult to colour and finish cotton fabrics satisfactorily, so it is always a first step in the art of dyeing and finishing to purify the cotton as completely as possible.

Fibre characteristics   Each cotton fibre is from $\frac{3}{4}$ to $1\frac{1}{2}$ in (20 to 40 mm) in length and its diameter varies slightly from one end to the other but averages 1/1000 in (0·025 mm). These dimensions vary with the type of cotton, the Indian and American fibres being shorter and the Egyptian and Sea Island fibres longer. Usually a long fibre is correspondingly finer. From the viewpoint of cotton manufacture, the finer and longer the fibre the more suitable it is for making high-quality materials. Long, fine fibres allow the production of threads which, though fine, have high strength.

Each cotton fibre weighs about $1/10000000$ oz (28×10^{-7} g), so that in 1 lb (450 g) of cotton there are about 160000000 cotton fibres. If these fibres were placed end to end they would extend for about 2500 miles (4100 km). The finer cotton fibres have more twists per inch than the coarser fibres. Thus there may be upwards of 250 twists (both right- and left-handed) per inch (10 per mm) in an Egyptian cotton fibre but only 150 twists per inch (6 per mm) in an Indian cotton fibre. All these points about the shape and character of cotton fibre are of importance in the manufacture of cotton threads and fabrics and they will be mentioned again later.

Chemical properties The cellulose of which cotton is made is a very tough and durable substance. Some proof of this is afforded by the fact that cotton fabrics of Indian origin have been preserved to us for over 3000 years. Yet it is a simple substance from the

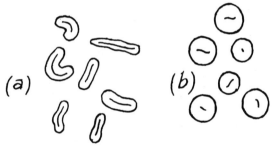

Figure 1.6 Cross-sections of cotton fibres showing the effect of mercerisation. (a) Before mercerisation, (b) mercerised. The cylindrical form which the fibres have before ripening and collapse (in the cotton boll) is restored by the treatment

chemist's point of view. It consists solely of carbon, hydrogen and oxygen, and the following formula, $C_6H_{10}O_5$, indicates that in each molecule of cellulose are combined six carbon, ten hydrogen and five oxygen atoms. So permanent and stable is cellulose that it has universally been accepted as the most suitable raw material for making the highest qualities of viscose rayon and cellulose acetate fibres now so much used. However, although cellulose (and therefore cotton) is highly resistant to sunlight, heat and alkalis, it is deteriorated by the action of acids, and oxidising agents such as hydrogen peroxide and chlorine bleaching compounds. These properties will be considered later, but at this point it is very convenient to refer to the reaction of cotton with caustic soda which is used to give it increased lustre in the process known as *mercerisation*.

Mercerised cotton Over 120 years ago (in 1850) John Mercer of Accrington noticed that when cotton fibres were immersed for a few minutes in a strong solution of caustic soda they swelled so as to become cylindrical, and at the same time contracted in length. He also found that after removal of the alkali by thorough washing the cotton fibres were chemically unchanged and still consisted of pure cellulose. By this treatment a piece of muslin fabric will become as compact and thick as closely woven material. This alkaline treatment brings about a permanent change in the cotton so that each fibre thereafter always retains cylindrical form dimensionally very similar to its original state in the cotton boll before its drying and collapsing as described earlier. It can never again be restored to the twisted ribbon-like form. The cellulose also appears to have gained a somewhat permanently swollen state so that the cotton is generally more reactive towards all those influences which affect cellulose. It is more hydrophile and thus in air absorbs about 12 per cent instead of the 6 per cent of moisture as absorbed by ordinary cotton. The cotton in its new form, now always referred to as *mercerised* cotton, has a much greater affinity for most dyes and is stronger. Some forty years later (in 1890) another cotton experimenter, H. A. Lowe, noticed that if the cotton yarn treated with caustic soda was held stretched to about its original length whilst the alkali was being washed out then the yarn acquired a much higher lustre. Mercer had missed this point about the stretching.

So, mercerised cotton as we know it today is produced by treating the yarn or fabric with caustic soda under conditions such that the shrinkage is largely prevented during the treatment and is entirely counteracted by strong stretching during the period in which the alkali is washed out. Only strong caustic soda solutions of about 50°Tw produce the maximum mercerising effect so there soon comes a stage in the washing when the residual alkali is too weak to have any appreciable effect.

Mercerised cotton fibres are cylindrical and have about the same length as before treatment; their dimensions are permanent. The important point, however, is that each cotton fibre has a smooth surface, and materials made of mercerised cotton have a more silk-like lustre than those made of untreated cotton. Mercerisation is today largely applied to cotton yarns and fabrics to give them a greater lustre and a higher quality appearance and increased strength. The mercerisation of cotton can now be effected by treatment with liquid ammonia at $-33°C$ instead of caustic soda (p. 249).

Modified and improved cottons The hydrophobic nature of synthetic fibres (p. 113) has given nylon, Terylene and such fibres certain advantages over cotton in the production of textile materials. For example, since synthetic fibre materials imbibe so little water (with accompanying small fibre swelling) when wetted they dry very rapidly, and at the same time they suffer negligible distortion and shrinkage as a result of such wetting (washing) and natural drying so that they need not be ironed before use. Cotton producers have therefore sought treatments for ordinary cotton which could give this fibre some of the desirable properties possessed by synthetic fibres.

Considerable success has been obtained in these efforts so that now it is possible to treat cotton with certain urea- or melamine-formaldehyde resins or with formaldehyde itself, or with substances able to cross-link the chain-like cellulose molecules of which cotton fibres are composed (p. 391) and reduce appreciably the hydrophile properties of the cotton. It is now possible to use these modifying treatments in the production of cotton fabrics which can be used instead of synthetic fibre fabrics in the manufacture of 'drip-dry' and 'wash-and-wear' shirts and other such articles.

Synthetic fibres are also immune to attack by micro-organisms of all kinds and by various types of insects. Cotton can be given somewhat similar properties by synthetic resins or partial acetylation (treatment with acetic anhydride and acetic acid) or by partial cyanoethylation (treatment with acrylonitrile in the presence of an alkali) and these special processes are now being commercially exploited.

Linen

Linen materials are made from fibres found in flax. Before 1939 Russia was the greatest producer of flax. In recent years research carried out by the Linen Industry Research Association on producing and processing flax in Great Britain to give linen fibres equal to those produced on the continent, particularly at Courtrai, has met with a limited degree of success.

Flax is grown from seed sown annually on land which has to be ploughed before seed-time. The seed is sown rather thickly so that the flax stalks grow up tall and straight. In due course heads of bright blue flowers crown each flax plant and these later give way to seeds. It is customary, however, to harvest the flax somewhat before actual maturity since under these conditions a better quality of linen fibre is ultimately obtained.

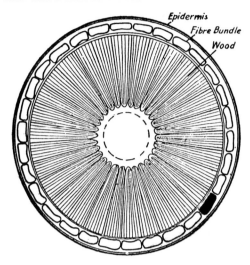

Figure 1.7 Diagrammatic cross-section of flax stalk showing the distribution of the bundles of linen fibres. (Courtesy the Textile Institute)

Figure 1.8 (left) Flax stalks showing the flower or seed head, the long straight stem with small leaves, and the roots. (Courtesy the Textile Institute)

We have seen in the case of cotton that the fibres are produced as a seed covering, but this is not the case with linen although these fibres consist of the same substance, cellulose. Linen fibres form part of the stalks, so that in harvesting it is necessary to collect these rather than the seeds. Nevertheless, the seeds are usefully employed as a source of linseed oil and after pressing out the oil the residual seed 'cake' may be used as a cattle food.

The linen fibres extend from the top of the stalk right down to the root so in harvesting flax it is the practice to pull up the plants rather than to cut them as in the case of wheat, oats and barley. If flax were cut then, a fair proportion of fibre would be left in the stubble and so be wasted.

Rippling   After pulling, which in the past has been carried out by hand but which in later years has been assisted by machinery, bundles of the flax are stacked to dry in the sun. Then the flax is passed through a rippling process by which the seeds are combed away to leave the stalks ready for extraction of the linen fibres.

To understand the processes which follow rippling it is necessary to be aware of the manner in which linen fibres are distributed in the flax stalks. A cross-section of a single stalk shows that first there is the outer bark, then a soft gummy portion, and then the cortex at the centre. The linen fibres are in the form of long bundles of finer shorter fibres which lie end to end or slightly overlapping each other. These bundles constitute an annular layer extending throughout the length of the flax stalk.

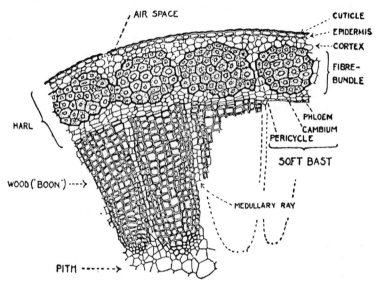

Figure 1.9 Cross-sectional view of a flax stalk, showing the position of the bundles of linen fibres which have to be separated from the remainder of the stalk during manu-facture. (Courtesy the Linen Industry Research Association)

Retting The fibre bundles are most serviceable for the manu-facture of linen goods when they are long, so that in the processes for isolating them in a pure state care is taken to avoid breaking them down into their finer shorter fibres.

For extracting the long bundles of linen fibres the traditional method, which is still used, is that of exposing the flax stalks, after rippling, to the action of weather or water. The conditions are such that bacteria and other organisms act on the stalks and disintegrate their substance with the exception of the linen fibres. At a suitable stage of this decomposition the stalk substance is so loosened that the linen fibres can be separated mechanically from the other parts of the stalks which have been softened and solubilised.

This bacterial decomposition is accomplished in the next stage of processing which is known as *retting*. In this the bundles of flax are allowed to lie for some weeks exposed to sun and rain in the fields or they are steeped in running water or pools of stagnant water. The changes which take place within the flax stalks are much the same, no matter how this retting is carried out. However, the better the control the higher can be the quality of the linen flax fibres obtained, for if the retting process is too prolonged or uneven the linen fibres also undergo a degree of decomposition.

Scutching   After retting, the flax is dried and then passed through a cleaning process in which most of the impurities are removed. Firstly, the stalks are led through pairs of fluted rollers to break down any hard parts in them without breaking the long linen fibres. Then the stalks are passed through a scutching machine in which they are subject to a kind of combined beating and combing treatment. This separates the long fibre bundles, substantially unbroken, from various impurities and shorter fibres. These short fibres are known as *tow* and can be used in the manufacture of linen materials but they are of less value than the long straight fibre bundles. Hand scutching is also practised.

Scutched flax is linen in a state more or less comparable to that of ginned cotton in so far as it is a marketable textile raw material ready for the later processes involved in spinning it into yarn.

Characteristics of linen fibres   It has already been mentioned that linen fibres resemble cotton in so far as they consist of cellulose. But, since the long linen fibre bundles are made up of an agglomerate of finer short fibres, each about 1 in (25 mm) in length, held together with natural adhesive substances, they have a lower cellulose content. On an average the linen fibres contain only about 75 per cent of pure cellulose, the remaining matter being a gummy pectic substance. The fine short fibres are almost completely pure cellulose. Each of these short fibres is straight but it is not round or rod-like. Its cross-section is polygonal and it appears to be divided into closely packed cells. Along the length of each of these

Figure 1.10 A single linen fibre and its cellular cross-section

short fibres are found notches or transverse markings and from end to end extends a central canal or lumen. The surface of each fibre is smooth and this helps to give linen materials their characteristic high lustre.

Since the long fibre bundles are those best adapted to the production of the highest quality linen yarns and fabrics, it is desirable to guard against breaking down these bundles into the finer shorter fibres of which they are built up. But, because these long linen fibres are in fact fibre bundles, they are naturally somewhat irregular in length and thickness. Linen fibres such as those resulting from Linron processing (pages 155 and 249) are more prized the more uniform they are as regards their form and dimensions.

In many of its chemical properties linen closely resembles cotton. Thus it is resistant to alkalis and is easily deteriorated by acids. Actually linen is stronger than cotton. It is extremely durable and evidence for this is to be found in the numerous linen mummy cloths which have been taken from Egyptian tombs and which even today have a fair degree of strength.

Linen materials may be mercerised along the same lines as cotton, but since linen is naturally highly lustrous there is much less need for subjecting it to this process. However, after mercerisation with strong caustic soda solutions, linen has a more useful affinity for dyes.

Linen goods absorb about 12 per cent of moisture from the surrounding air and in this way are somewhat more hydrophile and cold to handle than cotton materials, which absorb only about 6 per cent.

Wool

Wool, the first of the so-called animal fibres to be considered here, is perhaps the most useful of all fibres. Not only is it obtainable in very large quantities, but it is a warm soft-handling fibre and is thus particularly suitable for outer clothing and underwear.

Since sheep farming is carried on in most parts of the world with a view to providing food, it is obvious that sheeps' wool becomes available there at the same time. In a few countries, however, notably Australia, New Zealand and South Africa, sheep farming is carried on more with a view of producing wool than mutton. These countries are the main wool-producing areas and it is from them that the best types of wool are obtained.

Wool can vary a great deal in quality. Some types consist of short coarse fibres whilst others are very fine and long. Wool fibres may also vary considerably in respect of lustre. These variations depend on differences in the breed of sheep and on whether the sheep is shorn alive or dead.

The merino sheep, first associated with Spain, have been introduced with great success into Australia and South Africa. Merino wool is, in general, most prized by wool manufacturers. Wool obtained from crossbred sheep, which are largely farmed in New Zealand, is also very useful, but crossbred wool is usually firmer in handle than merino wool and cannot be used where the finest wool yarn and fabric is required. Other types of sheep are to be found throughout the world, but on the whole they yield coarser and less valuable wool than merino and crossbred.

It is not necessary to give any details of the shearing of wool from sheep but perhaps some mention should be made of the fact that wool taken from the slaughtered sheep is generally pulled off after the skins have been treated with lime or other depilatory agent to facilitate this. The conditions for obtaining this pulled wool generally make it inferior in quality to shorn wool.

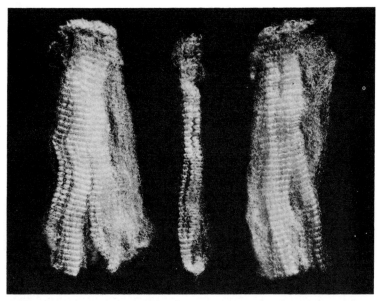

Figure 1.11 Flocks of superfine merino wool showing the highly crimped form as taken from a sheep's back. (Courtesy the International Wool Secretariat)

Wool fibres are quite different in form and composition from cotton and linen. As we shall see later, raw wool also contains a high proportion of impurities, mainly of a greasy character, which have to be removed by a scouring process before spinning and weaving operations can be carried out.

The nature of wool fibres   In its purified condition each wool fibre is composed of the protein known as *keratin,* a substance which consists of carbon, hydrogen, oxygen, nitrogen and sulphur. Keratin differs from cellulose in containing nitrogen and sulphur, and it is the presence of these two additional elements in the wool molecule which gives wool fibres properties profoundly different from those of the vegetable fibres. Notably greater warmth of handle is associated with a protein fibre.

A single wool fibre is rod-like and tapers from the root end to its tip. An important characteristic is that each fibre is closely covered with overlapping scales, known as *epithelial* scales, which all point in the direction of the tip. These scales are like those of a fish and generally act as a protective covering for the fibre. If the scales are removed in wear or processing, then the wool becomes much less resistant to deteriorating influences.

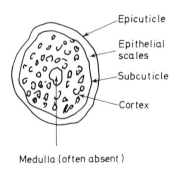

(a)

Figure 1.12 Structure of a wool fibre. (a) Cross-section of a normal fibre showing its structual components; (b) longitudinal section of a bicomponent fibre showing its spiral bilateral structure, in which A is the ortho-cortex and B the para-cortex; (c) cross-section of bicomponent fibre showing the combination of A ortho- and B para-cortexes; (d) general view of wool fibre covered by overlapping epithelial scales to protect the less tough core of ortho- and para-cortical cells—the core has the spiral bilateral structure shown in (b)

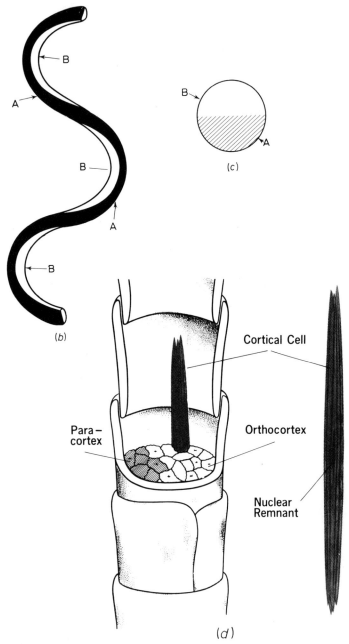

(b)

(c)

Cortical Cell

Para-
cortex

Orthocortex

Nuclear
Remnant

(d)

Figure 1.12 (continued)

The epithelial scales also play an important part in the shrinkage and thickening which wool fabrics show when repeatedly washed. Most wool fabrics and garments shrink a moderate degree when washed for the first time; this is to be expected since it results from the relaxation of fibre and yarn stretching forces which cannot be avoided in manufacturing operations. But, with additional washing, further shrinkage takes place and this is accompanied by a matting or closing up of the fibres in the fabric. This latter type of shrinkage is known as *felting* and continues with further washing until the fibres become packed to a maximum degree in the fabric.

The electron microscope and new chemical methods have revealed the presence of two very fine membranes which cover the epithelial scales. These earlier unrecognised and unsuspected parts of the wool fibre influence the way in which dyes are absorbed and also the role played by the epithelial scales in the felting and shrinkage of wool material (Figure 1.12).

Felting   This property of felting as it exists in manufactured wool materials can be both an advantage and a disadvantage. In recent years special treatments have been devised to reduce or eliminate it. These developments in wool processing are most interesting and will be considered later. Here it is only necessary to indicate that felting is largely dependent on the special roughness of the wool fibre surface due to the projecting edges of the overlapping scales. If this scale roughness is reduced the wool no longer felts.

Physical characteristics of wool fibres   Wool fibres have none of the twists which characterise cotton but they are unique in having 'crimp' or waviness. The finest fibres have most crimps or waves. Thus a merino wool fibre may have up to 30 crimps per inch (12 crimps/cm) whereas a low-quality wool may have only 5 or 10 crimps per inch (2 or 4 crimps/cm). This crimp is of importance in the spinning of wool yarns for it favours adhesion between the fibres as these lie in the yarn twisted about each other. Also it imparts a certain loftiness or fullness of handle in wool materials. The importance attached to this crimp may be seen by the fact that man-made fibre manufacturers have given much attention to possible methods for giving their fibres an artificial crimp.

The crimp in wool fibres has recently been found to be due to a spiral bilateral structure of each wool fibre which in its turn gives each fibre a degree of instability sufficient (when conditions are favourable) to cause its movement into a crimped form. Thus each wool fibre appears to be divided longitudinally into equal halves which stain differently when treated with the basic dye Janus Green—

the *orthocortex* half stains deeper and in a crimped fibre is always to be found on the outside of the curve, while the *paracortex* half is less accessible to the staining dye, stains less and is always to be found on the inside of the curve of the fibre. This special bilateral structure actually gives each wool fibre a tendency to coil spirally, but when the fibres are closely packed as in a wool flock on the back of a sheep there is a restriction of fibre movement so that the coiling is constrained into a linear crimp.

Wool fibres vary considerably in length. A merino wool fibre is about 3 to 5 in (75 to 125 mm) long, but other types may extend to 12 or 15 in (0·3 or 0·4 m). The thickness of a fibre varies a great deal, too. Generally, wool fibres are around 1/1000 in (0·025 mm) in diameter but they are frequently twice this thickness. Wool fibres are thus often thicker than cotton fibres.

Fibre thickness plays a large part in deciding the fineness of yarn which can be spun from any given type of wool. The highest quality wools are those which can be spun into the finest yarns, and the qualities are commonly graded downwards from 100. Generally, under this system, a wool is assigned a quality number corresponding to the finest counts of yarn which can be spun from it. Thus a 70s quality wool is one which can be spun under the most favourable circumstances of manufacture into yarn of 70s count. This, in turn, implies that 1 lb (0·45 kg) of this particular quality of wool can be spun into 70 skeins of yarn each of 560 yd (512 m), that is, into a total length of 39200 yd (35·8 km). The quality of a merino wool may vary between 60s and 100s, and a crossbred wool is usually 36s to 60s quality. Coarse wools, such as are used for carpet manufacture, are generally below 36s quality.

However, the quality of a wool is not determined by its source or type alone. It varies according to the position of growth on the sheep. The best wool grows on the shoulders, whilst the flanks carry the best average quality wool. Sorting wool into qualities is a highly skilled art. Moreover, even in wool lots of the same quality the individual fibres differ from each other to a fair degree; perhaps this is to be expected when it is remembered that on the sheep there may be 60000 fibres /in² (92/mm²). Present day research is steadily revealing how the quality of wool can usefully be influenced by attention to the pasture feeding of sheep, e.g. the presence of traces of cobalt in the feed (grass) can improve wool quality.

It is interesting to know how Professor A. Johnson has shown in Australia that the felting properties of wool can be influenced to be greater or less by dietary control of the living sheep. Trials have been made in which immediately after shearing, sodium molybdate and sodium sulphate were (*a*) added to the basic chaff diet and then after growth of the fibres to two thirds of their full length the diet was changed by omitting (*b*) sodium molybdate, when the fibres increased to their full length. The resulting fibres were then found to be modified especially with regard to their tip and root ends and to have much increased felting power. When the order of giving these additives to the diet was reversed the felting power of the fibres was reduced. In one trial the felting shrinkages measured under comparable conditions of wools (*a*), (*b*) and (*c*)—the last with fibres obtained with non-modified diet—were 24 per cent, 6 per cent and 15 per cent, respectively. Both of the modified diets reduced the power of the fibres to recover by relaxation from stretching.

Returning again to the characteristics of a single wool fibre, it is to be noted that below the outer layer of scales is a central core made up of so-called cortical cells which are probably embedded in a glutinous medium but of which very little is known. These cells are spindle-shaped and each is about 1/250 in (0·10 mm) long and 1/10000 in (0·0025 mm) thick. In any treatment which seriously deteriorates the wool fibre, e.g. steeping for about a month in concentrated ammonia solution, the agglomeration of these cells breaks down and they separate so as to be clearly visible under a microscope. Running through the centre of each wool fibre is a fine lumen or canal which is known as the *medulla*. In the fine merino wool fibre this medulla is, in most cases, so fine as to be scarcely observable, but in coarse, low-quality fibres it constitutes a fair proportion of the wool fibre. Actually this medulla is not empty but is filled with a honeycomb of cells considerably smaller than the spindle cells of the cortex.

Wool fibres are extremely resilient and so wool materials are highly resistant to crushing and do not form permanent creases under ordinary conditions. They are also very elastic (arising from the wool molecules being long, folded instead of straight and also crosslinked) and when the tension is released after stretching (wet or dry) they quickly spring back to their original length. No other natural textile fibre is the equal of wool in this respect.

A further property of wool is that it handles warm. This is largely because wool materials are spongy in structure and so occlude innumerable extremely small pockets of air. Still air is a very bad conductor of heat and it is therefore easy to understand that wool is a good heat insulator and gives the impression of warmth when handled. Associated with the warm handle of wool is the fact that on wetting (say by perspiration) it generates heat. Rayon and synthetic fibre manufacturers can partially impart the warmth of wool to their products by cutting their long fibres into shorter ones having a length comparable to that of the natural fibres to allow the production of more porous spongy yarns.

In many respects wool has chemical properties opposite to those of cotton and linen. Thus, whilst these cellulose fibres are highly resistant to alkalis but easily attacked by acids, wool is resistant to acids but easily deteriorated by alkalis. Even quite weak solutions of sodium carbonate (ordinary washing soda) are able to impoverish wool materials when used hot or for a prolonged period. On the other hand, it is useful to know that ammonia does not damage wool unless special steps are taken to make it do so. As a protein, wool has, in general, less stability than the cellulose fibres. For instance, whereas the prolonged boiling of cotton and linen in water causes them no appreciable damage, wool is harmed by such treatment. Prolonged immersion in boiling water causes a gradual breakdown of the wool keratin so that the fibres lose strength and become discoloured. If only a trace of alkali is also present in the boiling water then the resulting decomposition of the wool is so much the more rapid.

When air dry, wool contains about 18 per cent of absorbed moisture. Wool can hold a surprisingly large amount of moisture without feeling wet and it is this property which makes wool underwear especially serviceable since it absorbs perspiration readily. There is yet another useful point about this. In absorbing moisture or perspiration the wool becomes warmer. Thus the wearer of a wool garment does not feel it to be cold and clammy during periods of intense perspiration.

On the whole, wool has more lustre than cotton and some types are especially lustrous. By chlorination (followed by alkali treatment) wool can be made very lustrous and this method is sometimes used to give carpets a highly polished appearance. Such treatment, however, causes a definite loss in strength and durability by the wool and would not be recommended for ordinary wool fabrics and garments.

Recovered wool fibres   In normal times not enough wool is obtained from sheep to meet the requirements of the wool industry. This deficiency is made up by collecting waste and discarded wool materials such as rags, garments and yarns and then passing these through special machines which 'scratch up' the material into loose fibre. The product may pass under different names according to the care with which the waste materials have been selected. Thus *shoddy* is made from all-wool materials which have not been previously felted and in which any vegetable fibres have been destroyed by carbonising, whilst *mungo* is the product from milled woollen rags. *Alpaca* (not the high-quality goat's hair of the same name) is another type of recovered wool; it may contain a fair proportion of vegetable fibres.

All these recovered wool materials can be used again in the manufacture of yarns and fabrics either by themselves or, more frequently, in admixture with new wool. To distinguish such materials fron new wool, to which they are inferior, it is customary to refer to new wool goods as being made from virgin wool.

It is thus easy to understand that a label on a garment to indicate that it is all-wool does not necessarily guarantee that it is of the highest quality. It is only when such a label also shows virgin wool to have been used that such an inference can be made. Usually recovered wool materials readily absorb water when wetted, whilst a new wool material will resist wetting.

Recovered wool is quite hygienic to use. Its main drawback is its inferior durability. Obviously, the treatments involved in freeing it from vegetable fibres and dyes, as well as the mechanical treatment, involve the wool in a certain degree of degradation and fibre breakage. However, by mixing this wool with virgin wool these defects can be largely mitigated.

Goat and rabbit fibres

Allied to wool as obtained from sheep are a few types of fibres which are secured from other animals, particularly goats and rabbits.

The goat fibres include *mohair*, which is a silky material obtained from the Angora goat, and *cashmere*, which is a softer fibre obtained from another species of goat. These goat fibres resemble wool in being covered with epithelial scales and in consisting of keratin. But the scales are more smoothly arranged and so these goat fibres do not felt so readily as does sheep's wool. Mohair is used to a considerable extent for making plush fabrics.

Rabbit fur is widely used for manufacturing felt hats of the highest quality and for this purpose is much superior to sheep's wool. In recent years the supply of rabbit fur has been much reduced by large-scale extermination of rabbits by the disease of myxomatosis.

Silk

Silk as produced by the cultivated silkworm was at one time the most prized of all the textile fibres. According to tradition it originated in China about 2500 BC. Thence the silkworm was probably introduced into India, and much later (about the sixth century AD) the Romans brought silk to Europe. At the present time Japan is the most important producer of silk for export to other countries, but in India and China very large amounts of silk are produced and used locally.

The silkworm which is responsible for this silk is prone to disease and, in the past, strong French and Italian silk industries have largely disappeared as a result of disease which could not be checked. The Italian silk industry alone appears to have survived these calamities among the European countries.

There are two main types of silkworm, the cultivated and the wild ones. It is the former type which is under strict control that produces the highest quality of silk. Wild silkworms usually give a stronger, more alkali-resistant silk, but it has the disadvantage of lacking uniformity.

Cultivated silk   In speaking of silk, most people have in mind the silkworm which feeds on mulberry leaves. This is the cultivated silkworm known as *Bombyx mori*. In silk rearing establishments such as are to be found in Italy and Japan the first stage in the production of silk is that of hatching out the silkworms from tiny

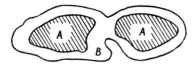

Figure 1.13 Cross-section of raw silk thread as taken from a cocoon, showing the two separate fibres A cemented together side by side with silk-gum or sericin B

eggs which have previously been carefully selected and shown by test to be free from disease. The eggs are kept in warm rooms on trays and towards the end of a month tiny silkworms emerge, each about $\frac{1}{4}$ in (6 mm) long. These silkworms, somewhat resembling caterpillars, are ravenous eaters of the cut-up mulberry leaves with which they are fed. In the course of the next month each silkworm grows rapidly and attains a length of about 3 in (75 mm). During this period it moults or sheds its skin four times. After the fourth moult the silkworm becomes restless and finally climbs upon one or other of the twigs which have been specially placed near it for this purpose and there spins around itself a silken cocoon.

In this operation the silkworm generates within itself a quantity of a viscous substance known as *fibroin*. This passes through two glands near its mouth so that actually two separate silk fibres are formed at this point. Immediately afterwards the two fibres pass out of the silkworm just under its mouth as a double thread cemented together and coated with another substance known as *sericin*. Thus the cocoon is spun with a double silk thread. During the formation of the cocoon the silkworm moves its head continuously so as to direct the silk thread into the form of an ellipsoidal or egg-shaped envelope. In this way, the silkworm spins a continuous double thread up to 2000 yd (1·8 km) in length until it reaches a stage at which it is completely and snugly enclosed. One end of the cocoon is made somewhat thinner than the other.

In the natural course of events the silkworm inside the cocoon changes into a chrysalis which subsequently awakes to life in the form of a moth. This, as a first act in its existence, moistens the thinner end of the cocoon adjacent to its head and so softens the silk thread. In this way it is ultimately able to force a gap in the end of the cocoon and make its escape. These moths live only for a few days and then die, but in this short life the female óf the species lays eggs which are collected. They are most important to the silkworm rearer for it is from these that he later hatches out fresh silkworms.

This natural cycle of changes is only allowed to take place with a small proportion of the silkworms handled in the industry. Only those cocoons which are required to provide the necessary silk-worm eggs are allowed to develop into moths. The remainder of the cocoons are steamed or otherwise treated so that all life within them is destroyed. The reason for this is easy to understand. If the moth is allowed to come to life and make its way through the cocoon then it disarranges the silk thread so that considerable difficulty is met with in unwinding it free from entanglement and waste at a later stage.

The cocoons vary from white to yellow in colour depending on the nature of the colouring matter formed by the silkworm at the time of spinning.

Reeling of silk It is now necessary to reel the silk thread from the cocoons. They are first sorted and those which appear likely to unreel easily are placed in bowls of hot water so as to soften all glutinous matter external to the thread. Loose fibre is then removed and the cocoons are placed in groups, the threads from them being brought together as they are drawn through a guiding eye positioned immediately over the bowl of water. Thus two, three, four or more threads can be formed into a silk yarn which is wound into skeins. The unwinding cocoons can be kept loosely distributed on the water by hand and broken ends of thread joined as occasion demands. The yarn which results is dried and suitably packed according to its quality. This constitutes raw silk ready for marketing and as it will later be received by manufacturers of silk goods.

Since World War II machinery for the reeling of silk from cocoons has been devised in Japan to take the place of hand labour. It is claimed that this type of machinery is efficient and makes the production of raw silk much cheaper.

Naturally there is a good deal of entanglement and breaking of the silk in reeling and many of the cocoons become damaged so that they cannot be reeled. Consequently there is a fair quantity

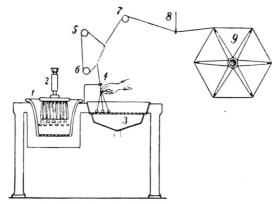

Figure 1.14 Hand silk reeling apparatus. The cocoons are softened in the hot-water trough 1 and brushed free from impurities by brush 2 before transference to the bowl of warm water 3. Single filaments are drawn from the cocoons and combined to form yarn whilst being drawn under tension through the guides 4, 5, 6, 7 and 8. The yarn is wound on the reel 9

of silk waste which is collected and ultimately brought into the form of comparatively short fibres, which can be spun into yarns by the methods similar to those used for spinning cotton. This *schappe* or spun silk is quite different in character from reeled silk since the individual fibres of reeled silk are many yards in length.

Raw silk fibres, as previously mentioned, are double, the fibres consisting of fibroin and the silk-gum or cementing substance being of sericin. These two substances are similar in chemical composition in so far as they both consist of carbon, hydrogen, oxygen and nitrogen. An important difference is that the sericin alone is soluble in warm soap solution, the fibroin being insoluble in all common substances. Thus, by boiling raw silk in a 1 per cent soap solution, the silk-gum is removed so as to leave the two fibroin fibres quite separate.

Weighting of silk   The amount of silk-gum in raw silk varies from 20 to 30 per cent so that in degumming there is a considerable loss of weight. This adds to the cost of the silk. For this reason the practice of silk weighting has arisen. This is a treatment by which certain cheap chemicals are deposited within the degummed silk so as to give it increased weight and also to increase the thickness of each fibre. The first idea of using a tin–phosphate–silicate compound for this purpose was just to counterbalance the loss of silk-gum but in later years the degree of weighting has been increased even to double the weight of the silk and so cheapen it. These treatments will be dealt with later, but at this stage it is sufficient to point out that if the weighting is kept within moderate limits the silk retains its strength and softness; if the weighting is carried to extreme limits then the silk is impoverished.

The degummed silk fibres are very fine, being only 1/2000 in (0·010 mm) in diameter, and in the cocoon are 1000 yd (0·9 km) or more long. They are smooth and lustrous. Contrary to general belief a silk fibre is not cylindrical but its cross-section is triangular with the corners rounded.

Silk characteristics   Characteristics of silk are its great strength, its softness, pleasing lustre and excellent elasticity. In addition, silk is a bad conductor of heat and thus handles warmly. All these desirable properties united in the one fibre make silk the most valuable of all the fibres used in textile manufacture. It is no wonder that the rayon industry has arisen with the initial idea of producing artificial fibres at least the equal of silk.

Raw silk has a specific gravity of 1·33 but degumming gives pure fibroin having the lower specific gravity of 1·25. With this change brought about by degumming, the silk fibres become nearly white or transparent and gain in softness.

The silk-gum affords some protection to the fibroin fibres which it coats and cements together so that, where circumstances permit, it is better to weave or knit with raw silk and then degum the fabric so produced. However, to allow this the raw silk threads must first be softened with an oil emulsion so that they are more supple.

The silkworm is not quite perfect in its spinning of silk for the thread changes somewhat in diameter from the beginning to the end of the cocoon. The general tendency is for the fibre to become thinner towards the end. The fibres also vary from cocoon to cocoon. This defect reveals itself particularly in the knitting of ladies' silk hose when 'barry' effects are sometimes noticeable after dyeing.

Since silk is made of a protein similar to wool keratin (it contains no sulphur, however) it is to be anticipated that it will resemble wool in being resistant to acids but easily damaged by alkaline substances. This is true, but on the whole silk is less adversely affected by alkalis than is wool. Tussah silk, the most important wild silk obtained from India, is indeed quite resistant to alkalis. Although Tussah silk has this particular advantage, it is inferior in other respects in so far as it is much more irregular in quality and is harsher in handle.

Degummed silk is hydrophile and under ordinary conditions retains about 11 per cent of moisture.

Miscellaneous fibres

In addition to the main natural fibres, cotton, linen, wool and silk, which are used for the manufacture of garments and fabrics used personally or in the home, there are a number of other types which can receive brief mention here, although they are employed for more menial purposes. The most important of these are ramie, jute, hemp and nettle fibres.

All of these are obtained from the stalks of plants which, like flax stalks, are built up of various soft ingredients strengthened by tenacious cellulose fibres mainly aligned in the lengthwise direction of the stalk. To separate the fibres it is generally necessary to take the same steps as with flax and decompose the non-fibre ingredients before scutching, hackling or otherwise combing to straighten the fibres and elimate residual impurities.

Hemp is extremely valuable for the manufacture of cord, rope and string, whilst jute is much used for making sacks, mats and wrapping-cloths. Ramie is highly lustrous and closely resembles linen; it is obtainable from a species of nettle and there are other types of nettle fibre in limited use.

These natural fibres are today being replaced to an increasing degree by stronger and more durable man-made fibres, notably by polypropylene.

Kapok fibre

Kapok is a fibre obtained from the seed pods of the kapok tree. It is like cotton, but owing to the presence in each fibre of innumerable small air pockets kapok is very light. Furthermore, it is not easily penetrated by water. Thus during World War II this fibre attained considerable importance as the most satisfactory filler for life-saving jackets and other equipment used to reduce loss of life at sea by drowning. The synthetic fibre made from polypropylene is more durable but not so light as kapok (p. 83).

Sisal

This is a vegetable fibre which in recent years has acquired increased usefulness mainly for the manufacture of twines, ropes and cordages. In the form of raffia it is well known to gardeners. Sisal is obtained from the leaves of the plant *Agave sisalana* which is cultivated on large plantations in Java and in east and west Africa. The leaves are fleshy and require mechanical decortication for separation of the fibres. Sisal fibres, unlike hemp, have only moderate resistance to deterioration in sea-water.

Asbestos fibre

Asbestos fibres (hydrated magnesium silicates) occur as deposits in the earth. The fibres are coarse and lack tensile strength, but they have the important properties of being fire-proof and good heat insulators. Thus they are much used in the form of soft string, rope and fabric for engineering purposes. As is well known, asbestos is also used for the production of fire-proof curtains such as are compulsory in places of public entertainment. It is an important constituent of motorcar brake linings. Asbestos slowly rots in sea-water and is decomposed by hot acids.

THE MAN-MADE FIBRES

Regenerated cellulose. Cellulose acetate. Synthetic glass and fibres made by fibrillation

The conception of man-made fibres

The possibility of making fibres artificially was predicted about 300 years ago by Robert Hooke but it is only within the past sixty-five years that this prophecy has been fulfilled. Anyone taking the trouble to watch a silkworm spinning its silken cocoon would become aware of the first principle which must underlie the production of man-made fibres. The silkworm forces a thick viscous liquid, fibroin, through small glands (orifices) and as the liquid is thus extruded it at once solidifies to form a fibre or filament.

If Man wishes to make similar fibres he must prepare the fibre-forming substance as a suitably viscous liquid either by melting it or by dissolving it in an organic or aqueous solvent, and then he must extrude it through a small orifice or group of orifices, with cooling to effect solidification (or by evaporation or other means remove the solvent) to leave a single fibre or a group of fibres in the form of a thread.

The shape of the orifice can be much varied say from square, circular, triangular or cross-shaped so as to obtain fibres having special cross-sectional shapes and corresponding special properties such as fibre-to-fibre adhesion, smoothness, lustre (including glitter), dye absorptive power, etc.

An entirely new and recent method for making fibres consists of making film from a fibre-forming substance and then (optionally after slitting it to form a sheet of narrow tapes) vibrating or otherwise subjecting this to stress to cause it to split or fibrillate lengthwise into fibres.

Definition of fibre terms

The nomenclature of fibres has changed in recent years. In this book the commonly used terms have the following meaning.

Rayon–fibres are composed of regenerated cellulose.

Acetate–fibres are composed of cellulose approximately di- or tri-acetate (otherwise known as *secondary* or *primary* acetates respectively).

Synthetic–fibres are composed of non-natural fibre-forming substances such as are manufactured by chemical methods independent of natural growth and climatic conditions.

Returning now to a consideration of the way in which these newer fibres are made, it can be noted that those rayons based on cellulose as a raw material are: (1) nitrocellulose or Chardonnet rayon, (2) viscose rayons (ordinary and polynosic), (3) cuprammonium rayon, (4) cellulose acetate.

Chardonnet rayon

Chardonnet rayon is no longer of commercial importance, but it is of much interest as representing the first successful rayon to be made on a large scale. This success was almost entirely due to the lifetime's persevering effort of Count Hilaire de Chardonnet. His process consisted of converting cotton into nitrocellulose for dis-

solving in a mixture of ether and alcohol to produce the necessary
viscous liquid, and then spinning this into air where immediate
evaporation of the ether and alcohol solvents took place to leave
solidified nitrocellulose fibres. It was soon found that nitrocellulose rayon was much too inflammable to be safe in everyday
use, so a de-nitrating treatment was used to split off the nitro
groups and so convert the fibres into regenerated cellulose. Chardonnet rayon was at one time very popular and much used. Having
served its purpose of showing that imitation silk fibres could be
produced, this type of rayon gradually became supplanted by
superior types, particularly cuprammonium and viscose rayons.
The production of cuprammonium rayon is now insignificant but
viscose rayon is very important and its manufacture exceeds that of
all the man-made fibres taken together.

Viscose rayon

This type of rayon is made from cellulose which can be obtained
either from cotton linters or, more frequently today, from the
cheaper wood pulp (preferably spruce). It is often referred to as a

*Figure 1.15 Sheets of wood pulp being placed into a large gate-end press ready for
steeping in strong caustic soda solution. (Courtesy Courtaulds Ltd)*

regenerated cellulose rayon. Present-day methods of manufacture are based on fundamental reactions of cellulose discovered in 1891 by three British chemists, Cross, Bevan and Beadle, and it is true to say that British effort has been most responsible for the development of this particular type of rayon.

The manufacture of viscose rayon is based on the following reactions of cellulose. By treatment with concentrated caustic soda solution (similar to that used in the mercerisation of cotton) cellulose is converted into alkali-cellulose. This is then capable of reacting with carbon disulphide to form cellulose xanthate, which desirably differs from the original cellulose in that it dissolves in a dilute solution of caustic soda to give a very viscous solution. This cellulose xanthate solution can then be extruded through orifices or spinnerets containing a group of orifices to be at once solidified to form fibres by passing through an acid coagulating liquor. The action of the acid is to decompose the cellulose xanthate with regeneration of the cellulose. Thus the resulting viscose rayon filaments consist of regenerated cellulose. It is necessary to make use of these intermediate cellulose compounds (alkali-cellulose and

Figure 1.16 Viscose solution suitable for spinning into viscose rayon fibres. (Courtesy Courtaulds Ltd)

cellulose xanthate) because only in this way can the cellulose be brought into solution suitable for extruding into filaments. The cellulose used as raw material has exactly the same chemical composition as the regenerated rayon but, nevertheless, there is a difference between these celluloses so that the regenerated cellulose is somewhat less durable than the original although the fibres have many compensating advantages.

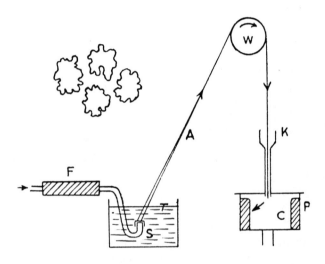

Figure 1.17 Essential features of viscose rayon spinning. Viscose is pumped through the candle filter F and the multi-holed spinneret S into the coagulating bath T. The resulting bundle of rayon filaments A forming a single thread is drawn upwards over Godet wheel W and then downwards into the funnel K of the Topham pot P which is rotating rapidly. Here the thread, thrown by centrifugal force outwards, becomes wound into a compact cake C. Typical cross-sections of individual viscose rayon filaments are also shown

Basic manufacturing processes

In the manufacture of ordinary viscose rayon the following essential chemical reactions are involved:

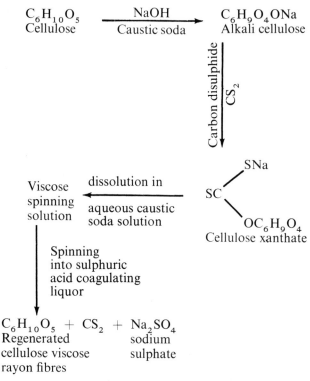

The raw material, cellulose, is utilised in the form of sheets since this is the most convenient form of handling it. These sheets are stacked in a frame so that they can be flooded with the caustic soda solution. After a suitable period has been allowed for the liquor to penetrate evenly and thoroughly, the sheets of alkali-cellulose are hydraulically pressed until they are about three times their original weight.

The compressed moist sheets of alkali-cellulose thus formed are then shredded into fibrous 'crumbs' and allowed to lie exposed to air in small containers for two to three days. In this period the cellulose undergoes a chemical change resulting in a shortening of the long molecules so that the final solution has a more desirable viscosity. When this shortening has proceeded to the desired extent the alkali cellulose is churned with carbon disulphide. The crumbs of soda-cellulose soon become yellow and then reddish orange. When this formation of the cellulose xanthate is complete the whole is dissolved in a dilute solution of caustic soda, adjustments being made so that the resulting viscous solution (generally known as *viscose*) contains 7 to 8 per cent of cellulose (as cellulose xanthate) and 6 to 7 per cent of caustic soda.

This freshly prepared viscose is not yet suitable for spinning since certain chemical changes are still taking place within it and it is only when these processes have reached an optimum stage that the best yarn properties are obtained. So it is allowed to stand at room temperature for three to four days until tests indicate that this stage has been reached. During this period the viscose is filtered and de-aerated since both suspended solids and occluded air bubbles would seriously interfere with spinning by blocking the orifices in the spinnerets.

The viscose is then forced by a metering pump (the forward flow of viscose must be maintained at a uniform rate to ensure the formation of filaments of a uniform thickness) through a small 'candle filter' and then through a spinneret which is immersed in the coagulating bath containing mainly sodium sulphate and sulphuric acid together with a small proportion of coagulation assistants such as zinc sulphate and glucose. The resulting solidified filaments are brought together in the form of yarn by passing through an 'eye', the coagulation being so advanced at this stage that the individual fibres do not stick together. The yarn is led upwards over a Godet wheel and it then drops downward to be collected in a rapidly rotating Topham centrifugal pot. This travel of the freshly formed yarn enables it to be stretched to a small degree. This stretching is found to be beneficial to the ultimate properties of the rayon more especially fibre strength and extensibility and elasticity. Inside the centrifugal pot the yarn is built up into a compact 'cake' and at the same time a small amount of twist is given to it.

It should be pointed out at this stage that the spinneret through which the viscose is forced has not one hole, but a number. The denier or thickness of the filaments depends primarily upon the rate at which the viscose is pumped through the spinneret and the surface speed of the rotating Godets, and the diameter of the number of holes in the spinneret.

The coagulating liquor is primarily designed to solidify the incoming viscose streams and to decompose them so that the cellulose xanthate is changed to regenerated cellulose. Sulphuric acid and sodium sulphate achieve this purpose, but other substances such as zinc sulphate and glucose are added for the purpose of modifying the character of the resulting filaments. Thus zinc sulphate gives the coagulating filaments useful ductility and an indented cross-section and glucose gives them greater softness and suppleness. A filament with an indented surface is more receptive to dyes. Surface

34

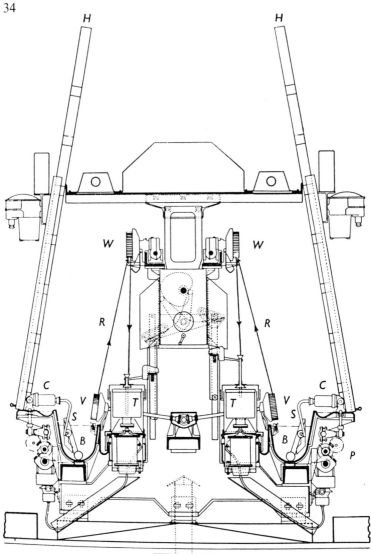

Figure 1.18 Sectional view of double-sided viscose rayon spinning frame. The viscose solution is pumped by pump P through candle filter C and spinneret S immersed in coagulating bath B. The resulting individual rayon filaments are brought together as a thread R which passes upwards round Godet wheels V and W. W is driven faster than V so that the thread R is stretched to increase its strength. The thread falls into a Topham pot T where it is cross-wound into cake form. H and H are sliding glass windows. (Courtesy Dobson & Barlow Ltd)

active agents are also added to promote break-down and dispersion of any coarsely precipitated substances which could encrust the orifices and cause fibre breakage or specky distortion.

Figure 1.19 Lacing up a box-type machine producing viscose continuous filament textile yarn showing the freshly coagulated viscose rayon thread passing around lower and upper Godet wheels thus being controllably stretched to give increased fibre strength.
(Courtesy Courtaulds Ltd)

Many properties of the viscose rayon, such as lustre, strength, softness and affinity for dyes, are influenced by the degree of stretch to which the freshly formed yarn is subjected in travelling from the spinneret to the centrifugal pot. So it is essential to maintain the conditions employed as constant as possible. Considerable trouble might occur in dyeing and finishing if in the same fabric there were present more than one quality of viscose rayon arising from differences in the manufacturing operations as just described.

Purification The freshly formed rayon in the centrifugal pot is impure and somewhat weak, so the next step is to pass it through a purification process. This involves thorough washing, treatment with a dilute solution of sodium sulphide to remove sulphur impurities, bleaching to remove a slight yellowness and secure a good white colour, and a final washing.

In the early days of viscose rayon manufacture it was a general practice to carry out this purification on the rayon after it had first

36

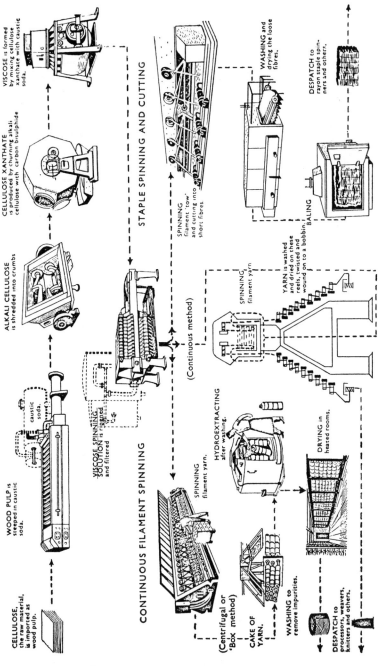

VISCOSE is formed by mixing cellulose xanthate with caustic soda.

CELLULOSE XANTHATE is produced by churning alkali cellulose with carbon bisulphide.

ALKALI CELLULOSE is shredded into crumbs.

WOOD PULP is steeped in caustic soda.

caustic soda.

CELLULOSE, the raw material, is imported as wood pulp.

STAPLE SPINNING AND CUTTING

SPINNING filament 'tow' and cutting into short fibres.

WASHING and drying the loose fibres.

DESPATCH to rayon staple spinners and others.

BALING

CONTINUOUS FILAMENT SPINNING

VISCOSE SPINNING SOLUTION is ripened and filtered

(Continuous method)

SPINNING filament yarn.

YARN is washed and dried on these reels, twisted and wound on to a bobbin.

SPINNING filament yarn.

HYDROEXTRACTING after washing.

DRYING in heated rooms.

(Centrifugal or 'Box method)

CAKE OF YARN.

WASHING to remove impurities.

DESPATCH to processors, weavers, knitters and others.

Figure 1.20 Flow sheet for the manufacture of continuous filament and staple viscose rayon. (Courtesy Courtaulds Ltd)

been wound into skein form. In later years, however, with the design of suitable machinery, progress has been made in a method by which the rayon is treated in the 'cake' form. It is obviously desirable to use the latter method since it avoids all winding processes in which entanglement and damage to the rayon yarn can take place. The main difficulty encountered in dealing with the rayon in cake form is that of ensuring that it is uniformly processed and dried. The highly swollen filaments resist the passage of liquids and the outside of the cake is liable to be better purified than the inside and to dry before the inside layers.

Finally the rayon yarn in 'cake' form is hydro-extracted to remove surplus moisture and then dried in hot-air rooms.

Continuous methods of viscose rayon manufacture   The above is a description of the essential features of the manufacture of ordinary viscose rayon, but there are several variations either to secure certain manipulative advantages or to produce special types of rayon. More economical continuous (one-run) processes for manufacturing viscose rayon (regenerated cellulose) fibres are now in general use (Figure 1.20).

In the modern method of rayon production by continuous spinning, the freshly formed thread is led downwards over a number of pairs of reels arranged in tiers towards the final drying reels; the thread is treated with washing, purifying and fibre-lubricating liquors as it passes over these reels. This use of a number of pairs of thread-advancing reels instead of the previously used multi-stage methods results in a more completely desulphurised and bleached yarn.

In recent years it has been found practicable to make viscose rayon manufacture more economical by also carrying out continuously the preparation and ageing of the alkali-cellulose, the conversion of this to cellulose xanthate and the preparation of the spinning solution from it.

Special treatments   For certain classes of goods the public has shown a preference for rayon having a subdued lustre or even a full matt appearance. In early days bright lustre rayon was favoured. To reduce the lustre of the viscose rayon the most satisfactory method is that of adding a small proportion of opaque pigment (usually titanium dioxide) to the viscose solution before spinning it into filaments. Thus the pigment particles are trapped permanently within each filament and the loss of lustre is roughly proportional to the amount of titanium dioxide pigment added. This particular pigment is used because it is highly resistant to all chemical influences to which rayon is likely to be exposed.

A modification of the manufacture of viscose rayon is to add certain fast colouring matters to the viscose solution before spinning so that coloured rayon is thus produced directly. Weavers and knitters of rayon goods can thus produce coloured fabrics without having to send them to the dyer for colouring after manufacture. Obviously there are limitations to this idea (it involves the holding of much stock of coloured yarn by the textile manufacturer), but rayon manufacturers are finding it acceptable to the trade to produce a proportion of their output in about twenty or more standard shades.

Viscose rayon is colder in handle than wool and so many efforts have been made to change this. One method (used with rayon staple fibre) is that of ensuring the presence of a large number of tiny air bubbles evenly distributed throughout the length of each filament. Air being a bad conductor of heat, these bubbles do make the rayon somewhat more wool-like. At the same time the rayon filaments become less dense and so the rayon yarn gives more 'cover' when woven into fabric.

Another method is to add a protein such as casein to the viscose spinning solution since the casein is a poor heat conductor. Neither of the above methods is much used.

As we shall see later, a fair amount of viscose rayon staple is used in the manufacture of yarns and fabrics consisting of a mixture of wool and rayon. Now these two fibres have very different dyeing properties and to secure a good dyeing of the mixture material it is often necessary to dye first one fibre and then the other. This makes dyeing expensive and complicated, so attempts have been made to add substances to the viscose spinning solution so that the presence of these in the resulting rayon gives it a special affinity for wool dyes. Another, and at the moment apparently better, method is to impregnate the fully formed viscose rayon with these substances. At least one of these special types of viscose rayon contains a small proportion of a synthetic resin to enable it to be dyed with wool dyes.

It is interesting to note that the presence of a certain proportion of viscose rayon staple in wool materials can cause them to felt more rapidly and to a greater degree. Synthetic fibre carpets can be made anti-static by evenly distributing in them 3 to 5 per cent of viscose rayon threads.

Special types of viscose rayon

As ordinarily manufactured viscose rayon has some disadvantages, e.g. the fibres lose about 30 per cent of their dry strength and also much of their resistance to shearing forces (threads and fabrics can much more easily cut when wet than dry), but these losses are completely recovered by drying.

Other disadvantages are that the fibres may have excessive lustre and lack of resistance to wear abrasion. Such disadvantages can be largely counteracted in the later stages of finishing to which fabrics and garments are subjected and which will be described later. Strength and durability deficiences can be largely corrected during fibre manufacture by stretching and other treatments.

Fine filament rayon   Softness of handle is a very desirable characteristic to impart to rayon materials and this aspect of rayon manufacture has in the past received much attention. It was soon recognised that greater softness was obtained as the rayon filaments were made finer, but manufacturing conditions set a limit to the fineness which can thus be obtained. To produce fine filaments, the holes in the spinnerets must be made smaller. These are more easily blocked by impurities in the viscose solution. Such blockages occasion great delay and reduce output considerably. So it is generally found that by simple spinning methods filaments of less than 1 denier cannot be satisfactorily produced.

High-tenacity rayon   Apart from producing a finer filament rayon the use of stretching treatments is most important for they can be the basis of methods for manufacturing very strong rayons. By highly stretching rayon fibres, at the time of their formation or during the process of their coagulation, it has become possible to produce fibres having a very high tensile strength.

Of all the various artificial fibres now being manufactured, viscose rayon is the one which is most difficult to stretch to a high degree in the course of its manufacture. It is very much easier to stretch, say, acetate fibres 1 000 per cent, nylon 250 per cent, or Orlon 300 per cent, than it is to stretch viscose rayon threads 50 per cent. Yet when viscose rayon threads became important for reinforcing motor tyres and it became necessary to make them as strong as possible, the best way found for achieving this was to use some form of stretching during manufacture. Nylon and polyester fibres have now largely replaced viscose rayon threads (*cords*).

In spinning viscose rayon there are two distinct but overlapping stages in the conversion of the cellulose xanthate to regenerated cellulose during coagulation. As the viscose solution is extruded through the holes in the spinneret so that fine streams of fluid viscose enter the acid coagulating bath, there is an immediate formation of solidified cellulose xanthate which then changes by acid decomposition to regenerated cellulose. In the normal production of viscose rayon fibres these two stages occur almost simultaneously but, by increasing the proportion of zinc sulphate and reducing the amount of acid in the coagulation bath, the two stages can be partially separated. The method of spinning viscose rayon using two coagulating baths containing suitable concentrations of salt and acid coagulants has in recent years been adapted to yield a strong stretched rayon. The success of such a process depends on the ductility of the freshly formed filaments during the stretching. Processes for producing high-tenacity viscose rayons usually, therefore, effect the necessary stretching while the freshly formed fibres are in an intermediate state between cellulose xanthate and regenerated cellulose. In other words, a two-stage or slow regeneration process is used. This type of spinning is used for the production of Tenasco (Courtaulds Ltd) which has a tensile strength nearly double that of ordinary viscose rayon.

An alternative spinning method is used for the production of Durafil high-tenacity rayon (Courtaulds Ltd) which has nearly three times the tensile strength of ordinary viscose rayon. In this method the viscose is extruded into a coagulating bath containing a high concentration of sulphuric acid so that the filaments are to some extent gelatinised or parchmentised. In this state it is possible to stretch the fibres to at least twice their original length.

Examination of these high-tenacity viscose rayon fibres has shown that they differ from ordinary fibres in possessing a large proportion of a tough, highly dense skin—it is now easy to ascertain the extent of this skin by use of X-ray and dye-staining methods and also by observation of the manner in which light is transmitted through the fibres. It is now recognised that fibre characteristics such as strength, elasticity, and extensibility are largely governed by the nature of this skin and that when the fibres are required for the purpose (in thread form) of reinforcing motor tyres it can be an advantage for each fibre to be almost all skin. Special manufacturing methods have been devised to produce viscose rayon fibres of this type.

All types of viscose rayon have a skin which arises from the outside of each fibre being coagulated in the spinning stage more rapidly than the inside and thus becoming more highly stretched. The thickness of the skin can thus be controlled by adjustment of the coagulation rate and degree of stretching.

In the skin of a viscose rayon fibre the long cellulose molecules are to a high degree aligned parallel to the length of the fibre and this parallelism allows the same molecules to lie closer together, that is, so that this skin cellulose is denser than the non-skin parts where the cellulose molecules are more randomly distributed and less capable of close packing.

Polynosic rayons   It has recently been recognised that in addition to the advantage that viscose rayon can be produced regular in quality and physical form (a handful of cotton, wool, etc. will contain a mixture of fibres of much varying length and thickness while a handful of comparable rayon staple fibres will have a uniform thickness and can have been all cut to any one required fibre length) it would be valuable for these fibres to be produced having something more of the toughness and resistance to chemical attack possessed by cotton. Following Japanese initiative, modifications of the ordinary viscose process have now been devised so that so-called polynosic fibres can be produced. Vincel (Courtaulds Ltd), American Zantrel and British Vincel are typical polynosic fibres.

Polynosic fibres are made from cotton by the viscose process but under modified conditions such that the original properties of the cotton are better preserved. They are made from cotton and not wood pulp because this latter cellulose raw material yields somewhat inferior ordinary viscose rayon to that obtainable from cotton.

It has been mentioned earlier (p. 5) that each cotton fibre is composed of long tenuous molecules largely aligned parallel to each other and to the length of the fibre and having a spiral twist. The high strength and toughness of such a fibre is dependent on this special structure and any treatment which disturbs it reduces these valuable properties and is generally termed *fibre degradation*. Thus any treatment of cotton which breaks the long molecules into shorter ones and/or which reduces their parallelism and closeness of packing degrades the fibres.

In manufacturing viscose rayon it is necessary to disintegrate the cotton used and separate the fibre molecules in order to prepare a viscose solution which can then be spun into fibres. The coagulation of the viscose while passing through the spinneret orifices into the acid coagulating liquor is a stage in which the separated molecules are brought together. In order that the resulting fibres may have maximum strength and toughness it is important that the molecules should become as closely packed and parallel as possible in the resulting rayon fibres. Stretching during coagulation can assist this.

It will be recalled that in preparing a viscose solution it is necessary to prepare alkali-cellulose and that this must be aged (an oxidation process) to aid its dissolution in caustic soda solution and in order that the resulting viscose solution shall have ultimately a viscosity which allows a satisfactory smooth flow of viscose solution through the spinnerets. A somewhat similar ageing of the freshly prepared viscose solution is also allowed for a similar purpose.

In these stages the alkali cellulose and cellulose xanthate molecules become shortened by a splitting process so that, although necessary for satisfactory spinning of the rayon fibres, this shortening does have the harmful effect of causing these fibres to consist of regenerated cellulose molecules much shorter than those in the cotton used as a raw cellulose material. This is unfortunate since the shorter, less closely packed cellulose molecules in the rayon fibres are responsible for the lower dry- and particularly the lower wet-strength than the cotton fibres.

The early Japanese research resulted in discovering how cellulose xanthate and the viscose solution made from it could be carried out with a reduced degradation of the cellulose cotton and, in effect, ensure the production of rayon fibres containing more closely packed and longer cellulose regenerated molecules and thus fibres desirably more akin to the original cotton fibres. Polynosic viscose rayon fibres can thus be regarded as having fewer of the disadvantages of ordinary viscose rayon and more of the advantages of cotton. In particular they are stronger both dry and wet, more durable and resistant to abrasion, less subject to easy stretching in both their dry and wet states, and less subject to swelling in alkaline solutions. Because of this last characteristic polynosic rayon fibres can withstand mercerising treatments as widely used for conferring high permanent lustre on cotton yarns and fabrics.

Because polynosic fibres swell and stretch less when wet, garments made from them are more dimensionally stable in laundering than garments made from ordinary viscose rayon. It is claimed that the presence of polynosic fibres in cotton fabric allows this to be given a wash-and-wear resin finish with less than the usual loss of strength and abrasion resistance.

Vincel (Courtaulds Ltd) and Zantrel (Du Pont Co) among the many brands of polynosic fibres available are produced in standard and high strength types. All of them in wetting pick up and retain less water than does ordinary viscose rayon while their moisture content (regain) in air is also considerably less than the 12 per cent for ordinary rayon.

Polynosic fibres can be regarded as being the man-made equivalent of mercerised cotton.

The close packing of the regenerated cellulose molecules in polynosic viscose rayon is most satisfactorily obtained by extreme stretching of the freshly coagulated fibres. Such stretching demands a high degree of plasticity in the fibres. In spinning ordinary viscose rayon a sufficient but lower degree of plasticity is obtained by having a small proportion of zinc sulphate in the coagulating liquor which essentially contains sulphuric acid and sodium sulphate. But if the higher proportion of zinc sulphate required to give the higher plasticity for producing polynosic fibres is present in the usual coagulation bath, irregular coagulation occurs. Hence, the expedient of having no zinc sulphate in a first cold coagulation liquor and then leading the partially coagulated fibres through a hot second liquor containing sulphuric acid and sodium sulphate and a high proportion of zinc sulphate (the relative high proportion of zinc xanthate now formed in the fibres makes them very plastic) so that on being led through a third hot liquor containing sodium sulphate and sulphuric acid only, they can be stretched 200 per cent with complete regeneration of their cellulose content. This gives them the desired improved properties expected in polynosic viscose rayon fibres.

Thus the resulting fibres can have dry and wet strengths of 4·9g and 3·5g/denier as compared with 2·5g and 1·6g/denier for normal viscose rayon. As might be expected these higher strengths are accompanied by lower extensibilities.

Darelle This recently introduced type of viscose rayon has in-built flameproof properties arising from bromine-containing compounds such as tris(2:3-dibromopropyl)-phosphate introduced into the viscose spinning solution. It is recommended for blending with wool fibres.

Cellulose acetate fibres

Although cotton is used as the raw material for making cellulose acetate fibres (usually briefly designated as *acetate* fibres) this type of fibre consists of cellulose chemically combined with acetic acid. Such a composition gives them properties uniquely different from those of viscose rayon which consists of regenerated cellulose.

Even before the discovery by Cross, Bevan and Beadle of viscose rayon it was known that cellulose acetate had potential fibre-forming properties. But owing to partial degradation of the cellulose unavoidably occuring in making this cellulose acetate, the fibres made from it were too weak to be acceptable.

However, during World War I cellulose acetate dope (a solution of cellulose acetate in organic solvent) was found to be the most satisfactory product for coating aeroplane wings, making them taught and weatherproof. To make this necessary product Drs Henri and Camille Dreyfus were engaged by the British Government to erect a cellulose acetate plant in the U.K. at Spondon. These Swiss chemists were chosen because they had previously discovered important new facts concerning the manufacture of cellulose acetate which enabled them to produce it in better quality.

During World War I the Spondon works manufactured huge amounts of cellulose acetate and when hostilities ceased there arose the question as to whether the product could be used in other directions than for aircraft. This was solved by turning it into acetate fibres, a very difficult problem at the time which, mainly through the perseverance of the Drs Dreyfus, was solved and a valuable new type of man-made fibre made available to the textile industry.

Cellulose di- and tri-acetate

Both di- and tri-acetate fibres are now being manufactured but in the early Spondon days attention was given solely to the di-acetate type since this was less difficult to manufacture and was more amenable (though nevertheless difficult) to dyeing with the dyes then available. In America a small production of tri-acetate fibres was achieved but because of being more difficult and costly to produce and also very difficult to dye it was, for a time, abandoned.

Both types are made by dry spinning an organic volatile solvent solution of one or the other cellulose acetates through a multi-hole spinneret into warm air which immediately evaporates the solvent to leave residual solid fibres. The di-acetate is soluble in relatively cheap acetone but the tri-acetate requires expensive chloroform or methylene chloride to dissolve it. This difference in solubility largely brought about the decision to make di-acetate fibres at Spondon and this decision was supported by the di-acetate fibres proving to be the more hydrophile type of fibre and to be the less difficult to dye with dyes then in common use for colouring the strongly hydrophile natural fibres such as cotton and wool.

Cellulose acetate manufacture

When cotton cellulose is reacted with a mixture of acetic anhydride and acetic acid and a relatively small proportion of sulphuric acid is present as catalyst, the reaction proceeds progressively with the formation of the mono-acetate and then the di-acetate and finally to the tri-acetate as indicated below. In this reaction the hydroxyl (—OH) groups of the cellulose molecules $[C_6H_7O_2(OH)_3]$ take part.

$$C_6H_7O_2(OH)_3 + 3CH_3COOH \rightarrow C_6H_7O_2(OH)_2(OCOCH_3)$$

Cellulose mono-acetate

$$\downarrow$$

$$C_6H_7O_2(OCOCH_3)_3 \leftarrow C_6H_7O_2(OH)(OCOCH_3)_2$$

Cellulose tri-acetate	Cellulose di-acetate
(primary)	(secondary)
Acetyl content	Acetyl content
61 per cent	53 per cent

Although it is usual to refer to the cellulose acetate employed for spinning the lower acetate type of fibre as being cellulose di-acetate, it actually is slightly more acetylated and as made commercially it has a composition which agrees with the formula $C_6H_7O_2(OH)_{0.667}(OCOCH_3)_{2.333}$ having an acetyl content of 53 to 54 per cent.

Another interesting feature concerning the formation of this particular di-acetate is that instead of arresting the acetylation re-action, when this particular acetyl content is reached, the reaction is continued up to the formation of the tri-acetate. The reaction mixture is then suitably diluted by addition of a small proportion of water and allowed to stand for several hours whereby hydrolysis of the tri-acetate takes place with a splitting-off of an amount of acetic acid and the desired commercial di-acetate is formed. By this hydrolysis the tri-acetate loses its solubility in chloroform and in-solubility in acetone to become easily soluble in acetone.

Di-acetate fibre manufacture

In the manufacturing process, cotton cellulose is kneaded with a mixture of acetic acid, acetic anhydride and a catalyst (usually sulphuric acid). The catalyst promotes reaction between the cellulose and the acetic acid but itself only enters into the reaction to a small extent. As a result of this reaction the cotton dissolves in the reaction mixture and it is then said to be fully acetylated. At this stage three molecules of acetic acid have combined with one molecule of cellulose to form cellulose tri-acetate. It is now usual to assist the acetylation process by first activating the cotton cellulose. This is achieved by impregnating the cotton (usually cotton linters) with a suitable amount of acetic acid, which may also contain a small proportion of sulphuric acid, and then allowing the product to lie at room temperature for a few hours. After this treatment the cotton reacts more easily when afterwards acetylated as described above. By such activation it has proved possible to effect economies in the

PRODUCTION OF ACETATE FLAKE

RAW MATERIALS (cellulose, derived from woodpulp and cotton linters, i.e. fibres of cotton too short to be spun; acetic acid and acetic anhydride, derived from petroleum)

ACETYLATION (in the presence of a catalyst and a solvent the cellulose and acetic anhydride react to give a viscous liquid—a solution of cellulose triacetate)

RIPENING (in acetic acid to produce a solution of cellulose diacetate)

DRYING (the flake is washed and dried before storing in bins or bags. Diacetate can be used to prepare film or plastic articles in addition to being spun into yarn)

CONVERSION OF FLAKE TO FIBRE

DISSOLVING of flake in acetone to form a spinning solution or 'dope'

FILTRATION of dope to remove any undissolved solids which would block the spinning jets

SPINNING. After filtration the dope is extruded by pressure through the fine holes of the spinning jet

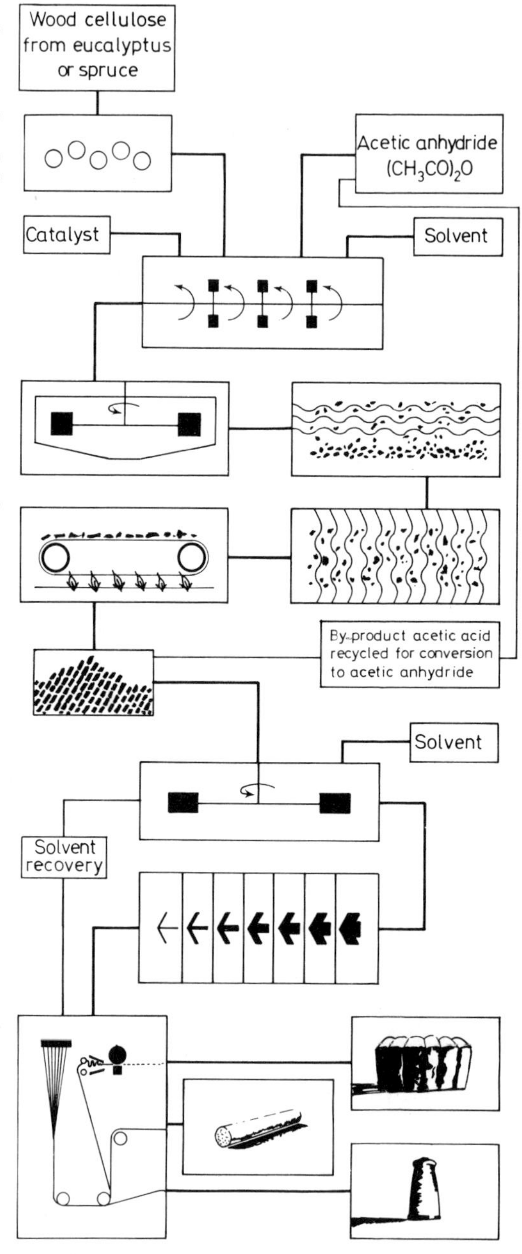

Figure 1.21 Essential manufacturing features of diacetate fibre. (Courtesy Courtaulds, Ltd)

manufacture of cellulose tri-acetate and also ensure that a higher quality of the cellulose di-acetate may ultimately be made from it.

Now follows a so-called hydrolysis treatment to change this cellulose tri-acetate into the corresponding secondary cellulose di-acetate. To the reaction mixture is now added a suitable amount of water and the mixture allowed to stand for a period during which one-half molecule of acetic acid is split off to produce the cellulose di-acetate in which two and a half molecules of acetic acid are combined with one molecule of cellulose. This particular cellulose acetate is insoluble in chloroform but readily soluble in acetone, a solvent which is comparatively cheap and has not the toxic properties of chloroform.

The reaction mixture is now diluted with water so that the cellulose acetate is precipitated in a finely divided fibrous form. After being thoroughly washed free from acetic acid and the by-products of the reaction, the fibrous product is boiled for a short time with a dilute solution of sulphuric acid. This splits off the small amount of sulphuric acid which has remained from the acetylation stage chemically combined with the cellulose acetate and which would ultimately cause discolouration and embrittlement of acetate rayon fibres made from it. Then the stabilised cellulose acetate is dried and dissolved in acetone to give a 25 per cent solution. This solution has just about the right viscosity to enable it to be spun into rayon filaments.

We have seen that in the case of viscose rayon the cellulose solution is spun into an aqueous coagulating liquor. It is quite different with acetate fibres. Acetone is a very volatile liquid and readily evaporates in warm air, so the spinning operation can be made very simple. It consists of forcing the cellulose acetate solution through spinnerets within small chambers or cells through which warm air is circulated. As the acetone evaporates, solidified filaments of cellulose acetate are left and these are collected through an 'eye' to form a thread or yarn which is then wound on to bobbins at the lower end of the cell. This type of spinning is usually known as 'dry spinning'. Moreover, since the cellulose acetate is so well purified before it is dissolved in the acetone to form a spinning solution, the resulting filaments are equally pure and require no further treatment except for grading according to quality. However, both acetic acid and acetone are sufficiently valuable to warrant the installation of recovery plant so that there is a minimum loss of these chemicals.

Properties of di-acetate   When acetate fibres were first introduced to textile manufacturers, considerable difficulty was experienced in dyeing materials made from them. Most of the dyes commonly used for cotton, wool and silk had but poor, if any, affinity for acetate fibres. Eventually new types of dyes (now known as *disperse* dyes) were discovered and so this dyeing problem was solved. In the intervening period, however, the difficulty of dyeing was indifferently solved by treating the acetate goods with alkalis for the purpose of changing the cellulose acetate to regenerated cellulose by splitting off the acetic acid residues. This involved a loss of weight, but after the treatment the fibres (now a mixture of cellulose acetate and regenerated cellulose) could be dyed with the usual cotton dyes. This conversion of cellulose acetate to regenerated cellulose fibres by alkali treatment is generally known as *saponification*. The process fell into disrepute as soon as the new acetate dyes became available.

When acetate fibres are strongly swelled say by treatment with an organic solvent such as acetone or acetic acid they can absorb acid dyes and give colourings fast to washing provided that after dyeing the fibres are collapsed by washing out the swelling agent.

Acetate fibres are much more plastic than those of regenerated cellulose and it was soon found that the fully formed yarns could be stretched up to ten times their original length under the influence of heat and softening agents. Stretching is easily carried out whilst the yarn is run through a steaming chamber. In this way very fine strong yarn can be produced but it suffers to some degree by having a low extensibility. But now a saponifying treatment can be utilised advantageously. If the yarn is stretched and also saponified so that practically the whole of the cellulose acetate is converted to regenerated cellulose, then an exceedingly strong and fine yarn is produced. Such a yarn is being used under the name Fortisan, and it is two or three times as strong as ordinary real silk and is perhaps the strongest textile fibre so far produced. During World War II it was largely used for the manufacture of parachute fabric, a sufficient testimonial to its high strength and durability. It is a matter of great interest that this roundabout method for producing a regenerated cellulose rayon should prove so useful.

Cellulose can be made to combine with acids other than acetic acid, for example, formic, butyric and palmitic acids. So far none of these alternative cellulose derivatives (termed cellulose *esters*) has proved so attractive as cellulose acetate for the manufacture of rayon.

Low lustre acetate fibres   Acetate fibres have a unique property of which much use is made. When material consisting of this fibre is boiled in water it rapidly loses its lustre. This effect is more pronounced in a soap liquor, or even more still in a soap liquor containing a softening or swelling agent for the rayon such as phenol. The loss of lustre can be controlled and by carrying the reaction to its limit a full matt appearance can be secured.

This loss of lustre can be prevented by holding the fibres stretched or by having a considerable amount of common salt or similar substance in the boiling liquor. Also, unless the liquor is quite near to the boil the degree of delustring is very low. Low lustre fibres produced in this way can also have their original lustre more or less restored by hot ironing, or by dyeing in a deep shade such as black or navy blue, or by stretching, especially with the aid of swelling agents. For this reason the titanium dioxide method (as described for viscose rayon) is generally preferred. The white opaque pigment is added to the cellulose acetate spinning solution so that the resulting fibres have just the degree of lustre desired. This delustring is permanent but, as with viscose rayon, it detracts slightly from the strength of the rayon.

Distinctive properties of di-acetate fibres The properties of acetate fibres are in general quite different from those of viscose rayon. Two important differences must be mentioned. Firstly, that acetate is thermoplastic and secondly that it is soluble (or at least, swellable) in many organic solvents. Cotton, linen, wool, silk and the viscose cuprammonium rayons are almost inert in these directions.

Acetate fibres are easily dissolved in acetone, acetic acid and other organic solvents. Fortunately they are not affected by petroleum products such as petrol, white spirit and the like, allowing them to be dry cleaned. Actually this solubility is not a serious disadvantage and may, in fact, be considered an advantage, for it enables acetate fibres to be employed in the 'Trubenised' process for the production of stiffened and laminated fabrics suitable for collars, cuffs and so on. This will be dealt with later (page 401).

When heated to about 210°C acetate fibres soften and, if the heating is continued, the filaments fuse together to form a solid brittle thread. So, in ironing acetate rayon materials, care must be taken to avoid the use of too hot an iron. As will be seen later, some of the newer synthetic fibres are even more liable to softening. At first this was considered a serious defect but today the public are sufficiently aware of it not to make the mistake of spoiling acetate goods by ironing too hot.

Acetate fibres can have their safe ironing temperature raised about 20°C by partial saponification, say by treating them with a boiling solution of sodium carbonate. This treatment hydrolyses the fibres so as to split off some of the acetyl groups (only a small proportion is necessary) and bring the fibres nearer to the regenerated cellulose type of fibre.

Owing to the fact that acetate fibres are composed of cellulose acetate they are less creasable than regenerated cellulose fibres and they handle warmer. It has been claimed in the past that acetate fibres are more wool-like and even more silk-like than other fibres. In fact the structure of a woven or knitted fabric can play a more important part than the composition of a fibre in deciding how closely it resembles wool and silk in warmth and handle.

Acetate fibres absorb considerably less moisture from the air than other textile fibres. Under ordinary circumstances their moisture content is about 4 per cent. Probably associated with this low moisture absorption is the fact that under the influence of heat and moisture acetate fibre materials can develop permanent creases during dyeing operations. This has given much trouble and will be referred to later. Viscose and cuprammonium rayons are not so troublesome in this direction in spite of the fact that under ordinary dry conditions they are more creasable than acetate fibres.

It may be pointed out that cellulose acetate is not only used for the production of textile fibres. It is a most important plastic and as such is widely used in the plastics industry for moulded articles. These uses all depend on the ease with which cellulose acetate can be pressed to any shape whilst held in a soft plastic condition by suitable heating. This thermoplastic behaviour of cellulose acetate makes it possible to emboss acetate rayon materials with pleasing patterns, during finishing and to produce washfast pleats in fabrics and garments. This holds especially for the tri-acetate.

Cellulose tri-acetate fibres

For the manufacture of this type of fibre cellulose tri-acetate is employed and this is made by the reaction of cellulose with acetic anhydride and acetic acid in the presence of sulphuric acid (a catalyst) as a first stage in the manufacture of secondary cellulose acetate as previously described (p. 45). But the required cellulose tri-acetate can also be made by acetylating the cellulose in an organic solvent medium so that the cellulose retains its fibrous nature although changed completely into cellulose tri-acetate. However this cellulose tri-acetate is made, it is soluble in chloroform but not in acetone so that cellulose tri-acetate fibres cannot be dry-spun from an acetone solution – instead it is usual to make use of a spinning solution prepared with methylene chloride containing a small proportion of alcohol.

At this point it will be useful to compare the properties of these two types of acetate fibres—di-acetate fibres typified by Dicel and cellulose tri-acetate fibres typified by Tricel (both products of British Celanese Ltd)—and to suggest reasons for manufacturing two such closely related fibres. To aid this it may be recalled that most of the properties of fibres which make them useful in the manufacture of textile materials depend on the way in which such fibres behave towards water and heat. The natural fibres such as cotton, wool, linen, silk, etc. and also the regenerated cellulose fibres typified by viscose rayon are all characterised in being *hydrophilic*, that is, they readily absorb water, and in doing so each fibre swells 20 to 40 per cent in diameter. In sharp contrast, the synthetic fibres represented by nylon, Orlon, Terylene, Dynel, etc. absorb very little, if any, water during wetting and thus do not swell when wetted so that they are termed *hydrophobic*. Further, hydrophilic fibres do not readily melt or soften at high temperatures and are thus not thermoplastic, whereas hydrophobic fibres do soften at high and sometimes at relatively low temperatures, thus being thermoplastic; the hydrophilic fibres are usually insoluble in organic solvents whereas the hydrophobic fibres are often soluble.

From these considerations it may be inferred that textile fibres are at present classified into two main groups according to whether they are hydrophilic or hydrophobic, and each group has distinctive properties which govern the types of textile material for which they are most suitable and also the characteristics of such materials. In general it may be stated that fabrics and garments made from hydrophilic fibres can be readily dyed with most of the water-soluble dyes which have over the past century been especially developed for the colouring of the natural fibres and that they are not made dimensionally unstable (liable to shrink and distort) when exposed to high temperatures; not being thermoplastic they are not amenable to heat-setting treatments for producing in them permanent pleats and special shapes. Fabrics and garments made from hydrophobic fibres have opposite properties for they are not easy to dye using ordinary dyes and conventional dyeing processes and equipment, while they are heat sensitive so as to distort and shrink badly at high temperatures. However, they have the advantage of being amenable to permanent shaping and pleating.

Experience with secondary acetate fibre materials has shown that these have properties lying about mid-way between those of hydrophilic and hydrophobic fibres (that is, between the natural and the synthetic fibres) and as such are able to satisfy many textile fibre requirements. But the gap between secondary acetate and synthetic fibres appeared large enough to justify the introduction of another

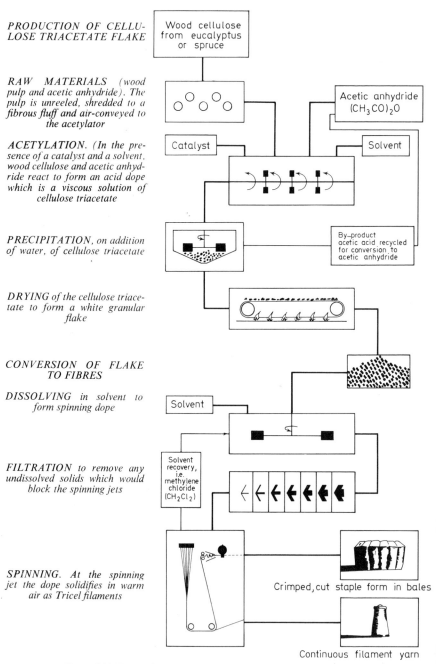

PRODUCTION OF CELLU-LOSE TRIACETATE FLAKE

Wood cellulose from eucalyptus or spruce

RAW MATERIALS (wood pulp and acetic anhydride). The pulp is unreeled, shredded to a fibrous fluff and air-conveyed to the acetylator

Acetic anhydride $(CH_3CO)_2O$

ACETYLATION. (In the presence of a catalyst and a solvent, wood cellulose and acetic anhydride react to form an acid dope which is a viscous solution of cellulose triacetate

Catalyst

Solvent

PRECIPITATION, on addition of water, of cellulose triacetate

By-product acetic acid recycled for conversion to acetic anhydride

DRYING of the cellulose triacetate to form a white granular flake

CONVERSION OF FLAKE TO FIBRES

DISSOLVING in solvent to form spinning dope

Solvent

FILTRATION to remove any undissolved solids which would block the spinning jets

Solvent recovery, i.e. methylene chloride (CH_2Cl_2)

SPINNING. At the spinning jet the dope solidifies in warm air as Tricel filaments

Crimped, cut staple form in bales

Continuous filament yarn

Figure 1.22 Essential manufactures of Tricel. (Courtesy Courtaulds, Ltd)

fibre having properties nearer to those of the synthetic fibres and so the cellulose tri-acetate fibres were introduced for this purpose. Obviously the many years of experience obtained in manufacturing secondary acetate fibres was found exceptionally useful in overcoming certain initial difficulties encountered in the manufacture of cellulose tri-acetate fibres. Experience obtained in solving the difficulties encountered in dyeing nylon has assisted devising satisfactory methods for dyeing tri-acetate fibres (mainly with disperse dyes).

One interesting feature of the tri-acetate fibres is that they are less sensitive to alkalis so that they do not so readily saponify when subjected to wet processing in liquors containing sodium carbonate, caustic soda, etc. Neither do the tri-acetate fibres delustre so readily when boiled in soap liquors. But perhaps more important is the fact that when heated to about 180°C or higher, cellulose tri-acetate fibres suffer internal structural changes so as to become more crystalline, become more amenable to heat-setting (including pleating) operations, lose some of their affinity for dyes, but if already dyed then the shade can acquire increased fastness to washing and sometimes to light also. The crease-resistance of the tri-acetate fibres is also increased by heat treatment. Tri-acetate fibres as ordinarily produced have about the same safe-ironing temperature as secondary acetate fibres (about 170 to 180°C) but this can be increased by heating them (preferably with a fibre-swelling agent present) at about 200°C or preferably by steaming them under pressure for about 5 min. Such treatment can raise the safe-ironing temperature by about 20°C. It is not effective in the same way with di-acetate fibres. Partial surface saponification can produce similar results (p. 48).

The dry heating of tri-acetate fibres can produce a movement of their molecules comparable to an annealing and produce a more stable fibre structure, and if fabric in this change is held in a distorted state, say with creases and pleats, then after cooling these are left permanently set and washfast so long as the washing temperature is lower than the annealing temperature. Heating and setting can lower the affinity of the fibres for dyes by inducing fibre crystallisation.

ALGINATE RAYON

Seaweed fibre

This type of rayon utilises the alginic acid found in seaweed as a raw material. This rayon must be considered to be a speciality fibre.

Seaweed contains a complex substance closely related to pectic acid and having a fairly large molecule. This is alginic acid which can be extracted in the form of its water-soluble sodium salt. A solution of this is very viscous and can be extruded through spinnerets in much the same manner as viscose or cellulose acetate solutions. By extruding this sodium alginate solution into a solution of calcium chloride there are formed coagulated filaments composed of calcium alginate. These can be brought together in the form of yarn.

Calcium alginate fibres are the most generally useful of seaweed rayons for, while they are not stable in alkaline soap solutions (the calcium alginate is converted into water-soluble sodium alginate) they can be treated with solutions of other metallic salts such as those of aluminium, chromium and beryllium to give the corresponding insoluble aluminium, chromium and beryllium alginate fibres. These are stable in soap solutions which are not too strongly alkaline. This property is useful, since it is generally less satisfactory to spin these last-named fibres directly by extruding a sodium alginate solution into a coagulating liquor containing the appropriate metal salt as the fibres thus formed are too highly swollen to handle without excessive breakage.

From the above it may be inferred that it is possible to manufacture a number of seaweed rayon fibres composed of alginic acid, or metallic alginates. Of these, the beryllium alginate fibres are most stable, but they cannot be generally used because beryllium is highly poisonous.

Alginate rayon fibres are quite strong in their dry state but because of their instability, particularly in alkaline soap liquors, it is unlikely that they can ever become as important as other types of rayon. Yet they can have special uses. Thus, the solubility of calcium alginate threads is utilised in the production of exceptionally light-weight wool fabrics. The calcium alginate threads are used to support the very fragile wool yarns through the weaving and other operations and are then removed by alkali treatment at a stage when the wool yarns no longer need this support. Calcium alginate threads can also be employed in linking strings of socks and imitation lace materials where it is convenient to be able to remove the alginate thread in a simple way and so separate the socks.

Metallic alginate rayon is flame-proof so that it can be used for the production of fire-proof curtains; the rayon is also stable to dry cleaning. This alginate as seaweed fibre is utilised to a small extent in the textile industry and it has other possibilities as a thickener in manufactured foods. Sodium alginate is a very important thickener for use in textile printing pastes especially those containing dyes reactive to cellulose.

THE SYNTHETIC FIBRES

As previously explained, all man-made fibres can be considered as being synthetic, but those to be dealt with here are, as it were, doubly synthetic, that is, they are made synthetically from synthetic raw materials. The rayons so far described have all been made from natural products such as cotton and wood cellulose, and animal and vegetable proteins.

The synthetic fibres are now very important and it is significant that they can be made practically independent of natural products unless we class oil and coal as such. It has been found that by using synthetic raw materials the manufacture of textile fibres can be brought under greater control and be made independent of climatic and other circumstances over which man has but little control.

The possibility for man to make synthetic fibres came after research had shown how natural fibres such as cotton, wool, linen and silk were built up with very long tenuous molecules usually aligned parallel to one another in the direction of the fibre length. It could then be seen that to make a satisfactory fibre it would be necessary to devise methods for making such long molecules and then means for bringing them together to lie side-by-side, closely locked laterally, in the form and with the dimensions of a natural fibre. The manufacture of viscose rayon fibres (p. 29) indicates partial success in this direction—only partial because nature-made long molecules of cotton cellulose are used, although bringing them together in fibre form is man's method.

In the viscose process, cotton fibres are broken down in a treatment with caustic soda and carbon disulphide. This gives a viscous alkaline liquor in which a high concentration of long molecules of cellulose xanthate are dispersed so that it is extruded into an acid coagulating liquor, thus forming a fibre in which these long cellulose xanthate molecules are simultaneously brought closely together and converted into regenerated cellulose molecules (usually considerably shorter than the natural cellulose molecules as present in the cotton fibres). The resulting highly swollen rayon fibre (the swelling is due to water present between the long cellulose molecules) then has, by stretching and drying, its molecules brought closer together and also more aligned in the length direction of the fibre, thus making it stronger and more rigid, with a structure more comparable to that of the original cotton fibres. In these latter, however, the molecules are somewhat spirally aligned, thus giving the fibres higher strength and resistance to wet swelling.

By suitable variation of the above process it is possible to make fibres of regenerated wool, silk and linen from the original natural fibres. All of these fibres are built up with very long molecules in which well-defined groups of atoms are, as such, joined end to end. For the above fibres these unit groups are represented by the following formulae:

$$(C_6H_{10}O_5)_n \quad (C_6H_{10}O_5)_n \quad (C_{42}H_{157}O_{15}N_5S)_n \quad (C_{24}H_{38}O_8N_8)_n$$
Cotton	Linen	Wool	Silk
(Cellulose)	(Cellulose)	(Keratin)	(Fibroin)

in which n can be a very large number, say 900 or more, thus making the molecules extremely long relative to their thickness. It can readily be understood how a natural fibre composed of a large number of these tenuous molecules, lying side by side and locked together laterally by physical and chemical forces can have a strength approaching that of steel wire.

Each of these long molecules has a kind of backbone throughout its length; this spine is made of some of the carbon, oxygen and nitrogen atoms suitably joined together (as shown later in Figure 2.3 for cotton and linen, wool, silk and nylon) and the remaining atoms are joined pendant or laterally to this backbone in a manner which is appropriate to the fibre composition. Figure 2.2 shows a short length of the backbone of a cellulose molecule with its laterally attached oxygen and hydrogen atoms.

From the above preliminary consideration of the nature and importance of long tenuous molecules as required for the build-up of textile fibres, it is convenient to turn to the truly synthetic long molecules which are now being used to manufacture synthetic fibres. The most important synthetic fibre is nylon (a polyamide fibre) but more recently Terylene and Dacron (similar polyester fibres) have gained nearly equal importance. The acrylic fibres, including Courtelle, Acrilan, Orlon, etc. are next in importance. The manner in which the required polyamide, polyester and acrylic long molecules are made and converted into the corresponding fibres, as described below, can be very convenient to illustrate the more interesting features of synthetic fibre production and use.

Since nylon has made the greatest appeal to the imagination of the public and is made in many countries throughout the world it will be described first.

POLYAMIDE FIBRES

Nylon

Early researches   The above-mentioned facts about long-fibre molecules became generally known in the late 1920s when Carothers was entrusted with research in the laboratories of the Du Pont de Nemours Co of America on the synthesis of long chain molecules. Carothers proved to be a brilliant chemist and he soon established a number of methods by which substances having long chain molecules could be built up in the laboratory from quite simple short chain molecules.

By 1935 Du Pont had acquired possession of much fundamental knowledge in this field of research and large numbers of substances having long chain molecules had been made. It became evident that the next stage was to select the best method and then further select the most suitable substance produced by this method as a raw material for fibre manufacture. This was done and one of the products known as No. 6.6 was selected.

The choice of No. 6.6 proved to be a good one. Very rapidly, the Du Pont Co erected a pilot plant and commenced the manufacture of a synthetic fibre which has since become known all over the world as nylon. It was a great venture and involved risk of a large amount of capital. But, as we now know, it was a venture well worth while. Nylon fibres have been made much stronger than any natural fibre and in addition these nylon fibres have other unique properties which commend them to the textile industry.

By 1939 the pilot plant had been superseded by a much larger plant and nylon yarns had been tried out on a large scale for the manufacture of ladies' hose and as bristles for tooth brushes. In each case the results were very encouraging. The widespread use of nylon for many purposes during World War II showed this fibre to be exceptionally useful so that now it is being manufactured in rapidly increasing quantities. It is now used not only for fabrics and garments but also for many industrial purposes, including ropes, parachutes, and army equipment, where its immunity to attack by mildew, bacteria, moth and dampness and other influences which rapidly deteriorate natural and rayon fibres is of great importance.

The nature of nylon Before considering the production and nature of nylon in more detail it will be useful to comment on the name. It has already been seen that nylon 6.6, the substance selected for manufacture of the nylon fibres now available, is but one chosen from a large number of substances. For this reason several types of nylon fibres are now being produced, each having modified properties to meet particular requirements, but all made from substances made according to the method elaborated and discovered by Carothers. So the term *nylon* is a generic one and refers to a class of fibres made by Carothers' methods. At the present time Nylon 6

(Caprolactam) and Nylon 11 (Rilsan) are among other nylon fibres now being made; the former is the more important. It is not possible to discuss in detail the methods devised by Carothers for making these special substances with long chain molecules but, nevertheless, it is possible to explain them in a fairly simple way.

Substances of the nylon type now being used for fibre making are composed of carbon, hydrogen, oxygen and nitrogen atoms, and in this respect resemble silk. They do not contain sulphur and thus differ from wool. The atoms may be represented by the letters C, H, O and N.

Carothers, for the purpose of nylon manufacture, selected two substances made from coal-tar products known as adipic acid and hexamethylene diamine and having the following molecular structures:

$$H_2N—CH_2—CH_2—CH_2—CH_2—CH_2—CH_2—NH_2$$
Hexamethylene diamine
$$HOOC—CH_2—CH_2—CH_2—CH_2—COOH$$
Adipic acid

The hexamethylene diamine molecule is made up of a straight backbone chain of six carbon atoms (to each of which are attached two hydrogen atoms) with a nitrogen atom (having two hydrogen atoms attached to it) at each end, and in a similar manner the adipic acid molecule consists of a straight backbone chain of six carbon atoms. The end groups of atoms in each of these chains are important. In the case of hexamethylene diamine the end groups are called amino groups (NH_2) and in adipic acid they are called carboxyl groups ($COOH$).

By heating these two substances together in equivalent proportions, one molecule of hexamethylene diamine is made to unite with one molecule of adipic acid. The end groups take part in this union and as a result one molecule of hexamethylene diamine joins end-to-end with one molecule of adipic acid as indicated below:

$$H_2N—(CH_2)_6—NH_2 \quad + \quad HOOC—(CH_2)_4—COOH \rightarrow$$
Hexamethylene diamine   Adipic acid

$$H_2N—(CH_2)_6—NH—CO—(CH_2)_4—COOH + H_2O$$
Hexamethylene adipamide   Water

Here we see the formation of a new substance which has a straight chain molecule made up of a total of two nitrogen atoms and twelve carbon atoms. The important feature of this new substance is that its molecule has an amino group at one end and a carboxyl group at the other.

Thus two molecules of this new substance on further heating are able to unite as before to form a more complex molecule with all the nitrogen and carbon atoms linked together linearly, that is, in the form of a straight chain as indicated below:

$$H_2N—(CH_2)_6—NH—CO—(CH_2)_4—CO—NH—(CH_2)_6$$
$$—NH—CO—(CH_2)_4—COOH$$

Again it is seen that by further heating further uniting can take place and in fact there appears to be no limit to the extent to which this joining end-to-end of the molecules can take place. In this way a long chain substance of the type indicated by the following formula can be formed:

$$H—[NH—(CH_2)_6—NH—CO—(CH_2)_4—CO]_n—OH$$

where n may be quite a large number.

In such a substance, which is a polymer known as polyhexamethylene adipamide, the hexamethylene diamine and the adipic acid residues are linked together by a special atomic grouping

$$—NH—CO—$$

which is known as an amide group. For this reason polymers of this particular type are known as *polyamides*, and it is this particular type of polyamide that Carothers selected as being most suitable for the manufacture of nylon.

Condensation and polymerisation In the early stages of this condensation of hexamethylene diamine with adipic acid the substances produced cannot be made into fibres, but there comes a stage when the product in its molten state can be drawn off with a rod into a coarse filament in much the same manner as treacle can be drawn off by a spoon first dipped into it. The difference between the molten nylon product and treacle is that the former solidifies on cooling whilst the treacle would remain liquid.

Union of hexamethylene diamine and adipic acid is generally known as condensation–polymerisation and the products are frequently referred to as condensation or polymerisation products. It is as well to remember these terms since they are being much used in connection with the synthesis of compounds for use in the textile and plastics industries. It is usually found that as the degree of polymerisation is increased so do the products become more and more inert to attack by chemicals or other influences. They usually also become less soluble in any solvent. If polymerisation is carried to extreme limits then the products are so hardened as to make it difficult to convert them into any shaped article or filament. So it is part of the art of making nylon substances to arrest the reaction at a suitable stage. If the nylon is required for making into textile fibres then condensation is allowed to continue until about 60 to 80 molecules each of hexamethylene diamine and adipic acid have joined together.

Now, instead of using hexamethylene diamine and adipic acid it is possible to employ other similar substances for condensation. For example, sebacic acid having the molecular formula

$$HOOC-CH_2-CH_2-CH_2-CH_2-CH_2-CH_2-CH_2-$$
$$-CH_2-COOH$$

may be used instead of adipic acid, and hexamethylene diamine can be replaced by pentamethylene diamine having the molecular formula

$$NH_2-CH_2-CH_2-CH_2-CH_2-CH_2-NH_2$$

It will be noticed that here the numbers of carbon atoms in the chains are increased and decreased respectively. Condensation of sebacic acid and pentamethylene diamine yield a slightly different nylon, but still a nylon which can be made into fibres.

Such then is the general method by which the raw material for present-day nylon fibres is made. Further, it is interesting to note that the moderately complex products are known as polyamides, whilst the products resulting in the final stages are sometimes termed super-polyamides. The term *polyamide* indicates to chemists that the constituent groups of atoms of the molecule are linked together by amide—NH·CO—groups in a certain order. Thus wool keratin is often referred to as a polyamide since it is a long chain molecule in which many groups of its constituent atoms are linked with amide groups.

Details of nylon manufacture   We can now deal with the essential features of nylon manufacture on the large scale and the production of textile filaments from it. Firstly, the equivalent proportions of hexamethylene diamine and adipic acid are mixed so as to form an initial combination, hexamethylene diammonium adipate, and this is placed within a large stainless steel cylinder provided with internal stirring gear where it is gradually heated until a first stage in the reaction is attained and quite complex molecules of nylon are formed. These molecules, however, have an insufficient degree of polymerisation. So the vessel is now completely exhausted and the heating continued. Under these new conditions, assisted by the simultaneous removal of the water formed, the reaction proceeds further until the required nylon is produced.

The molten charge of nylon is now run out on to a water-cooled revolving wheel where it at once solidifies into ribbon which is broken up into chips. In this way the nylon is obtained in small solid chips with drops of the cooling water adhering to them. It is now simply necessary to dry the chips previous to their conversion into filaments.

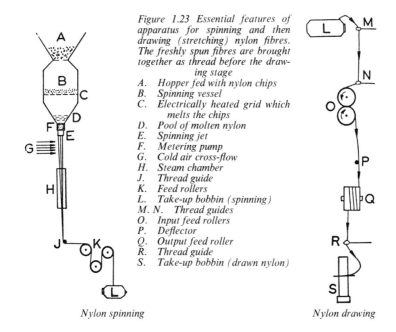

Figure 1.23 Essential features of apparatus for spinning and then drawing (stretching) nylon fibres. The freshly spun fibres are brought together as thread before the drawing stage

A. Hopper fed with nylon chips
B. Spinning vessel
C. Electrically heated grid which melts the chips
D. Pool of molten nylon
E. Spinning jet
F. Metering pump
G. Cold air cross-flow
H. Steam chamber
J. Thread guide
K. Feed rollers
L. Take-up bobbin (spinning)
M. N. Thread guides
O. Input feed rollers
P. Deflector
Q. Output feed roller
R. Thread guide
S. Take-up bobbin (drawn nylon)

Nylon spinning   Nylon drawing

It is interesting to note that a new feature in fibre manufacture is involved in the manufacture of nylon filaments. Hitherto it has been the practice to spin rayon filaments from solutions of the materials of which they are made. Thus viscose is spun from a solution of cellulose xanthate in dilute caustic soda solution, whilst acetate rayon is made from a solution of cellulose acetate in acetone. With nylon, however, the filaments are made by extruding the nylon in its molten condition through spinnerets. The streams of nylon which emerge from the holes in the spinnerets at once solidify as they cool and so form filaments. The spinning of the nylon filaments is thus reduced to its simplest terms as there is no use made of a solvent. As may be imagined, the Du Pont Co had to develop this special technique and the apparatus for it. Many difficulties were encountered but now they have been largely surmounted.

Use of stretching Now comes another new and interesting feature. The filaments and yarns thus produced are not completely satisfactory for use in textile manufacture. Firstly, they are extremely ductile and can be stretched easily to two or three times their original length, but on long standing they recover to their unstretched condition. Secondly, the nylon filaments are comparatively weak. It has been found that by stretching the filaments and then setting them in this stretched state they have their extensibility reduced to a suitable degree, comparable with that of viscose and other rayons, and at the same time are much strengthened. The process is known as *drawing* and it is optional to heat the filaments to plasticise them before stretching.

So the next step in the manufacture of nylon yarn is to stretch it up to four times its original length and then heat-set it by immersion for about 2 h in boiling water while held in this stretched condition. Thereafter the yarn keeps its stretched length except when treated with boiling aqueous liquors, when it may shrink anything up to a maximum of 12 per cent. By this stretching the nylon filaments are made much finer and ultimately approach the fineness of real silk. Their tensile strength may also exceed that of this fibre and this strength is associated with a much increased alignment (orientation) of the long chain molecules in the direction of the fibre length.

Nylon yarns may be made delustred by incorporating an opaque pigment (titanium dioxide) in the molten nylon at the time of spinning. Coloured nylon may also be made by using coloured pigments in the same manner.

Distinctive properties Considering now the properties of nylon filaments it is evident from the foregoing that these will be considerably different from those of the rayons already described. Some of these properties will be dealt with in more detail later in comparing the various fibres for manufacturing purposes, but here it is convenient to deal with the more distinctive ones.

Firstly, nylon is very resistant to many influences which have a deteriorating influence on other textile fibres. It is unaffected by bacteria, moulds, fungi and moth. It absorbs very little moisture from the air and in the ordinary air-dry state contains only 4 per cent of moisture. It does not swell appreciably in water and thus does not lose strength when wetted as do the cellulose rayons. Nylon is thus considered to be a hydrophobic fibre but other synthetic fibres such as Terylene and Orlon are much more hydrophobic.

Nylon filaments are very light and have the low specific gravity of 1·14. They are thus more voluminous than all other fibres except polythene and polypropylene (density 0·92). The elasticity of nylon is exceptionally high and is in fact superior to that of real silk. It is very resistant to creasing.

As regards attack by corrosive chemicals, nylon is quite inert except to hot mineral acids under the action of which it splits up into its original components, adipic acid and hexamethylene diamine. In weak acids it is stable. Boiling for several hours in quite strong solutions of caustic soda has scarcely any adverse effect on nylon materials.

As ordinarily produced (including a 'drawing' stage), nylon filaments have a dry tensile strength of about 5 g per denier and their normal extensibility at breaking point is 20 per cent. These values are only slightly affected by wetting the filaments.

Nylon has thermoplastic properties, and materials made of it are easily embossed under the influence of heat and pressure. Its softening temperature is higher than that of acetate rayon so that less trouble will be experienced in ironing nylon fabrics with a moderately hot iron. They should be ironed when dry but since they usually dry free from creases ironing can be unnecessary. The high elasticity and resistance to abrasion of nylon makes it very suitable for stockings and socks.

Most organic solvents have little action on nylon in the cold, but hot solvents can swell or dissolve it. Thus care has to be exercised in the dry cleaning of nylon garments.

When nylon yarns are rubbed, or otherwise are subject to friction, they readily accumulate static electricity, and this can cause trouble in winding or weaving or knitting such yarns. The electrically charged nylon fibres readily attract dust and other soil from the surrounding air (p. 121). Nylon garments can accumulate static electricity during ordinary wear and cause trouble—they can 'ride-up' and actually spark.

Nylon fibres which have been treated with formaldehyde under acid conditions, whereby the formaldehyde is caused to combine with the nylon molecules but not materially cross-link, have a much more elastic nature. Nylon filaments which have not been cold-drawn subsequent to their production by the ordinary melt spinning process are very extensible and in this state may have special uses but are not generally satisfactory as textile fibres.

Dyeing and finishing problems   When nylon was first used on a large scale for the manufacture of ladies' hose, it was then found to be subject to a curious defect. When the hose were scoured and dyed in the usual way they became permanently distorted and creased. Dyeing was carried out in a hot liquor and this was involved in the trouble. It seems that in hot aqueous liquors the nylon yarn in the hose contracted unevenly in length where the hose were tightly or loosely packed in the dyeing machine. A fairly simple remedy for this trouble was found. It consisted of placing each stocking on a metal shape to distend it slightly and evenly. These shapes were then placed in a steaming chamber (Figure 1.24) for a few minutes, and in this way the hose were set to the shape. Thereafter no contraction or distortion occurred. This high temperature pre-setting treatment has proved very useful in the wet processing of all kinds of nylon materials such as knitted garments, lace and fabrics. Modern dyeing machinery is available which allows nylon hose to be dyed while held to shape on metal 'shapes'.

At first it was found difficult to dye nylon materials, for it has only moderate affinity for most of the dyes used on other fibres. Fortunately, the discovery was made that nylon could be easily and satisfactorily dyed with that special (disperse) class of dye now

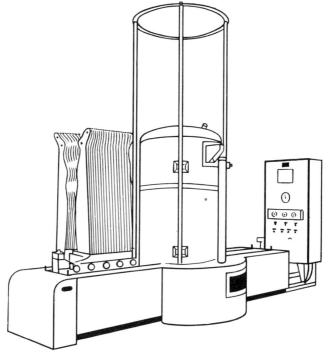

Figure 1.24 Unit for the high-temperature setting to shape of ladies' nylon stockings. They are placed on metal shapes and run into a cylindrical closed chamber where they are steamed for a few minutes. Thereafter they retain their shape in washing and resist creasing, as long as the washing temperature does not exceed 100°C

in general use for acetate fibres. All the nylon hose sold before World War II were coloured in this way. Since then, however, methods have been devised for assisting the absorption by nylon of many other dyes which were thought at first to be of little use for this fibre. Such dyes include the acid and chrome mordant wool dyes and the direct cotton dyes. So far vat dyes have proved less satisfactory, it being a surprising fact that they have poor fastness to light on nylon. The acid dyes are much used for nylon because they give bright fast colours and in these respects are superior to the disperse dyes. However, care is needed to ensure even dyeing.

Progress in developing improved methods for dyeing nylon have followed from a recognition of the fact that while nylon fibres may be considered as being comparable with acetate fibres in being amenable to dyeing with water-insoluble (usually termed 'disperse') dyes, in which process the dye dissolves in the fibre substance, such nylon fibres can also dye in a manner comparable with wool. Thus modern developments have come from discovering dyeing conditions such that nylon can be induced to absorb most types of wool dyes in addition to those employed for acetate fibres—a much wider range of dyes has thus become available for the colouring of nylon materials.

Wool owes much of its affinity for dyes to its content of basic groups which can enter into chemical combination with acidic dyes. Nylon fibres also contain basic groups, more especially the amide —NH.CO— groups which in the nylon molecule link hexamethylene diamine and adipic acid residues together end-to-end, and also the (—NH$_2$) groups attached to the ends of the nylon molecules which can function towards acid dyes just as do the basic groups in wool fibres. Unfortunately the number of these basic groups in nylon is much less than in wool and this explains how it is much more difficult to produce deep shades in nylon goods. But if during the dyeing process the acidity conditions are suitably adjusted to cause a slight breakdown of the long nylon molecules thereby liberating more basic groups and providing just that extra basicity to assist dye absorption, then deeper dyeings can be produced. It may be recalled that drastic hot acid treatment can be used to split up nylon into its components (adipic acid and hexamethylene diamine).

Part of the reluctance of nylon to absorb dyes is also due to a very close packing of the nylon molecules within the fibres so that dyes have great difficulty in penetrating the fibres. This has led to the adoption of methods of dyeing in which the dye liquor is applied within closed dyeing vessels so that high-pressure high-temperature conditions can be obtained—the effect is to loosen the fibre structure and thus facilitate dye penetration and fixation with the production of deeper shades than can be otherwise produced. An alternative method for loosening the nylon fibre structure during dyeing is to have present in the dye liquor a suitable proportion of a substance capable of producing a moderate degree of fibre swelling. It is important not to induce excessive swelling otherwise the fibres may be weakened as a result of the dyeing operation. Dye absorption can also be assisted by having present in the dye liquor a small proportion of a so-called 'dye-carrier' which swells the fibres and possibly also acts as a dye solvent. Typical dye carriers are benzoic acid, phenylphenol, chlorobenzene and diphenyl.

While the use of high-pressure high-temperature dyeing conditions and the use of fibre-swelling agents and of dye carriers has done much to help dyers of nylon materials, a number of snags have been encountered in making use of these expedients. The main snags are that the textile material may be weakened and the handle impaired while the fastness of the resulting colouring may be reduced due to the presence of fibre-swelling agents or dye carriers unavoidably left in the dyed material. It is generally better to use dyes which can be applied under more conventional conditions such as a type of pre-metallised dye at first introduced for the dyeing of wool. Under weak acid conditions these dyes can be used for the dyeing of nylon goods and indeed are much used. These so-called *pre-metallised* dyes contain chromium or cobalt as part of the dye.

More recently it has been found possible to modify the manufacture of the fibre-forming nylon so as to increase the number of amino groups and thus give the resulting fibres an increased affinity for acid dyes. Thus types of heavy-dyeing nylon yarns are now available and they are finding particular use in nylon carpets so that in a single dyeing operation two-tone (light and heavy) patterns can be produced.

Acid (more especially sulphonic acid) groups are also being introduced in a similar manner to ensure that the resulting fibres have an increased affinity for basic dyes and a decreased affinity for acid dyes.

OTHER POLYAMIDE FIBRES

Perlon L (Nylon 6 and Caprolactam)

This type of polyamide fibre, closely related to nylon 6·6, has been mainly developed in Germany as an alternative to nylon 6·6. In the early days of nylon 6·6 production in America, little attention was given to nylon 6 and outside of Germany it was considered to be a somewhat inferior polyamide fibre. But gradually its merits have become appreciated and its manufacture has steadily increased both within and outside Germany. In certain types of hosiery it is now preferred to nylon 6·6.

Nylon 6 (represented by Celon) is slightly more hydrophile than nylon 6·6 and this makes its more receptive to a wider range of the commonly available dyes. Its melting point of 215·6°C is 36·1°C below that of nylon 6·6 so that more care must be taken in exposing it to high temperature conditions.

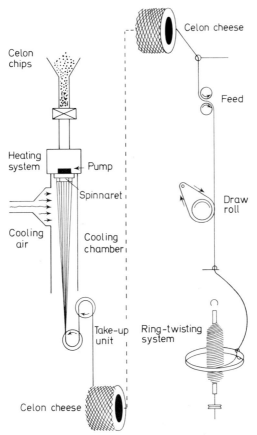

Figure 1.25 Production of Celon yarn. (Courtesy Courtaulds Ltd)

The fibre-forming polyamide nylon 6 is made similarly to nylon 6·6 by a polymerisation process in which caprolactam is the raw material used instead of hexamethylene adipamide. Its formation is indicated thus:

$$HN—CH_2CH_2CH_2CH_2CH_2CO$$

Caprolactam

$\downarrow$

$$H_2N—CH_2CH_2CH_2CH_2CH_2COOH$$
Caproic acid

$\downarrow$ condensation and polymerisation

$$H—[—HN—CH_2CH_2CH_2CH_2CH_2CO—]_n—OH$$
Fibre-forming polycaprolactam

It is to be noted that the caproic acid molecule carries both the required amino and carboxylic acid groups for condensation and polymerisation to give a fibre-forming polyamide.

Rilsan (Nylon 11)

This type of fibre-forming polyamide is made by polymerisation of 11-amino-undecanoic acid as indicated below:

$$H_2N—CH_2CH_2CH_2CH_2CH_2CH_2CH_2CH_2CH_2CH_2COOH$$
11-amino-undecanoic acid

$\downarrow$ polymerisation

$$H—[—HNCH_2CH_2CH_2CH_2CH_2CH_2CH_2CH_2CH_2CH_2CO—]_n$$
$$—OH$$

Fibre-forming polyamide
which can be melt spun into Rilsan fibres

The formula indicates that the recurring units which are joined end to end in Rilsan fibre molecules are much longer than those in nylon 6·6 and nylon 6 molecules and this difference is reflected in the more hydrophobic properties of Rilsan. At present Rilsan fibres are used in articles such as brushes where high water-repellency and stiffness are the properties most desired.

Quite a number of different fibre-forming polyamides have been made in a manner comparable to that for nylon 6 and nylon 6·6 and converted into polyamide fibres all having properties similar to nylon 6·6 but individually different. Included in these are nylons 5, 7, 8, 9 and 12 which are derived respectively from the following amino-acids:

$$NH_2CH_2CH_2CH_2CH_2COOH$$
Amino-valeric acid (for nylon 5)
$$NH_2CH_2CH_2CH_2CH_2CH_2CH_2COOH$$
Amino-heptylic acid (for nylon 7)
$$NH_2CH_2CH_2CH_2CH_2CH_2CH_2CH_2COOH$$
Amino-caprylic acid (for nylon 8)
$$NH_2CH_2CH_2CH_2CH_2CH_2CH_2CH_2CH_2COOH$$
Amino-nonanoic acid (for nylon 9)
$$NH_2CH_2CH_2CH_2CH_2CH_2CH_2CH_2CH_2CH_2CH_2CH_2COOH$$
Amino-lauric acid (for nylon 12)

The commercial possibilities of these newer nylons are now being assessed.

Success has followed a search for new nylon fibres which will have higher melting points and be more resistant to deterioration when exposed to high temperatures and to be superior to the nylons 6·6 and 6 in this respect. It has come from the introduction of benzene and naphthalene aromatic residues into the nylon molecule.

The formulae given above and also those previously for nylon 6·6 and 6 show that the nylon molecule is linear consisting of a straight chain of linked —CH_2— groups. Aliphatic compounds are characterised in consisting of such linear molecules. By contrast aromatic compounds contain some of their chains of carbon atoms in the form of one or more rings and because of this structure they can have properties quite different from those of aliphatic compounds. A typical aromatic compound (containing a phenyl ring structure) is phenyl acetone having the formula

It has been found that polyamide molecules containing phenyl groups among the aliphatic groups can have superior resistance to high temperatures even exceeding 500°C as against a temperature somewhat above 350°C for nylon 6.

One such special heat resistant polyamide fibre now in commercial production (Du Pont) is Nomex. It retains 96 per cent of its strength when exposed to 177°C for 3000 h and in the form of continuous filament or staple fibre yarns and fabric it is flame self-extinguishing during wear. These properties make it especially useful for aircraft upholstery. It is very good for the production of flameproof carpets. The fibre-forming polyamide from which Nomex is made is poly(meta-phenylene isophthalamide) and its long chain molecules have the formula

$$H_2N \left[\begin{array}{c} CH \\ C \diagup \diagdown C - NH - OC - C \diagup \diagdown C \\ \| \quad | \quad \quad \| \quad | \\ CH \quad CH \quad \quad CH \quad CH \\ CH \diagdown \diagup \quad \quad CH \diagdown \diagup \end{array} \right]_n COOH$$

Other heat resistant polyamide fibres have been prepared but so far Nomex appears to be most satisfactory. Incidentally it may be mentioned that Nomex fibres are also used to make paper and non-woven fabrics. As paper it can be used as an excellent electrical insulating wrapping.

Acrylic and modacrylic fibres

This type of synthetic fibre is characterised by being made from a polymer which may consist wholly of polyacrylonitrile or a co-polymer of a mixture of acrylonitrile with another vinyl compound; acrylonitrile is otherwise known as vinyl cyanide. It is usual for an acrylic fibre to contain at least 85 per cent of acrylonitrile, while a modacrylic fibre can contain less, even as little as 35 per cent.

Orlon is an example of a 100 per cent polyacrylonitrile fibre, while Dynel (the staple fibre form of Vinyon N) and Verel are modacrylic fibres containing less than 85 per cent of acrylonitrile. Creslan, Acrilan and Courtelle are grouped in the same class as Orlon but contain between 85 and 100 per cent of acrylonitrile.

Quite a large number of acrylic and modacrylic fibres other than Orlon are available and most of them have come from attempts to improve upon Orlon, which in the early years of synthetic fibres proved to be very difficult to manufacture. Also, while possessing certain attractive textile properties, Orlon was found to be very difficult to dye. Many of these difficulties have now been overcome and in various modified forms (especially those which are highly and permanently crimped—for example, Dynel and Verel) poly-acrylonitrile fibres find considerable use.

Polyacrylonitrile, of which Orlon fibres are composed, consists of long molecules in which a large number of vinyl cyanide molecules are joined end to end thus:

$$n\mathrm{CH_2} = \mathrm{CH(CN)}$$

Vinyl cyanide

↓ polymerisation

$$- - - -\mathrm{CH_2}-\mathrm{CH}-\mathrm{CH_2}-\mathrm{CH}-\mathrm{CH_2}-\mathrm{CH}- - - -$$

| | |

CN   CN   CN

Polyacrylonitrile

For the manufacture of the other fibre-forming long molecules, which arc copolymers of a mixture of acrylonitrile and another vinyl compound, it has been found most useful to use as the other vinyl compound vinyl acetate, vinyl chloride, vinylidene chloride, methyl methacrylate, etc. The resulting long molecules then have the individual molecules of these compounds distributed regularly or irregularly within each polyacrylonitrile molecule. Such long molecules may be represented in type by the following formulae:

$$- - - -\mathrm{CH_2}-\mathrm{CH}-\mathrm{CH_2}-\mathrm{CH}-\mathrm{CH_2}-\mathrm{CH}-\mathrm{CH_2}-\mathrm{CH}- - - -$$

| | | |

CN   Cl   CN   Cl

Copolymer of acrylonitrile and vinyl chloride

$$- - - -\mathrm{CH_2}-\mathrm{CH}-\mathrm{CH_2}-\mathrm{CH}-\mathrm{CH_2}-\mathrm{CH}-\mathrm{CH_2}-\mathrm{CH}- - - -$$

| | | |

O.COCH$_3$ O.COCH$_3$ CN   CN

Copolymer of acrylonitrile and vinyl acetate

The textile properties of such acrylonitrile copolymers will, of course, be dependent on their components and the proportions of these. Much research has been involved in discovering those relatively few fibre-forming acrylonitrile copolymers which are now being used for making commercially useful acrylic fibres.

It is interesting here to mention the early Vinyon and Dynel (Vinyon N) acrylic fibres and the improved types which followed since the process described illustrates how some early difficulties were surmounted in synthetic fibre manufacture.

Vinyon

This type of fibre is made from a copolymer which results from copolymerisation of a mixture of 88 per cent of vinyl chloride and 12 per cent of vinyl acetate. Neither of these vinyl compounds alone is a useful fibre-forming polymer.

Polymerisation is brought about by heating the mixture of vinyl chloride and vinyl acetate, together with some benzoyl peroxide as a catalyst in a closed reaction vessel, the reaction being arrested when the product is found by test to be suitable for fibre making.

It is possible that this vinyl chloride–acetate copolymer could be made into fibres somewhat after the manner of nylon production, but it is found much better to use the method adopted for acetate fibres. For this purpose, a 25 per cent solution of the resin in acetone is first prepared and this, after the usual purification, de-aeration and filtering treatments, is then spun through spinnerets into cells through which warm air flows. The acetone evaporates to leave solidified Vinyon filaments and these are collected in the form of yarn.

Now, as is the case with nylon, this Vinyon yarn has to undergo a drawing treatment during which it is stretched about 800 per cent in length before it can be considered to have the strength and extensibility to make them suitable for textile manufacture. After stretching, the Vinyon yarn is set by hot water treatment and it then shows little tendency to contract in length unless it is placed in water hotter than that used for the setting treatment.

Special properties of Vinyon   Only limited progress has been made in utilising Vinyon for the production of fabrics and garments. It has a very low softening temperature which hinders its usefulness along these lines. It should never be heated to 100°C (the boiling point of water) for under these conditions it fuses or melts. Generally, it should not be heated above 65°C.

Nevertheless it has certain other very valuable properties. It is unattacked by moth, bacteria, mildew and fungi. It is resistant to hot strong alkalis and to cold concentrated mineral acids such as hydrochloric and nitric acids. Its strength is absolutely unaffected by wetting. With such highly resistant properties it is little to be wondered at that Vinyon fabrics have proved extremely useful for filtering operations. They can obviously withstand the corrosive action of liquors which would quickly ruin cotton and wool fabrics.

Vinyon filaments are thermoplastic and in this way resemble nylon. This property can be advantageously used in textile finishing treatments such as embossing and calendering.

Vinyon materials have a very low moisture absorption of not more than 3 per cent.

The average quality of Vinyon being made has a tensile strength of 4 g per denier either wet or dry. Its extensibility is about 18 per cent. These values do not apply to some special types of yarn now being made and which have been left with a much higher degree of extensibility, say up to 25 per cent or more. Such yarns have a lower tensile strength but they can be used for special purposes where variable stresses have to be withstood. As a matter of fact, developments both with nylon and Vinyon have envisaged the production of yarns as extensible as rubber threads, which recover their original length very slowly.

Vinyon is soluble or swellable in various organic solvents, so care must be taken in dry cleaning Vinyon goods. Fortunately it is unaffected by petroleum products. The dyeing of Vinyon has proved difficult because of its poor affinity for most dyes but, as with nylon, the use of the disperse dyes has solved the problem.

The specific gravity of this fibre is 1·3, similar to that of acetate fibres. It has a high degree of elasticity. Vinyon is useful as a fibre binder in non-woven fabrics.

Vinyon N and Dynel

The description of Vinyon refers to one property, its low softening temperature, which, from the textile viewpoint, must be considered as a disadvantage and one that would seriously limit the use of this synthetic fibre for the manufacture of fabrics and garments. Attempts to raise this softening temperature to at least the boiling point of water by varying the proportions of vinyl chloride and vinyl acetate in the resin from which the fibres are made have proved completely unsuccessful. Fortunately, however, it was found that if the vinyl acetate is replaced by a closely related substance acrylonitrile (otherwise known as vinyl cyanide), and a 60/40 mixture of vinyl chloride and the acrylonitrile is copolymerised, then the resulting copolymer gives fibres which have the much higher softening temperature of about 158°C. This type of synthetic fibre, known as Vinyon N, is much more useful in the textile industry. The long continuous filaments of Vinyon N are cut up into shorter fibres comparable in length with those of cotton and wool for use in mixture yarns; in this cut-up form they are termed *dynel*. Dynel fibres which have been given a kind of crimp similar to that of wool fibres are now very popular and are being extensively used either by themselves or in admixture with other fibres for the production of all kinds of textile articles. The original Vinyon having a low softening temperature has now been almost completely replaced by Vinyon N and dynel, especially for textile purposes.

Vinyon N is highly stretched in the course of its manufacture to give it a high tensile strength of about 4·4 g per denier. Dynel is less stretched and so has a lower strength of 3·0 to 3·8 g per denier; on the other hand it is somewhat easier to dye.

Both Vinyon N and dynel are highly resistant to all deteriorating influences such as acids, alkalis, light, weathering, moth, insects, bacteria and damp. They absorb very little water when wetted and so retain their original high strength; in this respect they are quite unlike viscose and acetate fibres which temporarily lose about half their dry tensile strength when immersed in water. Both of these new synthetic fibres are fire-resistant and will not support combustion.

Dynel has a specific gravity of 1·28 and is thus considerably less dense than cotton and viscose rayon fibres although somewhat denser than nylon.

A peculiarity of Vinyon N and dynel is that these fibres lose much of their lustre in boiling water but it can be restored by boiling the fibres in a common salt solution or by exposing them to dry heat at about 116°C. Like other synthetic fibres, Vinyon N and dynel are thermoplastic.

Vinyon N and dynel can be fairly easily dyed with disperse dyes; they can also be dyed with some direct cotton and acid wool dyes, the absorption of these latter dyes being greatly assisted by the presence of copper salts (especially cuprous ions) in the dyebath. This aid has now been abandoned.

Vinyon N and dynel fibres have a compact structure similar to that of nylon and most other synthetic fibres so that it is usually difficult to induce sufficient dye to be absorbed under ordinary conditions of dyeing to produce really deep shades. To produce deep dyeings it is often necessary to use the high-pressure/high-temperature conditions, and the fibre-swelling agents and the dye-carriers mentioned in connection with the dyeing of nylon.

Basic dyes in their quaternary ammonium form have an excellent affinity for Vinyon N and dynel fibres. Such dyes are now extensively employed since they allow the production of fast deep shades using conventional dyeing methods and equipment.

Orlon

This synthetic fibre is made from highly polymerised acrylonitrile and is thus related to Vinyon N. Very many difficulties were encountered in the early production of Orlon and success came mainly when it was discovered that the polyacrylonitrile could be dissolved in a special volatile organic solvent such as dimethyl formamide to give a viscous solution and fibres be obtained from this solution by dry spinning as in the manufacture of acetate, or by wet spinning, as in the manufacture of viscose rayon, but using a hot coagulating bath containing glycerol. These fibres can be made very strong by stretching and are then suitable for use in the textile industry.

Orlon is manufactured in continuous filament form and also as staple fibre in cut-up form. It is proving very useful for many purposes on account of its high strength and resilience and its excellent resistance to deterioration during weathering. Orlon is now being used in men's suits which can be washed, rapidly dried and be at once ready for further wear.

Orlon fibres have a specific gravity of 1·17 and dry and wet strengths of 4·5 to 5·0 g per denier. In its air-dry state Orlon absorbs only about 0·9 per cent of moisture as compared with 15 to 18 per cent for wool. When immersed in water the fibres absorb only 12 per cent of water as compared with 95 per cent for viscose rayon and 45 per cent for cotton. It is thus easy to understand how wet Orlon fabrics dry rapidly.

Although Orlon is being manufactured by spinning solutions of polyacrylonitrile in dimethyl formamide it is possible to make use of spinning solutions prepared by dissolving the polyacrylonitrile in concentrated aqueous solutions of selected hydrated salts such as lithium and zinc halides; these spinning solutions are then extruded into water or aqueous solutions of the same hydrated salts but of lower concentration insufficient to maintain the solubility of the polyacrylonitrile.

Basic (cationic) dyes are now much used for dyeing Orlon and the affinity of the fibres for these dyes is increased by introducing sulphonic and carboxylic acid groups into the fibre-forming polyacrylonitrile. Similarly basic groups (amino) are alternatively introduced to give an affinity for acid dyes. By introducing organic acetate, acrylate and similar substances, the fibre structure can be extended to aid penetration of all dye types. A comparatively new type of Orlon is known as 'Orlon Sarille'. It is spun through orifices fed by two different solutions of polyacrylonitrile so that each fibre has a composite structure somewhat similar to that of wool which has a *para*- and an *ortho*-cortex (p. 17). This special structure causes the freshly spun fibres to acquire a permanent coiled type of crimp when entered into boiling water so that Orlon Sarille yarns are naturally bulky and soft, which can be an advantage for making some kinds of fibres and filaments. Although this crimp may be temporarily removed by stretching, it is regained when the fibres are entered into boiling water.

Acrilan

This acrylic fibre is closely related to Vinyon N and dynel, for it is made from a copolymer based on an 85/15 mixture of acrylonitrile and vinyl acetate. Its textile applications are now considerable; it is much used in carpets.

A feature of Acrilan is that it is much more amenable to ordinary dyeing processes than other synthetic fibres of similar type and contains a preponderance of acrylonitrile. Its good dyeing properties arise from the inclusion of a small proportion of a dye-attractive ingredient in the fibre molecules. Vinyl pyridine is one such basic ingredient and it is attractive for acid dyes particularly.

Creslan

This acrylic fibre is being manufactured by the American Cyanamid Company. It is made from a copolymer containing a preponderance of acrylonitrile and possibly some (basic) acrylamide. It has very good dyeing properties such that mixtures of Creslan and wool can be very easily and satisfactorily dyed. It has the usual physical properties possessed by hydrophobic synthetic fibres.

Verel

Verel is a modacrylic fibre based on a copolymer containing a small proportion of vinyl chloride or vinylidene chloride and a large proportion of acrylonitrile. It is being made by the Tennessee Eastman Company of America. Unlike some polyacrylonitrile fibres it is white and does not require bleaching. It has a useful affinity for neutral-dyeing and pre-metallised dyes, disperse dyes and the new cationic (basic) dyes so that it does not normally require the use of high-pressure/high-temperature methods to secure deep shades. Verel fibres are more hydrophile than usual for synthetic fibres in having a moisture regain of 3·75 per cent.

Courtelle

This is an acrylic fibre which has rapidly become very important, especially for the manufacture of knitted fabrics and garments required to have excellent softness and draping properties. First introduced by Courtaulds Ltd it is now being made in several countries.

The fibres are not so hydrophobic as many synthetic fibres, for example, Terylene, Kodel and Orlon, and for use in knitwear they

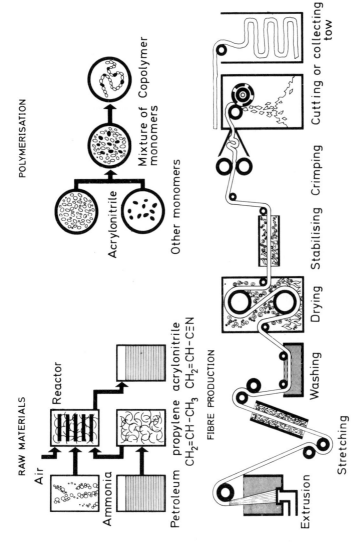

POLYMERISATION

Acrylonitrile

Other monomers

Mixture of Copolymer monomers

RAW MATERIALS

Air

Reactor

Ammonia

Petroleum   propylene   acrylonitrile
$CH_2=CH-CH_3$ $CH_2=CH-C\equiv N$

FIBRE PRODUCTION

Extrusion

Stretching

Washing   Drying   Stabilising   Crimping   Cutting or collecting tow

Figure 1.26 Stages in the manufacture of Courtelle, showing the manufacture of acrylonitrile and its mixture with other selected monomers. The mixture is then spun into fibres which are stretched to give increased strength, washed and dried, and is then either collected generally in cut-up staple form or as long continuous fibres collected as a thick tow. Stabilising is effected by a short high-temperature treatment and crimping by heating the stabilised fibres under compression while packed closely and unevenly. (Courtesy Courtaulds Ltd)

have the advantage of being voluminous with the low sp. gr. of 1·17. They can be satisfactorily dyed with basic (cationic) and disperse dyes. Courtelle fibres are mothproof and resistant to mildew while being only moderately susceptible to deterioration during exposure to sunlight. They do not appreciably weaken on wetting and have a relatively high extensibility.

Courtelle fibres are wet spun from solution in concentrated aqueous calcium thiocyanate, the fibre-forming polymer used being a copolymer containing a preponderance of acrylonitrile together with vinyl acetate and methacrylate. These latter ingredients confer the desirable good dyeing properties and also the other physical properties which make this type of fibre so useful.

Saran (Tygan)

This is now one of the older synthetic fibres and its uses in the textile industry have steadily expanded. Saran, which is also used under the name Velon, is closely related in chemical composition to the Vinyon HH made from the resin produced by copolymerising vinyl chloride with vinyl acetate.

Saran is made by copolymerising a mixture of vinyl chloride and vinylidene chloride to give a copolymer having the formula

$$---CH_2-\underset{\underset{Cl}{|}}{\overset{\overset{Cl}{|}}{C}}-CH_2-\underset{\underset{Cl}{|}}{\overset{\overset{Cl}{|}}{C}}-CH_2-\underset{\underset{Cl}{|}}{\overset{\overset{Cl}{|}}{C}}---$$

However, in contrast to Vinyon, Saran is made into textile fibres after the manner of nylon in so far as the resin is extruded as filaments in a molten state through orifices. The filaments, in the form of yarn, are cold stretched and set in a stable state by heat treatment.

Saran fibres are inert to water and retain their dry strength of about 4 to 6 g per denier even when wetted. Neither does wetting affect their normal extensibility of 20 to 30 per cent. As might be expected, Saran fibres are affected by heat. They soften at 115 to 137°C, and it is best to avoid exposing Saran to high temperatures since the tensile strength is lowered 50 per cent even at 100°C.

Saran has excellent resistance to most alkalis and acids but it is affected by strong ammonia. It is completely immune to attack by moth and bacteria or mildew.

From the viewpoint of dry cleaning it is useful to remember that Saran is insoluble in carbon tetrachloride, benzene and petroleum fractions but, nevertheless, other solvents can swell and dissolve it. Care must be taken in any treatment with an organic solvent especially if these are used hot. So far Saran has proved more useful for the manufacture of smallwares than for fine fabrics. Saran and Vinyon are thus not so far in competition with the older textile fibres and rayons except in these special directions. It is much used for covering deck chairs since it is highly resistant to weathering and in fabric form can be very strong (even when wet) and flameproof. It is difficult to dye and it is usual to add fast coloured pigments to the fibre-forming polymer before spinning into fibres.

POLYESTER FIBRES

Terylene and Dacron

Terylene is a particularly interesting synthetic fibre in that it was discovered, and its manufacture pioneered, in Britain. Actually it was first made by Dr J. R. Whinfield while he was working with J. T. Dickson in the research laboratories of The Calico Printers' Association Ltd in Manchester. Terylene, hitherto made on a relatively small scale, is now being manufactured on a greatly extended scale by ICI Ltd so as to be able to meet the present very large demand for this fibre. The same fibre is being made in America under the name of Dacron, for the right to do this was granted to Du Pont de Nemours and Co, who were responsible for the discovery of nylon.

When Dr Carothers was carrying out his fundamental research on methods for making long chain molecules such as those which constitute polymers and resins, he discovered not only the polyamides as used for the manufacture of nylon fibres but also another related type of fibre-forming polymer which is now known as a *polyester*. Just as a polyamide is made by condensing together a diamine and a dicarboxylic acid (hexamethylene diamine and adipic acid in the case of the most important nylon), so is a polyester made by condensing a glycol with a dicarboxylic acid. Polyamides and polyesters are similar in that they can both be melt spun into the form of fibres capable of being highly stretched to give them increased strength. When the Du Pont Co decided to proceed with the manufacture of synthetic fibres, a choice had to be made between using polyamides or polyesters. It was decided to use polyamides for the particular reason, among others, that all the polyesters which had been prepared at that time had low melting points and would thus yield fibres which would too readily soften when ironed with a hot iron. It was in this way that nylon came to be the first synthetic fibre to be manufactured.

The polyesters were neglected during the highly successful introduction of nylon to the textile industry, but in 1941 Dr Whinfield had the idea that a certain type of polyester not hitherto investigated might be capable of conversion into a new synthetic fibre at least the equal of nylon. Overcoming considerable difficulties he persevered and finally demonstrated that by condensing together ethylene glycol and terephthalic acid a particular polyester known as polyethylene terephthalate, having a high melting point, could be produced. He then showed that it could be melt spun into fibres which are now known to have exceptionally useful textile properties. Thus originated the polyester synthetic fibre known as Terylene.

Today the polyethylene terephthalate which is used for the spinning of Terylene is made from glycol and the dimethyl ester of terephthalic acid instead of the terephthalic acid itself since a smoother chemical reaction is thereby ensured—in the condensation, methyl alcohol instead of water is split off but the same polyethylene terephthalate is obtained as when terephthalic acid itself is employed.

The way in which long molecules of polyethylene terephthalate are made is comparable to that in which the long hexamethylene adipamide molecules for nylon are built up, and this is shown below:

$$HO—CH_2—CH_2—OH \quad + \quad HOOC.C_6H_4.COOH$$

Ethylene glycol   Terephthalic acid

$\downarrow$ condensation

$$HO.CH_2CH_2—O.OC—C_6H_4COOH$$

Ethylene terephthalate

$\downarrow$ polymerisation

$$HO—[—CH_2CH_2—O.OC—C_6H_4CO—]_n—OH$$

Polyethylene terephthalate
(a fibre-forming polyester)

In the polymerisation stage, which is effected by prolonged heating of the first formed ethylene terephthalate at a high temperature of 250°C and then higher under vacuum, the heating is continued until n is large and until the resulting polyethylene terephthalate has a viscosity which will allow it to be melt spun satisfactorily (that is, without too frequent breakage of the fibres thus being formed).

Like other synthetic fibres Terylene is 'drawn' following the spinning stage in order to increase its strength and reduce its

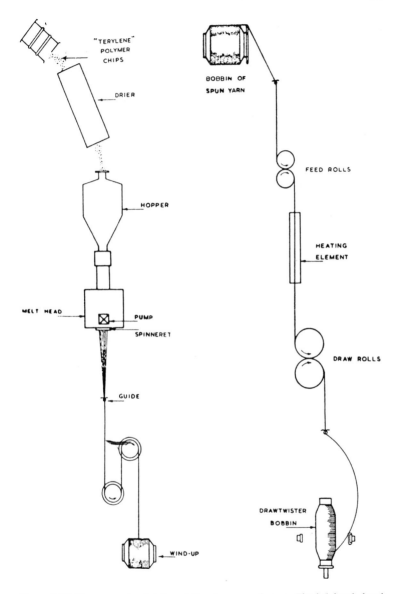

Figure 1.27 Diagrammatic sketches of Terylene manufacture. The left-hand sketch shows the extrusion of Terylene yarn and the right-hand sketch the stretching of the freshly formed filaments. The stretching takes place between the feed rolls and the draw rolls to give the yarn a high tenacity and an acceptable degree of extensibility. (Courtesy Imperial Chemical Industries Ltd)

extensibility to a useful degree—it is usually drawn or stretched to about four times its original length.

Terylene is a fibre difficult to dye and to some extent this factor has hindered its textile usefulness. Only by the use of the special dyeing techniques already mentioned in connection with the dyeing of nylon and polyacrylonitrile fibres is it possible to produce deep shades. At the present time the disperse dyes are of most importance for the colouring of this fibre. Terylene fabrics are admirably suited for permanent pleating and this property can also be used in fabrics consisting of mixtures with wool and especially cotton. Woven fabric made with a 2/1 mixture of Terylene and cotton or wool is much used for dress goods but especially the cotton+Terylene mixture which is crease-resist finished and made up into men's shirts and trousers having permanent washfast creases (p. 398). Terylene has wear properties somewhat inferior to those of nylon whose resistance to abrasion surpasses that of all other fibres.

Terylene is a remarkable synthetic fibre for it has a very high tensile strength, high resistance to attack by chemicals (except caustic alkalis), good durability when exposed to light and weathering, and immunity to attack by insects, fungi, bacteria and moth. It gives fabrics which are resilient and resistant to creasing. Terylene in its continuous filament and staple fibre forms has rapidly gained an importance and usefulness which is now comparable to that of nylon. Very large amounts of cotton–Terylene woven fabric are finished to have wash and wear properties for use in men's clothing.

Comparable polyester fibres are now being manufactured in many countries.

Kodel

This is an American polyester fibre spun from poly(cyclo-hexane-1:4-dimethylene terephthalate) instead of the polyethylene terephthalate used for Terylene and Dacron. It has the advantage of being more easily dyed than Terylene. Its polymeric molecules have the formula

$$HO-[-CH_2C_6H_{10}CH_2O.CO.C_6H_4CO-]_n-OH$$

POLYOLEFINE FIBRES

Polyethylene fibres

Ethylene is a gaseous substance which is capable of being polymerised under high pressures and temperatures and with the aid of suitable catalysts such as oxygen to give polymers, known as *polyethylenes*, which can be converted into synthetic fibres such as Courlene. The chemists of ICI Ltd must be considered as the pioneers in this field although polyethylene fibres are now being made in America and other countries.

The strongly hydrophobic properties of these fibres make them especially suitable in the electrical industry since they have excellent insulating and other related properties. Their textile uses are limited by their comparatively low softening temperatures (110 to 120°C), but such uses are being developed. The fibres are made by spinning the polyethylene substance in its molten state after the manner used for nylon production.

A so-called 'fibrillation' method has been devised for producing relatively coarse fibres which can then be made stronger and finer by drawing in the usual manner for synthetic fibres. It consists of extruding molten polyethylene as film and then vibrating or otherwise treating this to split laterally into fibres suitable for conversion into threads.

Much progress has recently been made in devising improved methods for manufacturing polyethylene. This arises from the discovery of new types of organo-metal catalysts which enable the polymerisation to be effected under much milder conditions and also to ensure that the polymer molecules are more truly linear— branched polymer molecules are not usually favourable to the production of the best textile fibres. Furthermore, the new polyethylenes have higher melting points so that the fibres made from them are more useful in the textile industry. But perhaps more important has been the discovery of harnessing these new polymerisation processes to the production of other polymers such as polypropylene. The raw materials for these new polymers can be cheaply obtained from the petroleum industry and the polymers themselves can be made into fibres having melting points above 250°C. The way has thus been opened for manufacturing cheaply several new fibreforming polymers from petroleum olefines—something hitherto impossible.

Polypropylene fibres

This very strong fibre, made by melt-spinning polypropylene has now assumed considerable importance and is not only employed for industrial purposes such as the manufacture of ropes, cordage and twines, etc. but is now being used in the textile industry for the manufacture of woven and knitted fabrics. The very low sp. gr. of

0·91 ensures that ropes made from these fibres float on water while it also allows the production of strong lightweight fabrics. Like polyethylene, polypropylene is extremely resistant to attack by most chemicals, organic solvents, bacteria and fungi. Ulstron is the British made representative of polypropylene fibres.

The discovery of satisfactory methods for polymerising ethylene to give a fibre-forming polymer, that is one having linear molecules substantially free from side chain groups attached along their length, was difficult, but more difficult proved to be the comparable discovery of a method for polymerising propylene (also a petroleum by-product) so that satisfactory fibres could be melt-spun from it. The tendency was for propylene to form chain molecules having an irregular structure not able to give strong fibres having the physical properties desired in a textile fibre.

Success in this field came from researches of Ziegler and Natti when they discovered that by using complex titanium and aluminium organic catalysts a so-called iso-tactic type of polypropylene having the formula

$$- - - -CH_2 - CH - CH_2 - CH - CH_2 - CH - - - -$$
$$\qquad\quad |\qquad\qquad |\qquad\qquad |$$
$$\qquad\quad CH_3\qquad\; CH_3\qquad\; CH_3$$

could be obtained instead of the useless atactic polypropylene of formula

$$\qquad\qquad\qquad\qquad CH_3$$
$$\qquad\qquad\qquad\qquad |$$
$$- - - -CH_2 - CH - CH_2 - CH - CH_2 - CH - - - -$$
$$\qquad\qquad |\qquad\qquad\qquad\qquad |$$
$$\qquad\qquad CH_3\qquad\qquad\qquad CH_3$$

characterised by an irregular spacing of the methyl groups.

This discovery was exploited by the Italian chemical manufacturing firm of Montecatani Soc Generale per l'Industria Mineraria e Chimica to the point at which the polymerising process could be licensed throughout the world. Polypropylene fibres are now being cheaply produced and although not a perfect textile fibre for making fabrics and garments it has very many industrial uses.

The fibres are made by a melt-spinning process similar to that employed for other man-made fibres including nylon and polyester fibres.

Polypropylene fibres are especially characterised by having the very low density of 0·91 and having a very compact hydrophobic structure. As might therefore be expected polypropylene fibres have proved very difficult to dye and they can only be dyed to pale shades by means of disperse dyes which have proved useful for dyeing other hydrophobic man-made fibres such as nylon. Steady but slow progress is being made in devising improved dyes and dyeing methods. For example the incorporation of nickel organic compounds in polypropylene fibres to give desirable increased resistance to deterioration during exposure to sunlight also improves their affinity for a range of mordant dyes. The colouring of polypropylene fibres is generally achieved by the addition of pigments to the fibre-spinning melt.

A most interesting development in the production of textile fibres has stemmed from the production of polypropylene fibres—it allows fibres to be made by melt-extruding polypropylene film flat or circular (tubular), stretching this to align the molecules in the length direction of the film and then vibrating this film as such or first slit into parallel narrow tapes under controlled conditions which cause it to split or fibrillate across its width thus giving extremely narrow tapes of approximately coarse fibre width so that they can be used as fibres. By such treatment the film becomes a sheet of polypropylene fibres which may be somewhat coarse and not perfectly uniform but nevertheless suitable for forming into yarn for weaving or knitting or for making ropes and twines. It is alternatively possible to effect the fibrillation so that the fibres are not completely laterally separated but are crosslinked periodically so that the sheet of fibres somewhat resembles wire netting or net curtain material. In this form the web of fibres can be superimposed on other webs and be fibre-bonded to give a non-woven fabric. This has found wide use for the backing of tufted carpet and is displacing the hitherto much used jute backing fabric (p. 125 and Figures 2.8 and 2.9).

Many modifications of the propylene polymerisation process and of treatments of the fibres to increase their usefulness are constantly being made. For example various acid and basic substances are added to the polypropylene spinning-melt so that the resulting fibres may have increased affinity for basic and acid dyes respectively. Also to improve the fastness of the fibres to light, selected substances such as hydroxybenzophenone compounds are added to the melt. The physical properties of the fibres can also be modified by spinning the molten polypropylene through special shaped orifices in the spinneret to give a special cross-sectional shape to the fibres, e.g. a clover leaf.

At present there is a need to modify polypropylene fibres (the British type is known as Ulstron) so that they are more amenable for the manufacture of high quality woven and knitted fabrics.

Vinylon

This fibre is mainly manufactured in Japan by spinning an aqueous solution of polyvinyl alcohol into a salt solution in which polyvinyl alcohol is insoluble. The fibres thus obtained are made permanently water-insoluble by treatment with formaldehyde under acid conditions and also by a high temperature treatment. A high stretching (drawing) stage assists this insolubilisation. Some types are also partially acetylated to increase their resistance to everyday use conditions. The fibres, unlike most synthetic fibres, can be readily dyed with a wide range of dyes.

Rhovyl

This type of synthetic fibre is made by dry and wet spinning processes which use polyvinyl chloride as a raw material. To this extent it is related to Vinyon HH and Saran fibres. It is manufactured in Germany and France and although it has useful properties it is much less used than such fibres as nylon and Terylene.

A feature of Rhovyl, Fibravyl, Thermovyl and Isovyl fibres is that they are all made by the dry spinning of a solution of polyvinyl chloride in a mixture of carbon disulphide and acetone. They are flameproof, extremely chemically inert and most difficult to dye although this problem is being tackled with some degree of success. Rhovyl and Fibravyl (continuous and staple fibre forms) are highly stretched during manufacture and are thus stronger and more stable than the Thermovyl and Isovyl types (both are staple fibre forms, but unstretched and therefore weaker and more extensible).

Zefran

This relatively new synthetic fibre is now being manufactured by The Dow Chemical Company in America, who term it a *nitrile alloy* fibre in view of the fact that it is made from a special type of copolymer derived from a 60/40 mixture of vinylidene chloride ($CH_2 = CCl_2$) and acrylonitrile ($CH_2 = CHCN$) and containing a small proportion of a hydrophilic component believed to be polyvinyl pyrrolidone to give the fibres improved dyeing properties. The fibres are dry spun from a solution of the fibre-forming copolymer in acetone. They have good dyeing properties.

The natural moisture regain of Zefran is 2·5 per cent at 65 per cent r.h. and this indicates the hydrophilic effectiveness of the special component since without it the moisture regain would probably be not more than 0·5 per cent. The fibres are given an artificial crimp to assist their processing with other fibres; they have a density of 1·19.

Darvan

This type of synthetic fibre is made by the B. F. Goodrich Company in America, who term it a *'dinitrile'* fibre by reason of its being made by the dry spinning of a solution in dimethyl formamide of a 50/50 copolymer of vinylidene cyanide and vinyl acetate. Vinylidene cyanide is a relatively new substance for use in fibre-forming polymeric substances so the introduction of Darvan made from this substance marked an innovation in synthetic fibre production.

At the present time the Darvan fibres now being manufactured have most of the properties of other synthetic fibres and which are commonly associated with a very strong hydrophobic character, but it is claimed that Darvan fibres have an especially soft handle, moderate dyeing properties, and also strength, extensibility, elasticity and other physical properties well within the range of those of other textile fibres.

The fibre-forming polymer from which Darvan is spun has the formula

$$----CH_2-CH-CH_2-\underset{\underset{CN}{|}}{\overset{\overset{CN}{|}}{C}}-CH_2-CH----$$

$$\quad\quad\quad O.COCH_3\quad\quad\quad O.COCH_3$$

Glass fibres

It may be surprising that glass fibres can be made and satisfactorily used in many kinds of textile materials manufactured by weaving, knitting and by fibre-bonding (in non-woven fabrics). This possibility stems from the essential feature that by extenuating a comparatively coarse, brittle fibre to one having a diameter below 0·01 mm it acquires a degree of pliability such that it does not readily break on bending and can accommodate all the stresses and strains to which it is likely to be exposed normally in a fabric or garment. Its additional properties including resistance to fungi, bacteria and high temperatures coupled with the possession of high strength and flameproof properties make glass fibres exceptionally useful in the particular fields where they can be employed.

Glass fibres are made by extruding molten glass through orifices in much the same way as man-made fibres (the process and apparatus

88

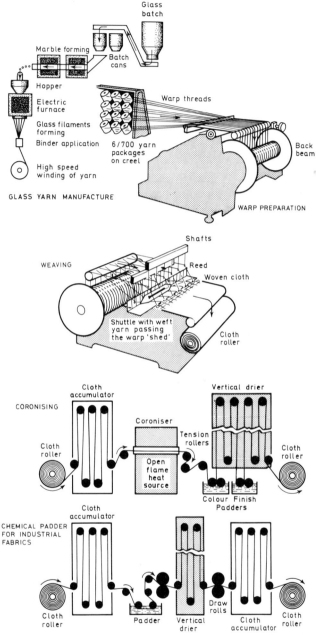

Figure 1.28 Manufacture of glass fibres, the winding of the lubricated yarn onto a warper's beam for weaving into fabric and the subsequent pigment colouring or protective coating of it. (Courtesy Marglass Ltd)

are adjusted to suit the very high temperature conditions of, say, above 1000°C involved) to produce yarn consisting of continuous fibres. Alternatively, entangled short fibres are made and brought together in lap or web form by turbulently propelling forward streams of molten glass which are made to issue from fine orifices by means of high pressure steam.

In manipulating freshly formed glass fibres and yarns there is a liability for much fibre breakage if the fibres rub against each other so it is necessary to coat them with a fibre-lubricating protective sizing composition at an early stage.

When the fibres in yarn form are later made into fabric the freedom of fibre movement must be restrained not only to stabilise the fabric dimensionally but also to guard against the fibre-to-fibre friction which could produce fibre breakage. This can be achieved by a much used 'coronisation' process which also allows the fabric to be coloured by application of a pigment and pigment binder. In this the glass fabric is padded with an aqueous dispersion of colloidal silica, dried and then rapidly led through a hot oven thus exposing the fabric in open width to a temperature which can burn off organic impurities and simultaneously fix the silica as a coating on the glass fibres and thus confer a degree of fibre-to-fibre adherence able to restrict fibre rubbing and stabilise the fabric dimensionally in use. The treatment also gives the fabric a soft handle. In the next stage the fabric is padded with a liquor containing the coloured pigment and a polymerisable resinous pigment binder so that on drying and curing the pigment plus polymerised resin become bonded to the fibres. This bond is substantially increased by drying into the fabric a complex chromium salt solution.

It could be thought that a faster and more easily obtained glass colouring could result from adding the pigment to the glass before spinning this into fibres but this is not so. As the glass fibres are made finer so does it become easier to remove the pigment by washing and wear—at the same time the fastness to light is reduced. By application of both resin and pigment superior fastness is obtained.

Board-like fibre plus glass material produced with the aid of a polyester resin is very strong and durable so that it is much used for the hulls of yachts and small boats.

Elastomeric (rubber-like) synthetic fibres

All synthetic fibres are very extensible as freshly spun and in fact they can be stretched from 3 to 15 times their spun length according to their type. From a moderate stretch the fibres will return almost completely to their original length but beyond that length recovery is incomplete since such stretching much disturbs the packing of the long chain molecules within each fibre. This stretching does not extend each fibre molecule except with certain fibres (exemplified by wool) whose molecules are wavy, folded or coiled; it causes the molecules to slide over each other and with high stretching they do not slide back again on release of the stretching force.

With such high stretching the fibres become not only longer but also thinner and by this latter change the molecules are brought together laterally which in its turn brings them within certain short range forces so that they adhere together more strongly to give the fibre increased strength.

This high stretching is generally applied to freshly spun synthetic fibres with the object of making them stronger. It also reduces their extensibility to the 20 to 30 per cent acceptable in textile fibres generally. The stretching effected with this aim is usually referred to as 'drawing' and can be carried out cold or at a suitably raised temperature. If the fibres are held in this highly stretched state at a fairly high temperature as, say, boiling water, then the stretch is permanently set and they will retain this state unless subjected to temperatures above that of boiling water. Such 'drawing of synthetic fibres' will be referred to later (p. 120) but it is here pointed out that whether freshly spun or in its drawn state synthetic fibres are not elastic as rubber is understood to be elastic. Yet recently it has been discovered how to make synthetic fibres similar to rubber. They are said to be *elastomeric* fibres and designated as Spandex fibres.

Spandex elastomeric synthetic fibres which can be stretched 600 per cent or more and yet return nearly completely to their original length under relaxed conditions are now being manufactured and are available under such brand names as Lycra, Vyrene, Blue C, Spandzelle and Glospan. They are dry or wet spun (not melt spun) from so-called segmented polyurethane polymers which can be made by various modifications of a common basic process. In one such process a polyethylene glycol is copolymerised with an aromatic di-isocyanate such as tolylene-2:4-di-isocyanate to form a prepolymer which ultimately forms a soft segment in the final fibre-forming polymer. Thereafter this prepolymer whose molecules are terminated by isocyanate groups are induced to react with glycol or diamine compounds to form hard segments. In the final polymer these two types of segment alternate and it is the soft segments which allow stretching as restrained (more particularly that due to irreversible plastic flow) by the hard segments. In an alternative process a polyether can replace the polyglycol as the starting component.

Spandex fibres are being used in the production of man-made elastic yarns and these latter find considerable use in woven and knitted fabrics and garments. The Spandex yarns have several advantages over natural rubber threads which can be made by cutting rubber sheet or by extruding rubber latex through orifices for immediate coagulation into threads.

Fabrics and shaped garments containing Spandex threads can be heat-set whereas this is not possible with comparable rubber-containing threads. Actually natural rubber is less resistant to high temperatures than is Spandex thread as is shown by the following data.

Physical property	Lycra	Vyrene	Rubber
Tenacity (g/denier)	0·6–0·8	1·7	0·4–0·6
Extensibility (%)	520–610	700	850–1 000
Recovery from 50% stretching (%)	95	98	—
Moisture regain (%)	1·3	1·5	
Density	1·21	1·28	1·00
Melting point (°C)	230	above 200	
Resistance to high temperatures	Decomposes above 150°C and yellows	Decomposes above 130°C	Decomposes above 100°C

The greater sensitivity to high temperatures is no doubt associated with the fact that the rubber threads have been vulcanised with sulphur compounds and Spandex fibres are less affected by such chemicals. However, to protect textile materials containing Spandex fibres against yellowing in hypochlorite bleaching it is advised to use only low active chlorine concentrations. Rubber threads are more sensitive to light and ozone (in weathering) than Spandex threads and this can account for the increasing use of the latter in sportswear of various kinds including swimsuits. As an advantage over rubber these elastomeric fibres can be dyed.

A single linear molecule as present in an elastomeric fibre consists of alternating soft and hard segments joined together end-to-end, and the fibre structure is shown in Figure 1.29.

As mentioned above the soft segments are highly extensible in contrast to the much less extensible hard segments. The function of the hard segments is not so much to restrict the reversible part of the extension of the soft segments but that part of the extension termed *plastic* flow and which tends to remain on release of the stretching force. The hard segments thus assist the fibres to be not only highly extensible but truly elastic and to recover after stretching and release by about 90–95 per cent.

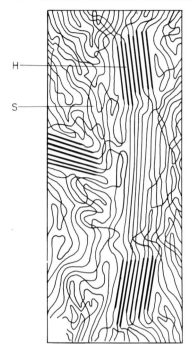

H

S

Figure 1.29 Spandex-type fibre with alternating interconnected hard (H) and soft (S) segments whose high overall fibre elasticity is due to that of the soft segments

STAPLE FIBRES

So far we have dealt mainly with rayon and the synthetic fibres as made in long lengths of at least several yards. In this respect they closely resemble real silk fibres. About 1920, interest was first taken in the production of these rayon fibres in lengths comparable to those of cotton and wool fibres. Today the production of short length fibres is very large and the importance of this staple fibre is great and is constantly increasing.

It is not difficult to understand how this demand for short length fibres has arisen. Firstly, it is difficult to mix fibres of widely different lengths so that for the production of yarns consisting of mixtures of artificial fibres with cotton or wool the artificial fibres must first be cut to lengths more or less the same as these natural fibres. Secondly, the handle and appearance of yarns made from continuous filaments and yarns made from the same filaments cut up into short lengths, are widely different. Generally the continuous filament yarns are thinner, firmer and more lustrous. Thirdly, by contrast yarns composed of cut-up short fibres are more spongy and softer and, because of their higher content of trapped air, are warmer.

It is advantageous to be able to spin all-artificial yarns on the spinning machinery universally available for the spinning of cotton and wool yarns.

Thus, by cutting the filaments into short lengths and spinning these either alone or in admixture with other fibres, it becomes possible to produce a wider range of yarns and thus a wider range of fabrics and garments. The usefulness of the man-made fibres is thus extended, as manufacturers of the various types have developed the production of staple or cut-up fibres.

For the manufacture of such staple fibre much the same methods and materials for ordinary continuous filament fibres can be used. It is towards the end of the process of manufacture that changes must be made. As the filaments are extruded from the spinnerets, which may each have 1000 or more orifices, they are brought together in the form of a bundle or tow and then, before or after purification or at some intermediate stage, they are led through a cutting machine which can be adjusted to deliver the fibres of any desired length. Thereafter the cut-up fibres are led on a travelling brattice or belt through purifying treatments, if necessary, and are finally dried in the form of a fairly thick continuous layer or lap, in which the fibres are considerably entangled. In this form the staple fibre is ready to pass through the various machines used for manufacturing cotton or wool yarns.

An important point about the manufacture of staple fibre in this manner is the large degree of control which the manufacturer has

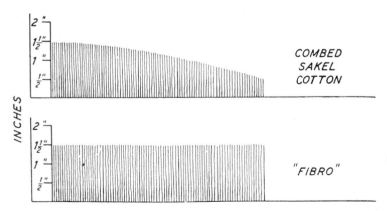

Figure 1.30 The variation in fibre length in similar samples of Egyptian Sakel cotton and viscose staple fibre (Fibro, Courtaulds Ltd). The cotton fibres vary from $\frac{1}{2}$ to $1\frac{1}{2}$ in (12·5 to 38 mm) in length, whilst the Fibro fibres have a uniform length of $1\frac{1}{2}$ in (38 mm)

over the physical form of the individual fibres. These cut-up fibres may be made so as to have any required thickness or length and this without variation. By contrast it is impossible to obtain a quantity of cotton or wool fibres all having the same characteristics; they vary widely when taken from the cotton boll or the back of the sheep.

Another method for producing staple fibre is that of leading the tow of continuous filaments through two pairs of rollers with the remote pair rotating faster than the near pair and spaced at a distance appropriate to the length of fibre ultimately required. Under these conditions the fibres are broken more or less regularly along the length of the tow but not in such a manner as to break the tow completely at any point. Thereafter the tow can be spun by the usual methods of drafting and twisting into a yarn which then consists of relatively short fibres.

Since the use of staple fibre allows warm lofty yarns to be produced, fabric manufacturers have today made considerable progress in their efforts to make wool-like fabrics out of artificial fibres.

In order that staple fibre may more closely resemble wool fibres, methods have been developed whereby the cut-up fibres are given a fairly permanent crinkle or crimp. This can be done by bringing the fibres into a plastic condition and then 'setting' them whilst in an entangled or twisted state or whilst passing between hot and fluted rollers. Another method for producing fibres having a crimp is to spin two different qualities of the fibre-forming polymer from pairs of adjacent orifices so as to form a composite fibre comparable to a wool fibre which has *ortho-* and *para-cortex* components (p. 17). These composite fibres can then be boiled in water when they assume a stable crimped form. These methods are being perfected to overcome the risk of the crinkle being lost when the fibres are later subjected to stretching by laundering and similar treatments.

It has already been mentioned that stretching can play a very important part in the manufacture of artificial fibres of all kinds since, by this operation, the fibres can be considerably strengthened and in some instances be made appreciably finer. The opportunity for effecting a high degree of stretching is less during the production of staple fibre and often there are opportunities for the fibres to contract and thus lose such stretch as may have been previously introduced. It is for this reason that the individual fibres in staple fibre are usually weaker than the corresponding continuous filaments. Since stretching reduces the affinity of most textile fibres for dyes, it is also found that staple fibre usually dyes more readily than the corresponding continuous filaments.

CHAPTER TWO

The properties of textile fibres

Anyone who considers the various textile materials now available and which have been used at different times in the past must be struck by the comparatively few types of fibres used. They could, until recent years when man-made fibres have appeared, be counted on the fingers of one's hands. It would seem that experience over many centuries has shown defects of one kind or another in many of the fibres which at first appeared useful, so that now only such fibres as cotton, linen, hemp, wool and silk are able to satisfy the requirements of our modern textile industry. These are now being supplemented by the various rayons and synthetic fibres, and present day indications are that a much larger number of these will be produced in comparison with the natural fibres.

A textile fibre to be really important and useful must, of course, be available in large quantity and be reasonably cheap. In addition, it must have at least a few of the several properties referred to below and which enable it to be converted into yarns and fabrics by means of machinery now commonly used in the textile industry. Further, it must be reasonably durable under everyday conditions and be capable of being dyed in a wide variety of fast shades.

Important properties of fibres

Considering the individual fibres, as distinct from yarns and fabrics made from them, the following properties are of importance:
 (1) shape—length, thickness, shape of cross-section, straightness;
 (2) strength;
 (3) extensibility and elasticity;
 (4) plasticity;
 (5) softness;
 (6) lustre;
 (7) general durability;

(8) density;

(9) solubility in aqueous and organic solvents.

Also important is the manner in which some of these properties are affected by wetting.

Useful properties of another kind desired in a textile fibre are indicated below:

(10) behaviour towards dyes;

(11) resistance to deteriorating influences, including:

  (a) light, particularly sunlight;

  (b) heat;

  (c) bacteria, mildew, fungi, moths and various destructive insects;

  (d) wet or damp conditions;

  (e) abrasion and wear;

  (f) corrosive chemicals;

  (g) creasing.

It is useful at this stage to discuss some of the properties enumerated above so that the usefulness of the various textile fibres may be better understood.

Fibre shape and strength of yarns

The present-day method for manufacturing textile materials is to bring the fibres together in more or less parallel form, draw these out into a tenuous state and twist them about each other into yarn or thread. This yarn can then be made into fabric by weaving, knitting or other means, all of which have the result of interlacing the yarn so that the resulting fabric holds together to a satisfactory degree.

Thus the basis of all fabric (excluding felts which will be dealt with later) is yarn. Obviously this must be strong enough to withstand the weaving and knitting operations and further to withstand everyday wear and tear in the fabric.

The strength of a yarn is governed partly by the strength of the individual fibres of which it is made up and partly by the degree to which these cling together and resist drawing or slipping over each other as the yarn is pulled lengthwise. From fibre to fibre the tensile strength varies considerably but the textile industry requires that it shall be at least 1 g per denier. The strongest fibres such as nylon and specially treated acetate can have a tensile strength up to 5 to 10 g per denier, this indicating that man-made fibres can thus be produced superior to the natural fibres in respect of this particular property. It must, however, be remembered that a strong fibre is useless if it is brittle; pliability must always accompany high strength.

The effect of twist on the strength of a yarn is considerable. It is to be noted that textile fibres are extremely long and tenuous. Cotton fibres are the shortest of textile fibres, yet it has been mentioned how these are about 1 000 times as long as they are thick. If a cotton fibre were magnified so that it was as thick as an ordinary pencil then this pencil would be about 20 ft (6 m) long. This analogy strikingly brings home the fact that a textile fibre is very tenuous. For a fibre to be useful for textiles it must be tenuous. Short thick fibres are useless.

Bearing in mind the long thin character of a fibre it becomes easier to understand how, when some fifty or a hundred of these

Figure 2.1 A cotton fibre drawn roughly to scale in length and thickness (but folded for ease of presentation). This illustrates its extremely tenuous form

are laid side by side as in yarn and this is then twisted, they become so closely interlocked that when the yarn is pulled they cannot slide over each other and thus allow the yarn to break. The greater the twist the more closely are the fibres interlocked and the more tightly do they cling to each other. Thus, up to a reasonable degree, yarn strength is increased by twisting the yarn tighter.

If a textile fibre is very smooth and perfectly straight it can generally be made into yarn and fabric in spite of this, but it will be necessary to place more reliance on the strengthening effect of increased twist. In contrast, fibres which have a natural roughness of surface or some special distortions are more amenable to manufacture of strong yarns. For instance, cotton fibres, which resemble lengths of twisted ribbon, readily and closely interlock when brought together in yarn form. Similarly, the crimp or waviness of wool fibres is favourable to the production of strong yarns.

Recently it has been found useful in the case of man-made fibres to manufacture these so that they are not round and rod-like but have special cross-section shapes. This can be readily achieved by extruding the solution or melt of the fibre-forming polymer through spinnerets whose orifices are specially shaped, say rectangular, elliptical, slot-like, star-shaped, etc. and drawing the fibres away from the orifices at a suitable rate. The cross-sectional shape can thus be modified so as to ensure that the fibres within the yarns made from them have a high or low degree of adhesion to each other as may be desired. For example, some shapes give such a high degree of adhesion that fabrics made from them have less tendency to acquire a hairy surface by the protrusion of fibre ends through it and are thus less liable to the defect of pilling (p. 410).

A further development is to extrude two different fibre-forming polymers side-by-side through special orifices so that composite fibres are formed in each of which one half has properties different from that of the other half throughout its length. The different polymers may in the composite fibres be aligned side-by-side or as a sheath-core. They must adhere very tenaciously to each other and resist separation in their subsequent use in fibres and garments. When such fibres, especially after stretching, are immersed in boiling water they contract irregularly and acquire (in the case of Orlon Sarille fibres which are of this type) a permanent coiled type of crimp which makes them more useful than simple round straight fibres for the production of soft spongy fabrics.

These methods for producing modified types of man-made fibres are now being actively developed so that fibres may be made which are specially 'tailored' to fit their intended use (p. 133).

Fibre extensibility

Most fabrics in everyday wear are subject to stretching or distortion from time to time. Generally, we expect a fabric or garment which has been stretched to return to its original size and shape after relaxing. To a considerable degree this elasticity or power to recover from stretching is decided by the structure of the fabric. For instance, everyone is familiar with the high elasticity of a knitted fabric compared with one which has been woven. The extensibility and elasticity of the individual fibres, however, plays a part and this is well brought out in the manufacture of yarns and fabrics.

The extent to which a fibre will stretch is roughly proportional to the stretching force. This property varies a great deal among fibres. For example, a wool fibre can be easily stretched 10 per cent, whilst a cotton fibre resists stretching more strongly and will break before stretching 5 per cent. Viscose rayon and acetate fibres stretch easily up to 20 per cent and some of the special forms of the synthetic fibres nylon and Vinyon can be stretched beyond 100 per cent without breaking.

There is no fixed standard of extensibility for a textile fibre since, in general, the manufacturing conditions can be adjusted to suit any particular fibre. But it is essential that a fibre must have a certain minimum amount of extensibility so as to withstand sudden strains placed upon it.

Usually fibres (especially hydrophile fibres such as cotton, wool and rayon) stretch more easily when wet because the absorbed water produces a lateral separation of the fibre molecules so that they can slide over each other more easily.

High stretching optionally at a high temperature to assist ductility followed by relaxation and shrinkage is to-day a useful method for crimping synthetic fibre yarns.

The properties of fibres (especially those which are hydrophobic) can be much modified (often beneficially) by suitable stretching and then relaxing optionally to produce a crimp.

Softness

Softness is today a very acceptable property in a textile fibre. For tapestry, upholstery and other purposes it is often an advantage that the textile fibres used shall be stiff and resilient, but for garments, and especially underwear, it is softness that is most desired. For babywear extreme softness is insisted upon.

For these reasons it is a good thing for a fibre to be naturally soft. But nowadays a number of softening agents (pp. 360, 372) are available and some of them are extremely efficient. Mechanical methods are also available to produce softness in fabrics (p. 334). Thus, if a textile fibre has other useful properties, it could still be utilised even if it naturally lacked softness.

Plasticity and thermoplasticity

It will be understood that when fibres are brought together in the form of yarn and fabric they have, as it were, to lie down and conform to the shape of the material. It is undesirable for the fibre ends to protrude, for this would make a yarn or fabric hairy when it was required to be sheer. Thus the fibres must have a fair degree of plasticity and not be too resistant to moulding or flattening influences. It is not necessary that a fibre shall be fine to be plastic but usually useful plasticity goes with fibre fineness. Wool fibres are very usefully plastic, especially under the influence of heat and moisture, and this is true for both fine and coarse fibres.

The introduction of synthetic fibres to the textile industry has drawn increased attention to the thermoplasticity of fibres, for most of these newer fibres soften under temperature conditions which have but little effect on the older natural fibres so that they can be usefully distorted or moulded by heat and pressure to acquire special and desirable finishes. Most synthetic fibre fabrics can, for example, be flattened, embossed and otherwise changed in appearance and handle by passing them between hot rollers. But the thermoplastic properties of synthetic fibres are now being most advantageously exploited for the production of permanent pleats in dress materials. Provided that a certain minimum proportion of the thermoplastic fibre is present in a mixture, fabric also containing non-thermoplastic fibres can be pressed under a suitable high temperature so as to acquire desired folds and pleats which will be fast to a reasonable amount of washing. At the present time the pleating of fabrics containing wool in admixture with say nylon, Terylene and similar fibres is largely carried out.

The thermoplastic properties of synthetic fibres has also made possible the production of special textured yarns of the 'high bulk' and 'stretch' types as described later (pp. 135 and 166).

The permanent pleating of fabric containing both polyester and cotton fibres assisted by the use of cellulose crosslinking agents is widely carried out (p. 397).

Lustre

The lustre of a fibre is important, for in these days the appearance of a fabric or garment is a deciding factor in its appeal to the public. At the present time it would seem that for clothing purposes a subdued lustre is generally preferred. In reviewing the progress made in rayon manufacture during the past twenty-five years it will be seen that lustre has received very careful consideration. The early types with their bright metallic lustre would not sell today, when dull and even matt textures are in demand.

Here again, as with softness of handle, various finishing processes are available for changing the natural lustre of a fibre. Opaque substances can be deposited either upon or within the fibres so that their lustre can be adjusted exactly as required. As we have seen previously, rayons can be made to comply with a lustre specification by the simple method of adding delustrants (especially the white pigment titanium dioxide) to the rayon spinning solution.

Generally it is more difficult to increase the lustre of a fibre than it is to lower it. An exception is cotton, which can be given a permanent silky lustre by the well-known process of mercerising.

Fibre density

The density of a fibre is an important characteristic since it is on this property that the so-called covering power of the fibre as present in a fabric depends. The lower the density the greater is the volume of fibre for any given weight. With yarns made of low-density fibres it is possible to produce woven and knitted fabrics having a full, solid appearance more cheaply than with more dense fibres. Cotton and viscose rayon fibres have the high density of about 1·50 as compared with wool fibres having a density of about 1·33 and nylon of about 1·14; polyethylene and polypropylene fibres have the very low density of about 0·92, so that they float on water, while the industrial fibre polytetrafluoroethylene (sold by Du Pont as Teflon) has the exceptionally high density of about 2·2.

Solubility in various solvents

A feature of the natural fibres is that they are practically insoluble in all organic solvents and, of course, are insoluble in water. It is desirable that this be so in a textile material, for then it can be repeatedly washed or dry cleaned without fear of damage or loss of weight. To some extent the newer synthetic fibres are inferior in this inertness. Care has to be taken in their dry cleaning, for unless an inert solvent is chosen, acetate, nylon and Vinyon, etc. materials will dissolve or swell so as ultimately to be left, after drying by heating, weak and brittle. An advantage of dry cleaning is that the organic solvent used does not swell the fibres and thus avoids the shrinkage of yarns and fabrics which can arise in wet cleaning when swelling occurs.

Affinity for dyes

While the physical properties of textile fibres must be such that they ultimately yield yarns and fabrics of good general durability, it is also necessary that they must be amenable to dyeing processes, so that these materials can be presented to the public in pleasing and attractive colourings. In the past there have been several instances of the introduction of a new fibre which has caused many difficulties to dyers and printers so that its utilisation has been much hindered. Acetate fibres may be cited as an example, but there are even more recent instances in nylon, polyester and the polyolefine fibres. Glass fibres require specialist colouring methods.

Not only must a textile fibre be capable of being coloured but it must be amenable to dyeing in really fast shades.

It is true that if a new textile fibre has very desirable physical properties then it will warrant special research to discover new dyes to give it colour. But all this research may take much effort and time. It is much to be preferred that a fibre shall, right from the commencement, be amenable to dyeing with at least a fair proportion of the many dyes already to hand.

It has long been recognised that colourings faster to washing might result if dyes were available which had the power to combine chemically with the fibres to which they are applied. To some degree acid wool-dyes combine thus with wool fibres, but the chemical bond between the dye and fibre is relatively weak so that the resulting colourings are not so fast as might be expected. Recently it has been discovered that cellulose fibres such as those of cotton and viscose rayon are sufficiently reactive to allow certain new dyes to be applied to them under alkaline conditions which induce the dye to combine chemically with the fibre substance— the resulting bond between dye and fibre is exceptionally strong so that it is only possible to remove the dye by removing some of the fibre with it. Several new ranges of dyes (they are termed 'reactive' dyes) are now available which allow this important method of dyeing to be carried out; they include the Procion (ICI), Cibacron (CIBA), Levafix (Bayer), Reactone (Geigy), Drimarene (Sandoz) and the Remazol (Hoechst) dyes. It would appear that the first two types were almost simultaneously but independently discovered by British and Swiss dye chemists, respectively; they contain cellulose reactive chlorine atoms.

The affinity of cellulose fibres for dyes is much influenced by the way in which the cellulose molecules are packed together within the fibres. Usually a cellulose fibre, say cotton or viscose rayon, contains both crystalline and amorphous regions—the crystalline regions are those in which the cellulose molecules are arranged in an orderly manner. As might be anticipated, where there is an orderly molecular structure the cellulose molecules are most densely packed together. When such fibres are wetted, as when present in an aqueous dye liquor, the water penetrates between the cellulose molecules and forces these apart—this separation allows the comparatively coarse dye molecules to follow the water into the fibre and there become fixed between the cellulose molecules. Now water finds great difficulty in penetrating the crystalline regions, whereas it readily moves in and between the cellulose molecules where these exist in a disorderly fashion as in the amorphous parts of the fibres. From this it will be understood that the wet swelling of a cellulose fibre can assist dye absorption and that a highly crystalline fibre is much more difficult to dye than one which is largely amorphous. Cotton fibres are much more crystalline than viscose ordinary rayon fibres, and in accordance with this, experience shows that the latter type of fibre is the more easily dyed. Most synthetic fibres are crystalline to some extent and thus consist of molecules very closely packed together. Polyester and polyolefine fibres are of this type. Thus synthetic fibres are usually difficult to dye in deep shades especially by use of water-soluble dyes. The dye affinity of fibres is largely modified by stretching and by heating whereby the fibre structure and particularly the closeness of the molecular packing can be changed.

It has been mentioned earlier that some new types of viscose rayon have a large proportion of skin in which the cellulose molecules are orderly arranged and which is crystalline. Although these fibres have special desirable physical properties, it is a defect that they are more difficult to dye in deep shades —this follows from the crystalline skin hindering dye penetration and fixation.

Resistance to deterioration

It is unfortunate that in ordinary wear and tear textile materials are exposed to all kinds of influences which can degrade or impoverish them. Sunlight, high temperatures, damp, mildew, rubbing and flexing are all influences which may shorten the life of a fabric or garment.

Exposure of textile materials to sunlight can be very harmful. The ultra-violet light rays particularly are destructive. They generally induce oxidation of the fibre and also of any dyes which may be present, so that progressive weakening of the fibres takes place. In some cases this action is accelerated by catalytic influence of the dyes. Thus, yellow and orange curtains especially may quickly become impoverished in sunlight if the dyer has not taken care to avoid the use of certain dyes which have this harmful influence. The effect of sunlight is often more marked in a humid atmosphere.

It is for this reason that in dry tropical climates textile materials can have a longer life than might be anticipated.

Prolonged exposure of fabrics and garments to damp conditions is always certain to shorten their life. The moisture assists attack by all kinds of bacteria and organisms. Thus a piece of dry cotton fabric can be stored for centuries almost without change, but it soon rots if left buried in damp earth for a few weeks. In the damp soil are all kinds of organisms which can attack the textile material; under dry conditions these organisms remain almost dormant (p. 412). When bacteria or related organisms thrive in a piece of fabric they either weaken this by living on its substance or excrete substances which are corrosive to the fabric. Thus mildewed fabric is often found to be acid in character, the acid resulting from the living fungi which cause the mildew. As previously pointed out, acids readily tender cellulose fibres.

Everyone is familiar with the harm caused to wool goods by attack by moth, although it is not really the moths which do the damage. The moths lay eggs in the wool material, and when these hatch out into grubs it is these which eat the wool and soon form holes. In tropical countries there are many kinds of insects and small life which are just as dangerous to textile materials as moths. It is interesting to note that American soldiers fighting in the South Pacific areas in World War II tested out bootlaces of various kinds and found that only nylon laces withstood the ravages of the harmful insects and organisms which thrive there under all conditions of climate.

It is a very wise precaution in storing textile materials of all descriptions to keep them cool and dry. All textile fibres are susceptible to deterioration at high temperatures. The heating causes their decomposition and this cannot fail to weaken them considerably. Some fibres are more adversely affected by heat than others. Acetate and Vinyon fibres may be noted as requiring special care in hot ironing. With cotton, linen and wool, evidence of overheating is found in the formation of a brown colour generally known as *scorching*. These fibres do not soften but they decompose into brown coloured products. In contrast, such fibres as acetate, Vinyon and nylon soften or melt and become brittle. The susceptibility to damage at a high temperature of silk, wool, acetate, viscose rayon, linen and cotton decreases in the order named. It must be remembered, however, that most fabrics are never perfectly pure and that the presence of quite small amounts of foreign substances can assist the harmful influence of heating. On the other hand, by suitable treatment textile materials can often be given a fair degree of resistance to fire. Fire-proofing will be dealt with later (p. 413).

Most fibres are susceptible to oxidation during exposure to air, ozone or nitrogen oxides and this can be accelerated by light, heat and moisture as during weathering. Oxidation is revealed by discolouration and loss of durability.

Repeated creasing or bending of a fabric must be considered definitely harmful, particularly if it has received any finishing treatment which has decreased its original flexibility or resilience. The weighting of real silk can, for example, so lower its resistance that the silk material soon cracks on repeated creasing under pressure.

Having in mind all the various properties which must be possessed by a textile fibre in order that it may have a high degree of serviceability, it may well be asked whether there are any general features about those fibres now in use which seem indispensable to a fibre. Is there something about the chemical composition or the build-up of a fibre which is characteristic and which naturally gives such desirable properties as strength, resilience, durability and resistance to wear?

Fibre structure

So far as our present knowledge goes there is a 'something' about all the natural and man-made fibres which seems essential. It is that these fibres are made up of long chain molecules running more or less in the direction of the length axis. Just as a durable flexible thread must be built up of fine tenuous fibres arranged more or less parallel to each other and aligned in the direction of the thread length, so does it seem necessary that each textile fibre shall be built up in a similar way with long chain molecules.

The ultimate units of a fibre are the molecules, but in a textile fibre these are never found singly or isolated; they are always present joined end-to-end in the form of long chains, and it is these which must be considered as the real building units. It appears essential that these chain molecules shall be long and preferably straight.

Considering the natural fibres, cotton, linen, wool and silk, it is to be noted that their molecules have the following formulae:

$$(C_6H_{10}O_5)_n \quad (C_6H_{10}O_5)_n \quad (C_{42}H_{157}O_{15}N_5S)_n \quad (C_{24}H_{38}O_8N_8)_n$$
$$\text{Cotton} \qquad \text{Linen} \qquad \text{Wool} \qquad \text{Silk}$$
$$\text{(Cellulose)} \quad \text{(Cellulose)} \qquad \text{(Keratin)} \qquad \text{(Fibroin)}$$

where n represents a high number, and C, H, O, N and S represent atoms of carbon, hydrogen, oxygen, nitrogen and sulphur. These atoms are in groups as indicated by the formulae and each long chain molecule is made up of these groups joined end to end. Thus the chain molecules for these typical fibres are seen from Figure 2.3 to be complicated and quite different from each other.

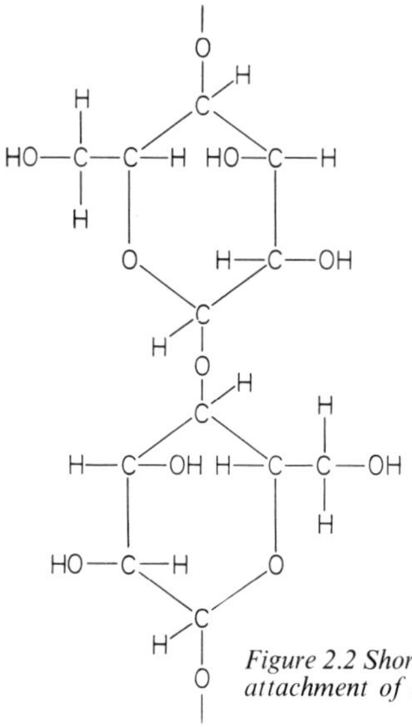

Figure 2.2 Short length of a cellulose long-chain molecule showing attachment of oxygen and hydrogen atoms to the backbone as shown in Figure 2.3

In some fibres, notably wool, the adjacent long chain molecules are also held together by lateral chemical forces and these play an important part in determining the properties of the fibres. It is believed that there are few, if any, cross-linkages of this kind in cotton, linen, silk, viscose, cuprammonium, acetate and the synthetic fibres. But cross-linkages (mainly produced by formaldehyde treatment) are purposely introduced into some fibres to give them increased stability and resistance to swelling during dyeing and other wet processing.

When a hydrophile fibre such as cotton or wool is placed in water it usually swells about 30 to 40 per cent in thickness but extends only about 3 per cent in length. This indicates that the forces by which the molecules are held together, end-to-end, are considerably stronger than those responsible for the lateral linkages. A hydrophobic fibre such as nylon or polypropylene swells only slightly. When a textile fibre undergoes deterioration it is often that the lateral linkages are broken first and then the fibre swells more easily and to a greater degree in water. It is thus easy to understand how such textile material has decreased wear especially in laundering.

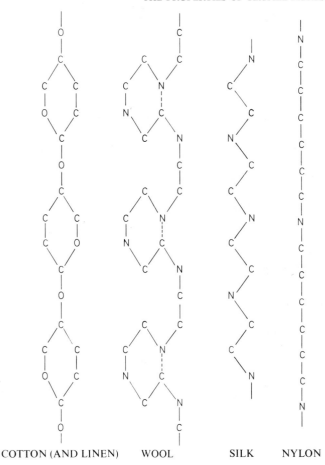

Figure 2.3 Backbone molecular chain structures present in cotton, wool, silk and nylon fibres. Such long chains are characteristic of most textile fibres. The chains are folded in the wool molecule thus giving it special extensibility

Effect of molecular length on fibre strength   A fibre is usually stronger and more durable in proportion to the length of its long chain molecules. When a strong fibre is overbleached, exposed to sunlight or damaged by chemical treatment its long chain molecules will be broken into shorter ones. There are various methods at the disposal of the textile chemist for determining the degree to which the long chain molecules are shortened and it is invariably also found that the fibres are weakened more or less in proportion to this shortening.

From these considerations it is easy to see that in modern developments concerning the manufacture of man-made fibres more progress will be made in the production of stronger and more durable fibres as it becomes possible to use for their manufacture substances which consist of very long molecules and which resist strongly all influences to break them down into shorter ones.

Alignment of molecules   There is just one further point about this structure of a textile fibre which is of interest. The fibres are stronger in proportion to the degree to which the long chain molecules are aligned parallel to each other and to the fibre length. In the case of natural fibres long chain molecules are mainly disposed in this way. But in the manufacture of man-made fibres it is sometimes difficult to ensure this in the freshly formed fibres. By subsequently stretching these, much better alignment can be obtained and so the strength of the fibres can be increased. Today some very tenacious tough rayon fibres are being made by application of stretching processes to the first-formed weaker fibres.

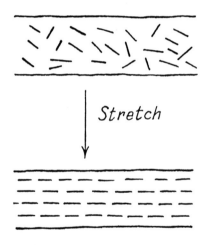

Figure 2.4 Diagrammatic sketch of the increased alignment of the molecular chains within a textile fibre which is produced by stretching. Such stretching gives the fibres increased tenacity but less extensibility

With this special alignment of the long chain molecules the fibres become less extensible and sometimes even brittle, so generally a compromise has to be reached. The stretching is carried out to a point at which a good increase of strength is secured and the extensibility of the fibre is not so lowered as to prevent the fibres passing satisfactorily through the manufacturing processes by which they are later made into yarn and fabric and during which they are often subject to considerable and varying stresses (Figures 2.4 and 2.5).

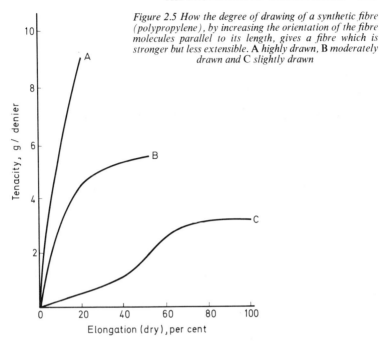

Figure 2.5 *How the degree of drawing of a synthetic fibre (polypropylene), by increasing the orientation of the fibre molecules parallel to its length, gives a fibre which is stronger but less extensible.* A *highly drawn,* B *moderately drawn and* C *slightly drawn*

The special properties of synthetic fibres

Although all the statements already made in this chapter are applicable in a general way to synthetic fibres, it has to be recognised that such fibres are very different in many respects from the older natural and rayon fibres. It thus seems appropriate to make some further observations which apply particularly to the synthetic fibres such as nylon, Orlon and Terylene.

The chemical compositions of the synthetic fibres vary considerably but they all resemble each other in respect of three very important properties:

(1) a tendency to soften at elevated temperatures and become thermoplastic;

(2) a lowering of internal structural stability so that the fibres tend to contract in length follows from heating them at high temperatures in either the wet or dry state or when exposed at ordinary temperatures to substances capable of swelling them;

(3) having a very low affinity for moisture which is correlated with a correspondingly small affinity for most of the water-soluble dyes which are in common use for dyeing other textile fibres.

It will be useful to discuss these important properties somewhat more fully.

(1) From the nature of the polymeric substances used in the manufacture of synthetic fibres and the way in which several of them are converted into fibres by extrusion of the molten polymer through spinnerets, it is easy to understand that these fibres must be liable to melt or at least soften when sufficiently heated. It may be recalled that acetate fibres are also liable to soften during hot ironing but the public is now so well aware of this that it is natural to take care and avoid the use of dangerous temperatures.

During the past few years some thousands of different polymers have been examined to test their suitability for conversion into synthetic fibres, and a very high proportion of these have been rejected almost solely on account of their low melting point. The first type of Vinyon was actually manufactured on a large scale but it is now being replaced by Vinyon N. The very low softening temperature of 65°C has proved too large an obstacle to the wide use of Vinyon H as a textile fibre. It is likely that for some years to come synthetic fibres will be liable to soften on heating. Of course there are types which are especially resistant to high temperatures such as Nomex polyamide fibres (p. 69) and which owe their heat stability to the presence in their molecules of aromatic ring structures.

A fairly low softening temperature is not altogether a defect in a synthetic fibre; the softening can be utilised by a textile finisher to produce permanent embossed patterns on fabrics made from synthetic fibres. As will later be described (p. 189), synthetic fibres are liable to become sticky or tacky when softened and this property can be employed in the manufacture of non-woven fabrics. But, in general, it is better for the synthetic fibre to have a high softening temperature, for when the fibres are heated to their softening point and then cooled they can become stiffer and more brittle so that the fabric or other article thereby loses some of its wear value.

(2) The internal instability of synthetic fibres is of considerable importance and has been the subject of much research on the part of synthetic fibre manufacturers. It has been noted that the freshly extruded filaments, whether they be melt spun or dry spun, are usually relatively weak and capable of being stretched to many times their original length. Generally these freshly made filaments are valueless for textile purposes on this account. So the filaments have to be highly stretched. All stretched materials tend to return to their original unstretched state when they are relaxed, and synthetic fibres are no exception to this rule. However, the manufacturers take steps to set the extended fibres in their stretched state. This is usually accomplished by holding them under tension in the form of yarn

while exposed, in either a wet or dry state, to a temperature which is higher than that to which the fibres will subsequently be exposed. The setting temperature will always be above the boiling point of water if the fibre is not to soften at such a temperature. It is found that when the synthetic fibres are set they can be handled without fear of shrinkage provided that the setting temperature is not approached too closely. Thus the stability of synthetic fibre materials is limited and if are hot ironed or treated in liquors near to the setting temperature they will commence to shrink and lose their shape. In all synthetic fibres there are latent forces which are awaiting the opportunity of inducing the fibres to attempt to return, at least partially, to their unstretched state and all users of synthetic fibre materials must fully appreciate the significance of this fact.

While dry and wet heat can free these latent contractive forces, it is possible for the synthetic fibres to shrink in another way. Many organic substances have the power to dissolve or swell synthetic fibres given suitable conditions. During the swelling the internal structure of the fibres is loosened, contraction is facilitated and the latent contractive forces are given the opportunity to exert their power. This type of shrinkage is of particular importance to dyers of synthetic fibre materials for the reason that such organic swelling substances are sometimes required to be used in dye liquors to assist dye absorption.

This internal instability of synthetic fibres has to be taken into account in various ways. It has to be counteracted in those instances where it can cause trouble in dyeing and finishing, and it can be utilised with advantage where it is desired to modify the form or shape of a synthetic fibre material. In dealing with nylon some mention was made of the fact that before nylon hose can be processed in hot liquors, say in scouring and dyeing, it is necessary to set them on shapes at a temperature which is somewhat below the softening point of nylon and yet appreciably above the highest temperature at which the hose will be processed (Figure 1.24). If this heat setting or pre-boarding is not carried out, the hose shrink and become distorted permanently during their treatment in the hot scouring or dyeing liquors and are thus made useless. However, the heat-set hose retain their shape and smooth texture perfectly, so long as the wet treatment is not allowed to reach the heat-setting temperature, which is usually about 122°C. Heat setting is also required for nylon fabrics and garments and, in this pretreatment, all that is required is to hold the nylon material to the size and shape desired while it is exposed to a dry or moist heat for a few minutes. With garments, as with hose, it is usually most convenient to have them stretched over suitable shapes. In the case of fabric there are two alternative methods of treatment.

The fabric can be held out in open width to the desired dimensions and exposed to hot air whilst travelling along a stenter frame such as is much used in finishing operations. The alternative (Figure 2.6) is to roll the fabric evenly on a perforated central tube, place into a closed chamber and blow steam through the fabric from inside to outside and then in the reverse direction. In all such setting operations it is necessary to prevent shrinkage during exposure to the high temperature. The setting of permanent pleats in nylon fabrics and garments is an instance of how the stability of a synthetic fibre can be modified beneficially.

Figure 2.6 Autosetter for setting the dimensions and surface characteristics of a synthetic fibre fabric by steaming it in roll form within a closed chamber. (Courtesy Andrews Engineering Co Ltd)

All that has been stated above in respect of nylon is equally applicable to all the other synthetic fibres. It is necessary to heat set them whenever they are to be processed at a high temperature, so that they will have no tendency to shrink or distort. The temperature conditions of such heat setting will vary from fibre to fibre and be governed by the softening point of the particular synthetic fibre being dealt with.

(3) The third common property of synthetic fibres distinguishes them from most other fibres such as cotton, wool, silk and the

cellulose rayons. All these last-named fibres readily pick up moisture from a damp atmosphere and imbibe much water when they are completely wetted. This is not the case with the synthetic fibres. In their air-dry state they may absorb less than 0·1 per cent of moisture from the surrounding air as compared with, say, 12 per cent for viscose rayon.

The power of a textile material to absorb water influences the kind and variety of uses to which it can be put. Thus, for outer-wear, it is desirable to employ materials which are moderately water-repellent; for underwear it is better that they should be absorbent since garments of this kind usually have to absorb perspiration. In the normal way synthetic fibre fabrics and garments with their hydrophobic or non-absorbent properties have been considered unsuitable for many types of garments, particularly underwear. However, now that these fibres have been put to use in their cut-up or staple form, it has been found that these restrictions on their use need not be observed so much as was anticipated. It is found that the porous nature of staple fibre materials can partly counteract the poor moisture-absorbent properties of the fibre substance.

In the case of synthetic fibres a small affinity for moisture is important in many directions but especially in their manipulation during weaving and knitting, and in dyeing and finishing operations. During the manipulation of synthetic fibre yarns in the weaving and knitting of fabrics and garments, they are normally subject to a considerable amount of friction as they rub against themselves and various parts of the machinery employed. This friction generates static electricity which is technically termed *static*. In the case of cotton, wool and other fibres, which normally retain 6 to 18 per cent of moisture in their air-dry state, the static leaks away to earth, just as fast as it is formed, via the surrounding air or the metal parts of the machinery used and so causes no trouble. On the other hand, owing to their extreme dryness, synthetic fibres are poor conductors of electricity, and this allows the static to accumulate on them. This static eventually causes the individual filaments and yarns to repel each other so that yarn and fabric manipulation can become very difficult. So considerable may be the difficulties caused by this static that, in the early days, much defective nylon fabric was produced and the elimination of static became quite a problem. Since then, methods for overcoming static have been devised but it still remains a disadvantage of synthetic fibres that they are so ready to accumulate static electricity (p. 121).

Synthetic fibre materials charged with electrostatic electricity readily attract dust, dirt particles, etc. to become soiled.

Three methods for overcoming static difficulties are now in use. In the first, the synthetic fibre yarns are sized with substances which make them much better conductors of electricity. The second method involves ionising the air around the machine where the synthetic fibre material is being wound, woven or otherwise processed, so that the static electricity formed on the fibres leaks away to earth via the ionised air. In the third method it is arranged that the synthetic fibre material touches one or more earthed conductors which carry away the static electricity.

When a textile fibre has a poor affinity for moisture, this is usually a sure indication that it will be difficult to dye with water-soluble dyes. In dyeing cotton, linen, wool and similar materials it is usual to apply the dyes from an aqueous dyebath, not only because water is the cheapest of all solvents for dyes but also because it is rapidly absorbed by the fibres. The imbibed water swells the fibres, loosens or opens out their internal structure, and so facilitates the entry of the dye particles into the fibre interior. Some dye particles are so large that they could not enter into, say, a cotton fibre unless this was first wetted so as to increase its porosity and thus make possible the entry of these large dye particles. Now, the synthetic fibres absorb so little water that their internal structure remains contracted and compact during dyeing, and a large proportion of the dyes commonly used for dyeing other fibres are unable to enter the synthetic fibres or only enter to such a limited degree that it is impossible to dye really deep shades. Thus with the large-scale production of synthetic fibres dyers have met a major problem, that of devising methods for dyeing these fibres in deep shades as easily as the older fibres.

It may be recalled that with the introduction of acetate fibres some thirty years ago a somewhat similar difficulty was encountered, for, as regards its affinity for moisture, acetate fibres lie about half-way between a hydrophilic (water-absorbent) fibre such as viscose rayon and a hydrophobic (water-repellent) fibre such as nylon. This difficulty was solved by the manufacture of entirely new types of dyes, generally known as *acetate* dyes, which differ from the usual dyes in being insoluble in water but soluble in organic solvents. It was fortunate for the early dyers of nylon that many of these acetate dyes were found to be applicable to the synthetic fibre, and it may be noted that all nylon hose are coloured with these dyes. Even with these acetate dyes, however, it has been found difficult to produce deep shades of very good fastness to washing. So new dyes, somewhat similar in type, have had to be discovered to supply this deficiency. Generally, the dyes which are applicable to acetate are useful for colouring all the synthetic fibres. Such dyes are now more usually referred to as disperse dyes.

In recent years, new methods have been devised for applying

ordinary cotton and wool dyes to synthetic fibres. They are based on the expedient of adding to the dye-bath a comparatively small proportion of an organic substance which is generally referred to as a dye *carrier* or *fibre-swelling* agent. It is probable that this substance splits the large dye particles into smaller ones and also opens out the internal structure of the synthetic fibres so that the dye particles can enter more freely. There is the alternative of dyeing at temperatures above the boiling point of water, say at 121°C, for such high-temperature conditions also have the effect of loosening the fibre structure. New types of high-pressure dyeing machines have had to be designed to allow dyeing at temperatures exceeding the boiling point of water. A very convenient new method for dyeing synthetic fabric with dyes for which it has but little affinity is known as the Thermofix method. It was devised in America and is now much used especially since it can be carried out rapidly by a continuous process. It consists of padding the fabric with an aqueous solution or suspension of the dye, drying and then passing it over hot rollers or through a heated chamber so as to expose it for about one minute to a high temperature somewhat below the softening temperature of the fibres. Under these conditions the dye externally adhering to the fibres rapidly passes into them by sublimation and solution and becomes uniformly distributed therein and fixed fast to repeated washing. This method of dyeing is especially useful in the dyeing of polyester (for example, Terylene and Dacron) fabrics. These expedients are proving very helpful in counteracting the generally poor affinity of synthetic fibres for dyes.

Much attention is now being given to producing coloured patterns on synthetic fibre fabrics by hot pressing these against transfer paper carrying the coloured pattern (p. 298).

So far the various textile fibres have been described in their normal forms as produced by nature and man. Over the years processes have been devised and machinery made to convert these fibres into yarns, fabrics and garments, etc. accepting their normal properties and especially their physical characteristics. This acceptance follows from the fact that nature's method for producing cotton, wool and other such fibres can only be modified by man within certain limits. By contrast in the manufacture of nylon, polyester, acrylic and other man-made fibres there is ample scope for giving these fibres modified characteristics to suit their use in textile manufacture. In recent years advantage has been taken of this valuable feature of man-made fibres as, for example, to produce them having special shape, lustre and appearance and other features so as to be more amenable to processing which allows the production of textile articles not possible with man-made fibres having their normal properties. It can now be useful to draw attention more specifically than hitherto to this line of progress with special reference to the man-made, so-called synthetic fibres.

Modification of fibre properties during manufacture   Some of the properties of synthetic fibres which can be usefully modified in their manufacture are concerned with: cross-sectional shape; close-packing of the molecules within a fibre; fibre tendency to shrink in length; affinity for dyes; lustre; tendency to become charged with electrostatic electricity; pick-up of soil and release of such soil in washing; resistance to deterioration during weathering, and exposure to light and heat.

Cross-sectional shape

Originally man-made fibres were universally extruded (spun) through spinnerets whose orifices were of circular cross-section and which would normally produce fibres of the same shape. By simple modi-fication of the spinning conditions without change of the orifice shape fibres could be obtained having an indented or other irregular cross section. However, more recently attention has been given to the effect of using orifices having various shapes and it has thus been found possible to produce differently shaped fibres, some of which have especially useful properties. However, the shape of the fibre does not always closely correspond with the shape of the orifice through which it is spun (p. 133).

Usually it is found that the round fibres have a firmer handle than those of irregular shape although softness can also be governed by fibre thickness (coarse fibres are harsher). An irregular shape which gives a fibre increased surface area favours an increased power to absorb dyes. Especially interesting and useful are those fibres which have a trilobar or clover-leaf cross-sectional shape, for if nylon fibres are thus shaped and conform to certain contour limits they reflect light to give a glitter or sparkle which can make knitwear and pile materials, such as carpets, more attractive.

It has been claimed that the glitter can have the useful effect of making less evident any soil (dirt) which may accumulate on a carpet containing fibres having a clover-leaf cross section. A similar claim has been made for a special porous type of fibre having innumerable very small air pockets or voids in them. Such fibres containing air bubbles (these may be partly collapsed) can be made by spinning a fibre-forming polymer containing a blowing agent such as is used in the manufacture of foamed materials.

Molecular packing within a fibre An important feature of man-made fibres is that their properties can be much influenced by suitably modifying the compactness of their structure—that is the degree of close packing of their fine tenuous molecules. In particular this compactness governs the rate and degree of entry into the fibres of any liquid or solid applied to them and most purifying, dyeing and finishing processes depend on the entry of applied agents. This internal structure of fibres is most important and is receiving close attention from all textile technologists.

In spinning synthetic fibres there is opportunity for withdrawing them from the orifices of the spinneret with or without stretching them. Further, after collecting these fibres into yarn form by winding them on a bobbin they can then be further stretched (optionally assisted by heating to a point suitably below their softening or melting point) by 200% to 1500%, according to the type of fibre. This stretching aligns the molecules more parallel to the fibre length and also to each other thus allowing them to pack more closely as the fibres are made correspondingly thinner. An interesting point is that whereas the molecules may be randomly arranged in the first place within the fibres they can, by the stretching and close packing, become so regularly packed in relation to each other (either throughout or in parts of the fibre) as to constitute crystalline regions.

The molecular packing within a crystalline region can be exceptionally tight so that penetration by liquids (even by water) can be much restricted and perhaps impossible. Thus in dyeing such fibres with an aqueous solution of a dye it may be impossible for the water to carry the dye into the crystalline regions so that dyeing may be restricted to the amorphous non-crystalline regions. Therefore the dyer would have to turn to the expedient of having present in the dye solution a substance able to swell the fibres and sufficiently loosen the fibre structure to allow dye penetration. Another expedient could consist of dyeing at a high temperature which may have to be up to 130°C, that is above the boiling point of water, for this also can loosen the fibre structure. However, it would be necessary to use an enclosed dyeing vessel able to withstand the high outward pressure involved. Obviously dyers would like the fibre manufacturer to produce fibres more easily dyeable, and this is now possible and is being done as described below.

In the early days of dyeing it was generally necessary to make dyes to suit the dye absorptive properties of the fibres such as cotton, wool, silk, etc. Thus because wool is basic and has a natural tendency to combine with acids, the dyes made for dyeing this fibre were made to be acidic. By contrast since cotton has no marked affinity for acid or alkali, dyes made for this fibre were largely neutral. Polyamide fibres such as nylon contain basic groups (—NH_2 and —NHCO—) in their molecules and so can be dyed with acid wool dyes but polyester fibres such as Terylene are neutral like cotton and so cannot be dyed with acid wool dyes. With these facts in mind synthetic fibre manufacturers hit on the idea of incorporating acid or basic substances in the fibre-forming polymers so that the fibres spun from them would have an affinity for basic or acid dyes respectively. Thus Orlon which is an acrylic fibre having an exceptionally poor affinity for dyes can now, by the above expedient, be made usefully receptive to acid and to basic dyes by incorporation in the fibres of the appropriate additive. For example, to make Orlon readily dyeable with acid wool dyes it is convenient to incorporate in the fibre-forming polyacrylonitrile polymer a small proportion of a basic additive such as a vinyl-pyridine. Basic dyeing properties can be similarly conferred on the Orlon fibres by arranging them to contain (in combination with the polyacrylonitrile) a small proportion of the acid additive 5-methyl-2-methylallyl-oxybenzene sulphonic acid.

Such additives serve a dual purpose. In one they change the fibre so that it has an acid or basic character and by the other they loosen the structure of the fibre. Both of these changes increase the power of the fibre to absorb and retain applied dyes. Loosening of the fibre structure comes from the fact that the acid or basic additive is combined with the fibre molecules—it is not present simply mixed with the fibre substance. The additive is present attached to the long fibre molecules at spaced intervals along their length and, according to the size and form of the additive, they make the fibre molecules more bulky and irregular in shape and prevent their close packing. Such a change can be appreciated by reference below to the molecular changes which are more particularly concerned with the use of the acid additive mentioned above, 5-methyl-2-methylallyl-oxy-benzene sulphonic acid.

$$CH_2 = CHCN$$

acrylonitrile
↓ Polymerisation

$$----CH_2—CH—CH_2—CH—CH_2—CH—CH_2—CH----$$
$$\quad\quad\quad | \quad\quad\quad\quad | \quad\quad\quad\quad | \quad\quad\quad\quad |$$
$$\quad\quad\quad CN \quad\quad\quad CN \quad\quad\quad CN \quad\quad\quad CN$$

(A) Polyacrylonitrile for spinning into Orlon fibres (allows close molecular packing in fibres)

$$\text{(ring structure)} \quad SO_3H \qquad CH_3$$

$$CH_3 - \langle ring \rangle - OCH_2 - C = CH_2$$

(B) Molecule of 5-methyl-2-methylallyl-oxybenzene sulphonic acid.

(A)+(B)
↓ copolymerisation

$$-----CH_2-CH-CH_2-\overset{CH_3}{\underset{CH_2}{C}}-CH_2-CH-----$$

$$\qquad\quad CN \qquad\qquad CN$$

$$O$$

$$C_6H_3CH_3SO_3H$$

Orlon fibre molecule made more bulky by pendant 5-methyl-2-methylallyl-oxybenzene sulphonic acid molecules which hinder close molecular packing within fibres.

Such additive molecules pendant on the fibre molecule give the fibre a permanent strong dye affinity proportional to their number. Since such separation of the fibre molecules must reduce to some degree the strength of the fibres there is a limit to the amount of additive which can be employed.

Of course it is not necessary for the additive chosen to have acid or basic properties. Quite neutral substances may be used provided that they are capable of combining with the fibre molecules (usually by copolymerisation with them) and effect a separation of the fibre molecules sufficiently to allow the desired easier dye penetration. Use is thus made with acrylic fibres of additives such as vinyl acetate, methyl acrylate and methyl methacrylate which it may be noted have the necessary property of copolymerising.

Experience has shown that such modification of a fibre-forming polymer even to but a moderate degree can radically influence the spinning of this polymer to give fibres. Frequently appreciable changes are simultaneously produced in the viscosity of the molten polymer such that the emerging fibres break more frequently than is acceptable and it is also possible for the physical properties of the resulting fibres to be adversely affected. Many difficulties of this kind have been encountered. For example in endeavouring to produce polycaprolactam fibres having a loosened structure more amenable to dye penetration by incorporating polypropylene, polystyrene and polyethylene teraphthalate in the polymer before spinning, unsurmountable difficulties were encountered until it was found that a small addition of calcium chloride could form a complex compound with the polycaprolactam (nylon 6) and the polypropylene (or other of the above mentioned polymers) and thus could promote satisfactory spinning and thus allow the production of modified nylon 6 fibres having a pronounced affinity for acid wool dyes such that deep dyeings could be readily obtained having the advantage of being fast to washing.

In spite of any difficulties or defects produced by fibre stretching it is almost always necessary to stretch highly the freshly spun synthetic fibres since it gives them a much increased strength and reduces their initial very high extensibility to a value of 15 to 30 per cent which enables them to pass satisfactorily through the various machines used to convert them into fabrics and garments. The stretching may be say 20 to 1500 per cent to produce a 5-fold increase of tensile strength. Nylon fibres are usually stretched about 400 per cent while acrylic fibres may be stretched as much as 1000 per cent. However, it is sometimes required that the fibres should be amenable to stretching in the textile articles made from them and in this case a lower strength fibre must be accepted.

Man-made fibres are available to textile manufacturers in two main forms—as long continuous fibres and as cut-up or staple fibre. Yarns made from the former are liable to have a wire-like handle and appearance whereas yarns made from staple fibre are likely to be softer and have a warmer handle. Almost invariably the staple fibres are somewhat weaker than the long continuous fibres owing to the latter having been more highly stretched. As will be seen later (p. 165) high-bulk yarns can be made by spinning a mixture of staple fibres, one type having been stretched and the other non-stretched, and then heat-treating these wet or dry to cause the stretched fibres to shrink much more than the others.

Synthetic fibres resistant to heat and light

Polyamide, polyester, acrylic and polypropylene synthetic fibres can all become discoloured by exposure to heat and light and in fact great care has to be taken to avoid this during those stages of polymer making and its spinning into fibres. The discolouration is usually due to oxidation and so a preventitive measure often adopted is to displace the air surrounding the polymer or fibres during a hot treatment by inert nitrogen gas. Another useful step is to incorporate in the polymer itself one or other substances (generally termed *additives*) which can counteract the harmful action of light and heat. Selected polyhydroxy-benzophenone compounds can give protection against light degradation while selected salts of manganese and copper and certain phosphites and phosphates can counteract the harmful effects of high temperatures. Fluorescent whitening agents and titanium dioxide additions to the fibre-forming polymer can also serve to give increased whiteness to the fibres. See Nomex polyamide fibre (p. 69).

Anti-static fibres

It is well known that when different types of materials are rubbed together as for example an ebonite rod against fur they become charged with electricity which is negative (−ve) on one and positive (+ve) on the other so that if the charge is sufficiently large a spark can pass from the material when brought near to any metal or other electrical conductor. If the electrically charged ebonite and fur are kept apart and suspended in air their charges will gradually be lost by leakage through the surrounding air to earth. This phenomenon arises from the fact that the ebonite and fur are non-conductors of electricity and so the charges which are formed on them as a result of repeatedly separating them (the rubbing is not essential except in so far as it causes their surfaces repeatedly to make and break intimate contact) cannot leak to earth and they therefore accumulate.

Textile fibres and especially synthetic fibres are not good conductors of electricity when they are dry. Cotton, wool, silk, rayon, linen fibres under normal air-dry conditions usually contain sufficient moisture absorbed from the surrounding air to make them moderately good electrical conductors. By contrast synthetic fibres such as nylon, Terylene, Courtelle and Dacron are all hydrophobic and in their air dry state contain so little moisture as to be non-conducting and so in normal wear and during their manufacture and conversion into yarns, fabrics and garments where they are liable to rub against non-conductors they accumulate high charges of static electricity. This special behaviour of synthetic fibres and materials made from them can be a nuisance and demand a preventitive treatment, which will result in making the fibres electrically conducting. Such treatment can be applied to articles made from the fibres usually in the form of yarn or fabric and thus be considered as a finishing treatment which is dealt with later (p. 362) or the fibre-forming polymers may in the first instance be made anti-static, that is electrically conducting so as to require no further treatment provided that the anti-static properties are not impaired by any of the subsequent treatments to which textile materials are normally subject including more particularly repeated washing.

The adverse effects of 'static' (a common abbreviation for static electricity among textile workers) are met with in many ways. Thus ladies removing synthetic fibre underwear frequently notice a crackle and sparking on removing slips which can be disturbing for the spark can be quite a strong one arising from perhaps an electrostatic charge on the garment of up to 5000 V. In the same manner a person walking on a synthetic fibre carpet can have generated on him a charge which may be even greater. With carpets the charge can be influenced by the type of shoe as for example whether it has a rubber or leather sole and of course also by the type of fibre. The following data are interesting and show how static voltages produced on the human body by walking on different types of carpet with different shoes and with the surrounding air at different relative humidities (this humidity will decide the moisture content of the carpet and influence leakage of static from the carpet) can vary widely (see also Figure 2.7).

| Type of carpet and fibre content | Static voltage produced on human body of walker | | | |
| | Rubber shoes | | Leather shoes | |
	15 per cent r.h.	40 per cent r.h.	15 per cent r.h.	40 per cent r.h.
Tufted: viscose rayon	− 600	− 205	− 382	− 23
Axminster: wool × nylon	− 6 300	− 2 900	− 6 560	− 1 000
,, acrylic	+ 10 140	+ 4 000	+ 10 560	+ 4 220
Tufted: wool	− 10 700	− 3 900	− 11 870	− 1 760
Axminster: nylon	− 6 660	− 4 320	− 4 220	− 1 900
Wilton: wool	− 9 020	− 4 180	− 12 040	− 2 080
,, nylon	− 3 240	− 2 870	− 2 280	− 2 160

The above voltages can of course be considerably reduced by spraying an anti-static agent over the carpet and many such agents are available. But it is widely considered better to introduce the agent into the fibre-forming polymer before it is spun into fibres. In general these agents have hydrophile properties arising from hydrophile groups in their molecules and can thus confer electrical conductivity

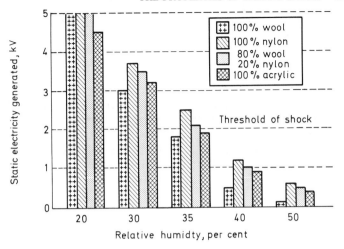

Figure 2.7 Comparison of the amount of static electricity developed by a person walking on a carpet composed of various fibres and fibre mixtures under varying humidity conditions; the threshold of shock to the walker is approximately 2kV

to the fibres. Thus piperazine compounds can for this purpose be introduced into fibre forming polyamides while various polyethoxylated compounds may similarly be introduced into other types of polymers before fibre spinning. But it is here that a difficulty is encountered for many such additives cannot be used since they would adversely interfere with the stage of fibre spinning. It is also necessary that the added anti-static agent should dissolve or disperse uniformly in the molten polymer or polymer spinning solution and not be decomposed by the high temperature spinning conditions.

Ultron is a nylon 6·6 fibre (Monsanto Textiles Ltd) having excellent wash-fast anti-static properties due to the addition of a selected agent to the spinning melt. It appears that this agent becomes concentrated in the fibre surface and perhaps this is a good feature since it is generally agreed that the static electricity produced in a fibre is largely situated in, and flows along, its surface. Courtaulds have recently introduced Celon Anti-stat which is a modified wash-fast nylon 6 fibre. Various fabric 'conditioners' such as Comfort, Glo-White, Soft Rinse, etc. can be added to wash-rinsing liquors to combat 'static'. They leave an electrical conducting film on the fibres (it is not wash-fast).

A new type of washfast antistatic synthetic fibre having the brand name of Epitropic (ICI Ltd) is, at the time of writing, now being developed. It consists of nylon or polyester distinguished by each fibre having up to 5% of carbon particles embedded in it; it can therefore be black or grey, and such fibres are electrical conductors. Those having the higher proportion of carbon can be used in fabrics and garments, etc. which can be maintained warm by an electric current passing through them. This greatly widens their potential use.

It has been found that most people can feel an electric shock from textile materials when the voltage of the static reaches about 3000 V. Such a shock is not lethal because the current flow that could arise from this is so small. However, the sparking can be a fire risk.

There are several methods for determining the liability of textile materials to accumulate static but a quite reliable, simple method involves rubbing the fabric and then holding it about 1 in (25 mm) above a little cigarette ash—if it fails to attract this ash to it the fabric can be judged to be sufficiently anti-static.

Flameproof fibres

The flameproofing of textile fibres is usually carried out when they are in the form of fabrics (p. 413) and for this a solution of a flame-proofing agent such as tetrakis hydroxy methyl phosphonium chloride $(HOCH_2)_4PCl$ is padded into the fabric and then dried. In one form of the treatment the padded fabric is further treated with ammonia before the drying for the purpose of better insolubilising the agent and thus making the flameproof finish fast to repeated washing. However, it is now found possible to produce flameproof fibres directly by adding suitable agents to the fibre-forming polymer before spinning it into such fibres as Darelle, Teklan and LoFlam (all of Courtaulds). The first is a viscose rayon containing bromine compounds such as tris (2:3-dibromopropyl) phosphate of formula $O=P(OCH_2CHBrCH_2Br)_3$. The second is an acrylic fibre made from a copolymer of acrylonitrile and flame-resistant vinylidine chloride and the last-named is a modified diacetate fibre. Teklan delustres in hot dyeing liquors to acquire a matt chalky appearance but can be relustred by steaming.

Recently the production of flameproof textile materials has received much attention and many countries have special legislation to ensure their wider use. This applies particularly to the USA where disastrous fires have resulted in appalling loss of life. It is found difficult to fireproof without harming the textile aesthetic properties.

Although all three fibres are proving useful in the production of fabrics and garments required to be flameproof the Du Pont flame-proof polyamide fibre Nomex has special interest because unlike ordinary nylon its polymer molecules are aromatic in so far as they are composed of poly (meta-phenylene isophthalamide) having the formula given on page 70 and made from metaphenylene diamine and isophthalic acid.

Nomex is especially heat resistant and in addition to being flame-proof it is able to withstand prolonged heating with but little loss of strength. For example even when held at the high melting point of nylon (260°C). Nomex fibres lose only about 20 per cent of their strength in 100 h and only 40 per cent in 300 h.

Nomex film is amenable to fibrillation (p. 69) to give fibres which separate or are crosslinked to form a network; in this latter form they are very useful for making industrial fabrics including filter-cloths. Now that Nomex fibres can be dyed with cationic dyes with the aid of a dye carrier (as for polyester fibres) without losing their flameproof properties there is a wide opening for them in furnishing and apparel fabrics and in aircraft interiors.

Fibrillated film fibres

Quite a novel and relatively simple method for producing fibres is based on the use of film as the raw material. This film can be made of almost any polymer say of the nylon, polyester and acrylic types, but the method originated with polypropylene film and this is the most widely used. Essentially the process commences with wide poly-propylene film which can have a content of an added substance favourable to its splitting at a later stage. Firstly the film may be of the chill-cast type produced by extruding the molten polymer through a horizontal slit so as to meet cold conditions such as a water-quench or contact with cold rollers to induce it immediately to solidify. Alternatively the molten polymer may be blown to solidify as a tube which can then be slit to form a sheet film.

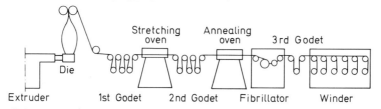

Figure 2.8 Extrusion of polypropylene film from the molten polymer and its fibrillation to produce fibres

Such film is then drawn in open width over a number of so-called 'Godet' rollers, whose peripheral speed steadily increases from the first to the last, in order to stretch the film lengthwise up to about nine times its original length with the object of aligning its long mol-

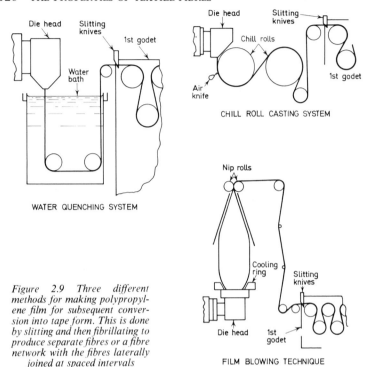

Figure 2.9 Three different methods for making polypropylene film for subsequent conversion into tape form. This is done by slitting and then fibrillating to produce separate fibres or a fibre network with the fibres laterally joined at spaced intervals

ecules parallel to each other and in the length direction of the film and thereby giving it much increased lengthwise strength, but not a width-wise increase, for this could make its splitting less easy. The stretched film passes through a hot annealing oven to eliminate latent strains which could later cause objectionable shrinkage.

At this stage the film is again run between rollers spaced about one or more yards apart and in this zone the film is vibrated or brushed or otherwise surface disturbed to cause it to split lengthwise and change into a sheet of closely spaced, parallel relatively coarse fibres thus much resembling a beam of warp yarn as prepared for weaving. Alternatively the film may be passed over suitably spaced cutting knives and then split to form tapes composed of the resulting coarse fibres.

The fibre tapes can be used as such or they may be twisted to give a twine or cord according to their width and these may then be made into rope if desired—such ropes are very suitable for marine use since the low density of polypropylene (0·92) ensures that they will float in water and their strong hydrophobicity will prevent their absorption of water. The twine is useful for agricultural purposes and the tape for packaging.

When a simple sheet of split fibres is produced this can, by introducing a weft thread into it, be made into fabric and this be employed for making sacking and bagging materials. By simply superimposing layers of the fibre sheets a suitable thick non-woven fabric can be made for backing carpets and other types of floor covering made by tufting or on a needle loom. But the splitting can be modified so that the resulting fibres are not completely separated but are left cross-linked at spaced intervals to form a sheet material somewhat resembling a length stretched piece of net. Obviously this material can be used as a type of fabric and form a basis for the making of a non-woven fabric with superimposed layers of loose cotton or other fibres ultimately bonded as a whole by means of an adhesive or by needling.

At the present time much effort is being directed towards improving the manufacture of split fibres by this fibrillation method and for diversifying the uses of the fibrous products obtained.

Self crimping fibrillated film fibres

An ingenious method for making fibrillated film fibres which can be readily induced to shrink and crimp by dry or wet heating (after the manner of bicomponent fibres, p. 133) consists of fibrillating a laminated or double film in which the two component films adhering together differ in respect of their physical properties but especially in the extent to which they shrink when heated.

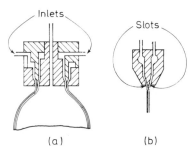

(a)   (b)

Figure 2.10 Extrusion of bicomponent (double) film for fibrillation to give self-crimping fibres. (a) Film extrusion head comprising two concentric annular extrusion slots which are fed under pump pressure through inlets with two different film-forming polymers. The extruded polymers together form a double-thickness tubular film which is maintained tubular by internal air pressure that causes the films to adhere to each other. Flat bicomponent film is then obtained by slitting lengthwise. (b) Extrusion of the two polymers through adjacent slots to form flat bicomponent film directly

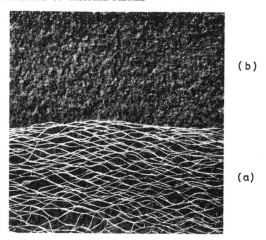

(b)

(a)

Figure 2.11 Net type of 'scrim' made by fibrillation of polypropylene or other film, (a) under special conditions that can be used as a backing to a non-woven fabric as shown at (b). (Courtesy Henschel Kunstoff-Maschinen Co)

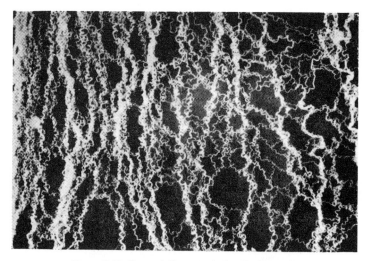

Figure 2.12 Crimped fibres made by fibrillating bi-component film. (Courtesy The Textile Manufacturer, July (1971))

Such double film may be made in various ways. For example the two films may be extruded separately in the usual manner and then hot pressed together after softening the surfaces with a polymer-swelling solvent or coating them with an adhesive. Alternatively the two films may be extruded side by side so that in the one run they are induced to coalesce.

Thereafter, the double film can, after the usual high stretching, slitting and fibrillation, give tapes of separate fibres which can then be converted by twisting into yarn and then optionally into woven or knitted fabric. In either of these forms high-bulk characteristics can be produced by dry or wet high temperature treatment to induce shrinkage and crimping.

Nylon fibres having modified dyeing properties

Mention has already been made of methods by which the dyeing properties of synthetic fibres can be modified so that they can be dyed with classes of dyes not normally applicable to them. These methods comprise a loosening of the fibre structure so as to make them accessible to dyes whose molecules are too large to penetrate them and by incorporating in the fibres substances usually of a basic or acid nature which will confer an attraction for dyes having an acid or basic character respectively, while at the same time effecting a loosening of the fibre structure. These methods are now being used for thus improving the dyeing properties of most types of synthetic fibres such as polyamide, polyester acrylic and polyolefine (polypropylene) fibres. With polyamide fibres, more particularly nylon (nylon 6·6 or nylon 6) a special method is used to modify their molecular structure and thus modify their dyeing properties so as to be more useful in producing colour patterned carpets (especially of the tufted type) or almost any type of nylon fabric having a pile surface.

Here it may be noted that the use of nylon fibres in a carpet can be specially advantageous since nylon possesses among the other textile fibres the highest resistance to abrasion and this property is noticeable when a carpet made say of wool or cotton contains only 20 per cent of nylon. In fact this proportion of nylon is often that used in carpet manufacture.

In making a woven fabric say of wool or other fibre and of say the Wilton or Axminster type it is possible to use coloured yarns and so produce very complex and attractive multi-coloured patterns which are quite beyond the scope possible in making the tufted carpets which are today so popular since they can be produced more cheaply and yet have quite good wearing properties even if not the equal of the woven carpet. This problem of colouring tufted carpets is, of course, attracting much effort to solve it and several different methods for producing more varied coloured tufted carpets are now being used and others are being investigated. As is later described (p. 305) tufted carpets can be printed with up to eight or more coloured patterns and they can be easily dyed solid to any one colour by using much the same methods and machines used for dyeing woven and knitted fabrics.

To help manufacturers of tufted carpets with this colouration problem producers of nylon (and acrylic) fibres are using special methods of manufacture so that while retaining their usual affinity for disperse dyes the fibres have an enhanced affinity for acid or basic dyes, according to whether the fibres are made more basic or more acid, respectively. Thus if these modified nylon fibres are introduced into a carpet together with ordinary nylon fibres according to any desired pattern then the carpet may be finally dyed, say in a winch machine with an appropriate dye mixture, thus allowing these different fibres to absorb the dyes preferentially to give a coloured pattern in which the colours may differ in depth of shade or be in contrasting shades.

An important feature of each molecule of nylon is that it has an amine group ($-NH_2$) at one end and a carboxylic acid group ($-COOH$) at the other end (see p. 58). The end amine group gives the nylon an affinity for acid dyes while the end carboxylic acid group only confers an affinity for basic dyes. Additionally the nylon has a very useful affinity for disperse dyes which it absorbs by a process of solution of the dye within each fibre and is not connected with either of the above two end groups.

Nylon manufacturers have now found how to increase the proportion of amine groups in the nylon molecule and thus give the nylon fibres a much increased affinity for acid dyes as compared with that of ordinary nylon. They have also been able to increase the affinity of nylon for basic dyes while at the same time very much reducing its affinity for acid dyes.

With the availability of three different types of nylon (deep acid dyeing, ordinary dyeing and basic dyeing) it is possible to use these in carpets and then dye in a single bath with a mixture of the appropriate dyes to secure a two- or three-coloured pattern. Thus when the carpet has a pattern in which ordinary nylon and deep dyeing nylon fibres are present and it is dyed with disperse and acid dyes in the same dye bath the deep-dyeing fibres acquire one colour (that of the acid dye) while the ordinary nylon fibres acquire the colour of the disperse dye.

Modification of fibre properties by irradiation

Increased interest is now being taken in the effect on textile fibres, such as cotton, nylon, polyester, wool and others, of high energy irradiation (for example gamma rays from the decay of Cobalt 60) and low energy irradiation (for example by bombarding electrons from a glow discharge) in the presence of gases and vapours of polymerisable monomers (expecially vinyl monomers) or while immersed in solutions of these. It has been found that according to the treatment employed, some polymer may be formed within the fibres, but more often a tenacious polymer film is formed on the fibre surface which can make the fibre more water repellent, flame-proof or hydrophile, in the last case giving the fibre useful soil-resist and soil-release properties and increased absorbency for dyes. Very interesting is the observation that irradiation can accelerate crease-resist and permanent press processes. For example, with the latter process, which is usually carried out on garments (with pleats and creases) on the premises of the garment maker, a washfast finish can be completed within 5 s. This applies when methylol acrylamide is used as the finishing agent.

It is difficult to summarise all the results of irradiation research so far carried out, for this is in its infancy, but the following selected observed features will indicate that several uses of irradiation of fibres and fabrics may ultimately be devised for commercial application.

Polyester fabric subjected to a glow irradiation in the presence of acrylic acid can be dyed with basic dyes for which it otherwise has no affinity.

Nylon and cotton irradiated while immersed in a solution in acetone of monomeric methylmethacrylate or in an alcoholic solution of monomeric styrene, acquires a substantial increase in strength. The cotton also acquires a very high resistance to attack by micro-organisms of various kinds so that irradiated cotton fabric could be made rot-proof for use in tropical countries. Very high amounts (up to 80 per cent calculated on the weight of fibre) of some reactive dyes can be induced to combine with viscose rayon with the aid of irradiation.

Cotton irradiated with polystyrene acquires thermoplastic properties. In the case of high energy irradiation of a roll of fabric about 50 per cent of its energy passes through at 100 mm depth so that by irradiation of only one thickness of fabric the efficiency of the treatment would be very low.

New fibre-forming polymers

For many purposes synthetic fibres are too hydrophobic and do not have the useful moisture absorbent properties of the natural fibres and thus are difficult to dye. Further, for this same reason, they accumulate static electricity and as a consequence more readily absorb soil to become dirty. Synthetic fibres also have low melting or softening temperatures and so become easily damaged by hot ironing and are subject to deterioration by exposure to light. In contrast, however, synthetic fibres have many very useful and unique properties not possessed by the natural fibres. Hence the incentive to discover new or modified fibre-forming polymers having these advantages but without the disadvantages. Perhaps it will never be possible to attain this object—the ideal fibre—but failure will not come from lack of effort.

Already there has been progress towards making synthetic fibres less hydrophobic (that is, more hydrophile) by forming them from polymers which contain a small proportion of water-attractive substances and in its turn this same modification can make the fibres more anti-static and easier to dye by the use of water-soluble dyes. Most of the fibre-forming polymers have a natural tendency to weaken by exposure to light (especially ultra-violet light) and so far the best remedy is that of introducing into the fibres, during their manufacture, ultra-violet light absorbers which may be organic or inorganic (manganese compounds, for example) as protectants.

It is not sufficient merely to make the synthetic fibres more hydrophile to give them easy-dyeing properties. This expedient must be assisted by incorporating in the fibre-forming polymer during its manufacture, or by adding to it at the spinneret, acid or basic substances having dye-attractive properties. Considerable success has been obtained in this way so that the dyeing of the fibres has now been much simplified. Additionally, it has by such methods become possible to enable carpets and the like pile materials to be dyed to different shades or in two-tone colours in a one-stage dyeing operation.

In quite another direction it has been found that by the use of special fibre-forming polymers, fibres somewhat weaker than usual can be produced and because of their lower strength they are less liable to cause the defect of pilling in fabrics and garments.

By addition of bromo-organic compounds containing phosphorus such as tris(2:3-dibromopropyl) phosphate to fibres during their manufacture it has been found possible to give them excellent resistance to catching fire as with Teklan and Darelle (Courtaulds).

An important line of progress which is now being exploited is that of using fibre-forming polyurethane polymers, since these allow the production of highly elastic fibres which are equal to rubber threads in many ways and are being used in swimsuits, corsets and foundation garments, etc.

Modified spinning techniques to produce fibres having special cross-sectional shape or a composite (heterofil) structure

A line of development which is now being actively followed and which is yielding valuable results is that of using special spinnerets so that it is now possible to produce fibres having special cross-sectional shape or to produce composite or conjugate fibres (also termed *heterofils*) or to produce hollow fibres.

It is obvious that the shape of the orifices in the spinneret through which the solution of the polymer in its molten state is extruded, can influence the cross-sectional shape of the resulting fibres. The normal shape of the orifice is circular, but it has now been found that advantages are obtainable in the fibres by using shapes which may be irregular or cross-like or of clover shape, etc. Clover and tri-lobal cross-sections give fibres which can have a special attractive sparkle type of lustre in garments and carpets. They can resist pilling. Such fibres also pack together well and support each other when they form a pile surface and thus can give increased resilience to the surface of tufted carpets. Various indented cross-sectional fibres can, in carpets, hide the soil picked up during use so that they appear less soiled.

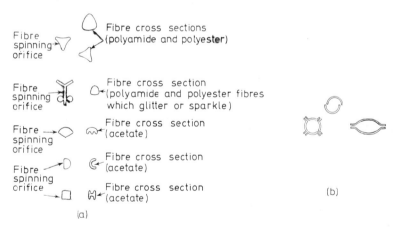

Figure 2.13 (a) Some special spinneret orifice shapes and the corresponding cross-sectional shapes of fibres spun through them. Double slotted orifices (b) give hollow fibres

Polyester fibres can be given specially enhanced brilliance by incorporating 15% of acrylamide in the fibre-forming polymer before spinning.

The use of orifices which allow two streams of fluid polymer to be extruded through them simultaneously are now being widely used for it has been found that the resulting composite fibres can have many advantages. In these composite fibres there may be two components each having a half-moon cross-sectional shape adhering tenaciously together to form a double fibre similar to that produced by the silkworm. But it is optional to produce such composite fibres with the two components lying side-by-side or having a sheath-core structure. The two components may be of the same polymer differing only in their physical properties or they may be different chemically and of course it is thus possible to make a wide variety of composite fibres having properties adjusted to suit their subsequent use. One new type of sheath-core composite fibre contains not one, but two or more cores.

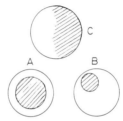

Figure 2.14 Sheath-core (A and B) and side-by-side bicomponent fibres (C)

At the present time it is useful to make a composite fibre having its components differing as regards their shrinkage in boiling water or at high temperatures. When such fibres are so treated they crimp (become wavy) as a result of the unequal shrinkage and thus they can be very useful for giving increased bulkiness to yarns and fabrics in which they may be present.

Composite fibres have rapidly acquired increased importance in so far as they allow the production of fabrics and garments by advantageous methods not otherwise possible. Characteristics much desired in garments and especially underwear are softness and warmth of handle. When the fabric of which a garment is made consists of long synthetic fibres, lying side-by-side parallel to each other in the constituent yarns, it usually has a firm and cold handle. Additionally, it can lack the useful moisture absorbent properties normally possessed by garments made from natural fibres such as wool and cotton. To give the garment these desirable properties a common expedient is to use yarns which consist of cut-up or staple synthetic fibres, since this ensures that the yarns have Hi-bulk. But if the fibres in these yarns are of the composite type and thus have strong self-crimping properties then the Hi-bulk which can be developed by hot water or dry heat treatment is much greater and the garment can be thus made much softer and warmer.

Composite fibres can be very useful for the manufacture of non-woven fabrics and especially fabrics having a pile surface, including carpets. It can be arranged that one of two finer fibres, which together form a composite fibre, has a lower melting point than the other. Then, when these composite fibres of suitable cut-length are carded and brought into the form of a web with the fibres lying randomly distributed or orientated in the length or width direction of the web, this can, in a needle loom, be converted into non-woven fabric with or without a pile surface. Then can follow a light hot pressing at a temperature sufficient to soften the fibre having the low melting point, so that it effects a bonding of the other fibre which has the high melting point and which is unaffected by the hot pressing conditions. This method (aided by steaming) avoids the necessity of using an adhesive to effect the fibre bonding.

The manufacture of Melded (ICI Ltd) non-woven fabrics is based on the use of sheath-core bicomponent fibres in which the sheath has a melting point lower than that of the core and is able to effect fibre bonding by becoming adhesive when heated to a suitable high temperature or steamed.

Non-woven fabrics produced by the above method have been found useful for backing tufted carpets. Previously, jute fabric has been almost exclusively employed as the backing fabric, but the non-woven fabrics have advantages which have now been recognised, one of which is that they do not shrink so much when the freshly made tufted carpet is dyed.

While developments are continuing along the lines outlined above there is always the possibility of fresh discoveries opening up other lines of progress. An example of this is the recent discovery that synthetic fibre yarns can be produced by a method termed 'spun bonding' in which the usual yarn spinning procedures are entirely by-passed—in this method the fibres within a continuous bundle of parallel fibres are spot-welded together at numerous points along its length to produce a yarn completely free from twist but nevertheless strong enough for making into fabrics.

'Islands-in-a-sea' fibres   A different type of composite synthetic called *islands-in-a-sea* fibre has now become available. Instead of consisting of two equal fibre halves joined throughout the length of composite fibre, the new fibre consists of a bundle of very fine fibres enclosed within a larger single fibre throughout its length (Figure 2.15). The fine fibres and the encapsulating fibre are made from

different fibre-forming polymers so that they have different physical, dyeing and lustre properties. Differences such as these can be utilised in many textile materials.

Figure 2.15 Alternative cross-sections showing two of the many different distributions of the fine fibres

A special variant of these fibres has been produced in which the fine fibres are of carbon while the encapsulating fibre has a thin carbon surface, but otherwise consists of a fibre-forming polymer. The special property of these fibres is that they are electrically conducting due to the carbon content. Thus by varying the amount and distribution of the carbon, the fibres can be made antistatic, but perhaps more interestingly, fabric such as a blanket or an article of clothing made electrically conducting by containing a sufficient number of these fibres can be made warm (and the heat is adjustable) by passing an electric current through it.

CHAPTER THREE

The conversion of fibres into yarns and fabrics

In most cases the raw material for the production of yarns and fabrics is a mass of entangled fibres which are extremely long in comparison with their thickness. The individual fibres also differ from each other greatly since, in general, they will be at different states of growth and maturity. The exceptions to this are real silk, the rayons and the synthetic fibres. As we have seen, the silkworm spins a continuous length of a double filament, and this is at once reeled into skeins of yarn, whilst manufacturers of continuous filament rayons and synthetic fibres also produce these directly in the form of yarn. Waste real silk and rayon and synthetic staple fibre have to be brought into yarn form by ordinary methods.

It is further to be noted that in order to make transport easier the natural raw fibres are usually highly compressed and contain most of their natural impurities.

General principles of yarn and fabric manufacture

The first stage in the production of a fabric from fibres is to clean and mix them thoroughly. The fibres are then generally straightened, but for the production of certain types of fabric they must be brought into a condition in which they are all parallel. The fibres are next drawn out into the form of sliver, which resembles a rope but with the fibres having no twist. Repeated drawing (extenuating) and twisting follows. This twisting is to give the resulting roving just sufficient strength to prevent breakage in its manipulation (extenuation). Thus a fine roving is produced which is finally twisted into yarn. The yarn is used to produce fabrics by either knitting or weaving.

An alternative method of producing a fabric from fibres, without making a yarn, is that of felt making. The wool or rabbit fur fibres which are employed can be felted or matted into the form of a fabric, the strength of which depends on the degree of entanglement or interlocking of the fibres. Yet another method now very popular is that of assembling a layer (web) of superimposed fibres of suitable width and containing an adhesive (this may be a special type of fibre which becomes tacky and adhesive at a raised temperature) and then 'needling' or hot compressing the web to produce a so-called non-woven fabric (p. 200).

It will be realised that for the carrying out of these manufacturing processes a wide range of different types of complicated machines and a great variety of methods are used. Such processes have taken more than two centuries to perfect and even now, partly owing to the increasing use of the rayons and synthetic fibres, modifications are constantly being introduced. Thus it is not proposed to deal here in any detail with yarn and fabric manufacture. All that is necessary is to outline the essential facts and principles so that the reader may have a better understanding of what a yarn or fabric is, and how it has been produced.

Preliminary processes

The first point to notice is that, in general, thorough purification of textile fibres is left to that stage at which the fibres are in the form of yarn or fabric. This is because it is more convenient to apply purification treatments with the fibres in a manufactured form such as yarn

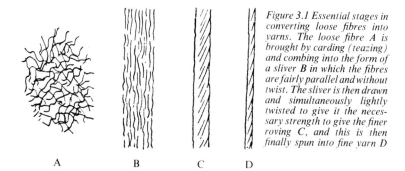

Figure 3.1 Essential stages in converting loose fibres into yarns. The loose fibre A is brought by carding (teazing) and combing into the form of a sliver B in which the fibres are fairly parallel and without twist. The sliver is then drawn and simultaneously lightly twisted to give it the necessary strength to give the finer roving C, and this is then finally spun into fine yarn D

A　　　　B　　　　C　　　D

and fabric as they then lend themselves better to mechanical handling. This does not apply to such solid impurities as sand, dirt and bits of stalk or leaf. These impurities are removed before spinning since they would interfere with these early processes. It is the natural fat, wax, colour and similar impurities which can remain until the later stages.

Exceptions to this rule are provided by wool and waste silk. Raw wool is generally so impure, perhaps containing only 50 per cent of wool fibre, that it is necessary to give it a scouring treatment quite early, in order to remove from it most of the natural grease and suint (perspiration) which it contains. Raw silk waste contains about 25 per cent silk-gum, and it is sometimes better to remove this before spinning for its removal leaves the fibres softer and more pliable.

The scouring of raw loose wool fibre is carried out continuously in long machines which comprise three or four shallow iron tanks. These are filled with a warm detergent liquor, usually soap solution, which may contain an alkali such as sodium carbonate, and the wool is gently propelled through these tanks and the liquor by means of swinging rakes. Between successive pairs of tanks are mangles so that as little as possible of the dirty liquor from one tank is carried forward by the wool into the next one. In this way the wool becomes progressively cleaner as it travels through the machine until, in the last tank, it has a final rinse in clean water and then passes forward for drying. The construction and operation of scouring machines of this kind (p. 212) are such as to disturb the wool as little as possible. If the action is too vigorous the wool fibres become entangled and felting occurs. Thereafter it is most difficult to remedy this and trouble is experienced in all the succeeding processes.

A relatively small proportion of raw wool is purified by extraction with organic solvents such as white spirit. It is claimed that the wool is less harmed than when alkaline soap scouring solutions are used.

In the case of purifying waste silk the removal of the silk-gum is best carried out by boiling the silk in a strong soap liquor. This softens the gum and soon dissolves it. The amount of silk-gum thus removed generally amounts to about 25 per cent of the weight of the raw silk. The spent liquor is frequently used by silk dyers since when added to a dye solution it can assist in producing even dyeing.

YARN MANUFACTURE

We can now deal in somewhat more detail with the methods and machines used for producing yarns from the different fibres, cotton, wool, linen and silk. It must be understood that only typical methods can be described since views on the best combination of these methods vary from one spinner to another. Moreover, since many of the machines are used with all these fibres, such machines will receive fuller mention in connection with cotton so that it will not be necessary to describe them again in dealing with other fibres.

Cotton

Mixing, cleaning and opening operations   When raw cotton arrives in bales at the spinning mill it is found that these are so highly compressed that when opened the cotton falls apart in large hard lumps which obviously must be thoroughly loosened and opened out before spinning into yarn. Furthermore, there is a considerable amount of impurities such as broken stalks, leaf, sand, grit, seeds and motes which must be separated and removed from the cotton. Each bale will differ from its neighbour to some extent as regards the quality and nature of the cotton which it contains, even when the bales are supposed to be of the same type. In many instances it is intentionally required to mix different qualities in order to secure a blend having desired properties. Thus, in the next stages of treatment designed to bring the cotton into a form suitable for spinning into yarn, arrangements are made so that purification and mixing take place in addition to the other changes.

Stages in the manufacture of cotton yarn

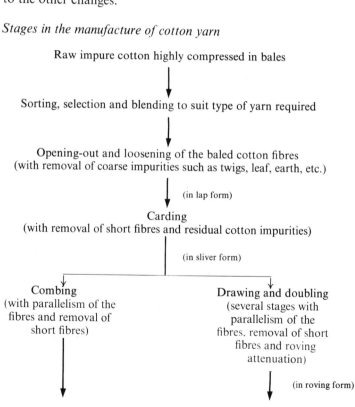

Raw impure cotton highly compressed in bales

Sorting, selection and blending to suit type of yarn required

Opening-out and loosening of the baled cotton fibres
(with removal of coarse impurities such as twigs, leaf, earth, etc.)

(in lap form)

Carding
(with removal of short fibres and residual cotton impurities)

(in sliver form)

Combing
(with parallelism of the
fibres and removal of
short fibres)

Drawing and doubling
(several stages with
parallelism of the
fibres, removal of short
fibres and roving
attenuation)

(in roving form)

↓   ↓

Drawing and doubling   Spinning into coarse
(several stages to   and medium-fine yarns
effect mixing and
further parallelism of
the fibres and roving
attenuation)

| (in roving form)
↓

Spinning into fine cotton
yarns by mule or ring
machines and finally winding
as cops or on bobbins

The first part of the treatment involves converting the highly compressed cotton bales into the form of an even uniform layer or lap of fibres neatly rolled so that it resembles a much-enlarged roll of surgical cotton wool. In such a lap the fibres are required to be only lightly packed, evenly distributed, as straight and parallel as is possible at this stage, reasonably free from tufts or hard matted lumps and with a good proportion of the original impurities removed.

A number of machines are used to produce this result, the chief ones being hopper bale breakers, hopper bale openers, hopper feeders, Crighton and porcupine openers and, finally, scutchers. These machines vary considerably in their construction, but their action is mainly based on progressively loosening the compressed cotton by the combing or teazing action of moving spiked rollers or belts, the beating action of rotating or striking devices and the distending action of air currents within the machine. It is arranged that the cotton is continuously moving forward during such treatment and that it is brought opposite or over grid surfaces to allow the loosened impurities to pass through. The mixing which necessarily takes place in such machines is an important aid in securing a uniform cotton lap.

The machines enumerated above are employed in series and in groups according to the types of cotton being dealt with and the character of the ultimate yarn required. Different mills also have different ideas as to the best combinations of machines. For these reasons it is not possible to describe any one arrangement or method as being the best but rather to indicate the nature of the machines and the manner in which they act on the cotton. However, one satisfactory layout of machines for dealing with medium qualities of American cotton is illustrated in Figure 3.3. It includes at least one of each type of machine.

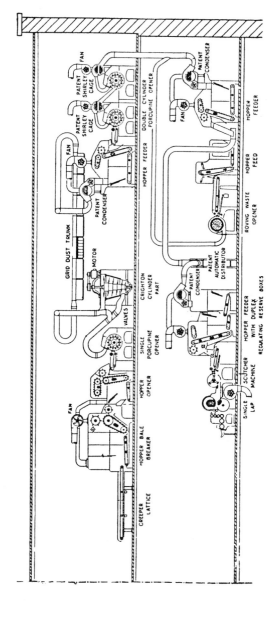

Figure 3.2 Typical layout of machines for processing American cotton. The cotton passes in turn through a series of bale breaking, opening and scutching machines in which it is opened, purified and mixed. At the end of this procedure the cotton is in lap form ready for carding. (Courtesy Platt Bros & Co Ltd)

It has already been mentioned that mixing of the cotton takes place whilst it passes through these machines. In the illustrated layout it is to be noted that such mixing is obtained entirely by this means. There is, however, the alternative of introducing a stage to secure further mixing in which the partly processed and loosened cotton is led to rooms where it can be stacked one layer above another. Then when the cotton is withdrawn to pass through the next machine vertical sections are taken so that a little from each layer becomes mixed with similar portions from the other layers. This method is used in producing fine yarns. A variant of this method of mixing is to use a large sealed chamber into which the loosened cotton is sucked so that it falls like snow-flakes, thus being thoroughly mixed. When either of these stacking methods is used it is convenient to stack the cotton after it has passed through a hopper bale breaker and a hopper opener and this system is illustrated in Figures 3.3 and 3.4. Alternative mechanical mixing is now favoured.

The sequence of treatments illustrated for American cotton can now be described.

After breaking the bale fastenings, the cotton is led through a hopper bale breaker with the object of commencing the opening-out treatment and at the same time removing some of the heavier impurities. In feeding the cotton into this machine it is easy to secure a fair degree of mixing by throwing in equal amounts of cotton taken in turn from a number of bales around the machine instead of simply throwing in cotton taken all from one bale and then all from the next bale and so on.

As the cotton passes through the bale breaker it is brought to the bottom of an upward-inclined endless brattice which carries numerous spikes. Small portions of the cotton are thus torn or teazed from the hard lumps and carried upwards on the spikes. At the top, excess of cotton is knocked back by a spiked roller rotating in the opposite direction. Then as the cotton is carried downwards on the brattice in another part of the machine it is beaten off by a so-called *stripping* roller so as to pass over a grid to be discharged out of the machine. The result of all this is that the cotton receives a good whipping-up and loosening whilst many of the impurities fall through the grid.

The cotton next passes into the hopper opener which is somewhat similar to the bale breaker in so far as the cotton is carried on spikes attached to an endless upward-inclined brattice with arrangements for combing off excess cotton on the upward path and beating off residual cotton on the downward path so that the loosened cotton passes over a grid and then falls down on to a brattice which carries it into the porcupine opener.

144

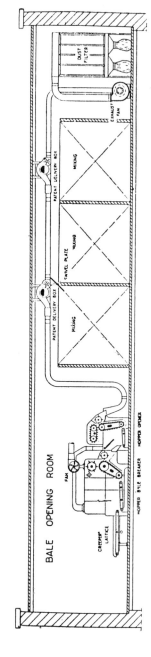

Figure 3.3 Layout of the bale-opening room for the preliminary opening and mixing of cotton. Raw cotton taken from several bales is led through the hopper bale breaker and the hopper opener and is then drawn pneumatically through delivery boxes to be evenly distributed in the mixing bins. Dust carried from the cotton is filtered out of the exhaust air. (Courtesy Platt Bros & Co Ltd)

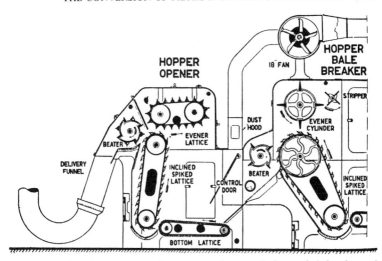

Figure 3.4 Sketch showing essential construction features of a hopper bale breaker and a hopper opener as used in the preliminary stages of processing raw cotton. (Courtesy Platt Bros & Co Ltd)

The porcupine opener consists essentially of two concentric cylinders, the outer one having a slotted or grid construction, whereas the inner one is built up of parallel metal discs spaced out on a central shaft. Small combing plates are secured to the periphery of each disc and there is sufficient clearance for cotton to move between these plates and the inside of the outer cylinder. The inner cylinder is rotated at a high speed. Cotton from the hopper opener is fed into the porcupine opener by means of a moving brattice and finally by a pair of feed rollers. At this point the 'teeth' of the inner cylinder quickly teaze or comb off small portions of the incoming cotton and these immediately fly against the grid bars. Loosened impurities pass through the grid whilst the purified cotton passes on around about three-quarters of the circumference and finally into the next machine, the Crighton opener. There is thus opportunity during the whole path of travel adjacent to the grid for the cotton to lose its impurities.

The Crighton opener consists of a vertical arrangement of strikers. The conical portion is built up of circular plates suitably spaced and each has a number of striking blades riveted to its periphery, the upper plates having more blades than the lower plates. This conical beater rotates on a vertical shaft within a cylindrical grid. Thus, as the cotton passes upwards between the beater and the inside of the grid, it is teazed so that flakes of cotton are detached and thrown against the grid which allows the loosened impurities to pass through.

The action is comparable to picking out small portions from a wad of cotton wool with thumb and finger.

The cotton, which is now in a much bulkier state, is drawn pneumatically through a trunk system where further impurities can fall out through a grid and then it enters a hopper feeder. Its action is similar to that of a bale breaker but considerably lighter. Further, it not only removes impurities and opens the cotton, but it evens it out in lap form so that it can be better presented to the next series of opening machines of the types previously described. The idea behind this hopper feeder is to pass the cotton on to further opening machines

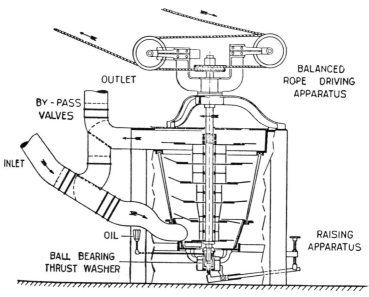

Figure 3.5 Sketch showing essential features of a Crighton opener for raw cotton. (Courtesy Platt Bros & Co Ltd)

at a regular rate and in the form of a wide lap of fairly even texture.

All the different types of opening machine involve a beating of the cotton under conditions such that the loosened dirt can pass through a suitably disposed grid. Possibly this idea of separating the impurities is ultimately based on an old-time method of beating or flaying the cotton whilst laid on open mesh material. The sand or grit would then fall through the meshes and leave the cotton substantially purified.

In opening machines the cotton is assisted in its travel by suction which also helps to loosen it. Passage of cotton through such opening

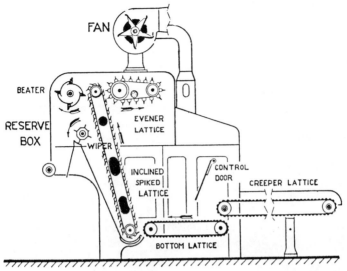

Figure 3.6 Sketch showing essential features of a hopper feeder. (Courtesy Platt Bros & Co Ltd)

machines makes it up to twenty times more bulky and this is evidence of their opening efficiency.

The cotton is delivered in the form of laps from the opening machines and it is convenient to make these laps into rolls.

Scutching   The next machine, which supplements the work of the opening machines just described, is known as a *scutcher*. This machine essentially consists of a rapidly rotating two- or three-bladed beater mounted over a grid. As the lap of cotton is fed into this machine it passes under the revolving beater and it is still further loosened by suction. Remaining impurities are removed through the grid. By feeding into the machine three or four laps of cotton superimposed instead of in succession a further degree of mixing is ensured.

Carding   There remains yet one more machine through which the cotton has to pass in order that it may be in a state suitable for making yarn. This is the carding machine which removes all the remaining impurities and at the same time removes very short and thin immature fibres. It breaks up any particularly hard tufts of fibre that have escaped the action of the previous opening and scutching machines and also attenuates the mass of fibres and makes them more nearly parallel to one another.

The carding machine is very interesting, and it has taken much ingenuity and perseverance to bring it to its present-day state of perfection. In many ways it may be looked upon as a combing machine, but this viewpoint must not be too strongly emphasised because there are other much more efficient machines for combing textile fibres when it is desired that they shall be aligned really parallel to each other.

In the carding machine use is made of card clothing. This consists of leather or composite fabric in which are anchored densely packed fine-pointed wires, their pointed ends being bent over.

Figure 3.7 Lap scutching machine in use showing three different laps passing simultaneously into the machine where they are combined into one lap to give a more uniform fibre mixture

The machine itself consists essentially of a large horizontal cylinder covered with card clothing, two small cylinders, one on each side, one being covered with card clothing and the other having cut teeth, and an endless brattice covered with card clothing which is disposed near to and around the top half of the large cylinder. All these parts move in relation to each other and the cotton passing through the machine.

The cotton from the scutching machine is fed in lap form on to the toothed small roller (the 'licker-in') which is rotating very rapidly. The teeth of this pierce the lap and carry it forward in attenuated form to meet the large cylinder, where it is taken off on to the wire points of this cylinder, again in more attenuated form. The cotton, held by the wire points in the form of a very thin film on the large cylinder, now passes under the moving wire points attached to the moving brattice ('flats') above. The cotton is thus combed and straightened whilst it is carried forward to the second slowly rotating smaller roller (the 'doffer') which picks it off the large cylinder. From

Figure 3.8 Cotton carding machine. The cotton is freed from impurities and made more even by a process somewhat similar to combing. Immature fibres are removed. Carding is important, since the entangled raw cotton is brought together in a more or less parallel sliver form which can then be progressively extenuated before twisting into yarn. (Courtesy Platt Bros & Co Ltd)

the doffer roller the straightened and purified cotton passes in the form of a lap through a vibrating comb and is then narrowed to the form of a thick sliver which is coiled into tall narrow cans.

It is to be noted that, apart from its cleansing action, the effect of carding on the cotton is to attenuate a comparatively thick lap of cotton to a gossamer-like film by drawing it out, comb the fibres in this film more or less parallel to each other, and then bunch this thin film or lap into rope form, which is known as a *sliver*. At the end of the carding operation the cotton has passed through the first stage of its conversion into yarn.

First drawing process   The cotton sliver is now ready to pass through a drawing process which has the objects of further mixing the fibres and making them more parallel. These objects are attained by leading several of the slivers through a machine known as a draw-frame, which comprises three or four pairs of rollers suitably spaced and in line; each succeeding pair of rollers is driven at a higher speed than the preceding pair. Thus, on leading a sliver through these rollers, it becomes drawn out or attenuated and made much longer. In order to secure mixing, several slivers are together led through at the same

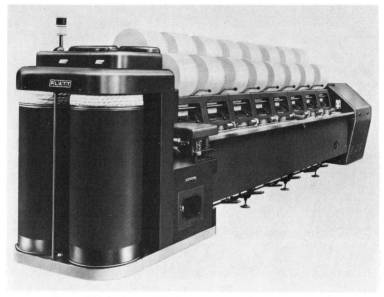

Figure 3.9 The Century Comber. (Courtesy Platt Bros & Co Ltd)

time and combined in the drawing-out so as to yield a corresponding longer single sliver. If the resulting sliver is then led through the machine again and again each time in combination with others, it is evident that the mixing and attenuation will be very thorough.

The rollers have to be spaced accurately so as not to lead to breakage of the cotton fibres, for it is obvious that if any single fibre were held gripped by its ends between two succeeding pairs of rollers then it would break under the strain. The necessary grip on the sliver whilst passing between the rollers is secured either by having the upper ones covered with leather and the lower ones fluted, or by having both upper and lower rollers fluted.

Cotton combing   If the cotton is to be made into very fine yarns, it is advisable to subject the sliver from the draw-frame to a combing operation. Two machines are used in the cotton industry for this purpose, the Heilman and the Nasmith combers, the latter being more popular. These combers are highly intricate machines. Not only do they comb the cotton sliver by means of fine needles but at the same time they remove the shorter fibres and leave the cotton much more uniform as regards the lengths of the fibres of which it is composed. The spinning of fine yarns would be difficult if the cotton were made up of a mixture of long and short fibres, and since fine yarns are better spun from long fibres it is arranged that the short ones are removed in combing.

Draw-frames   We can now return to the further preparation of the cotton for spinning, since the combing must be regarded as an extra process. At this stage the original cotton has now been converted into one of two distinct forms. One is the carded sliver and the other is the carded and combed sliver. The differences between them are important yet simple. The carded sliver contains fibres not uniform in length and not particularly parallel to each other. These fibres were made parallel whilst travelling over the large cylinder of the carding machine, but they were crossed again whilst being taken off by the doffing roller and brought from the form of a wide lap to that of a sliver. In contrast, a combed sliver consists mainly of long fibres of about the same length and they are all parallel to each other.

These two types of sliver are now led through special draw-frames, so that they may be steadily reduced to a thickness of sliver (later termed *roving*) suitable for spinning into yarn.

The first machine is a draw-frame provided with four pairs of rollers driven at increasing speeds, as described for the draw-frame used to prepare the carded sliver for combing. In this machine the slivers are mixed or combined and then attenuated, a process more usually spoken of as *drafting*. The draft is such that on emerging from the draw-frame the sliver is only just strong enough to withstand further handling.

This sliver is then led through a *slubbing frame* in which it is drafted by the usual arrangement of pairs of rotating rollers, but on emerging from the last pair of these rollers it is led through a flyer which slightly twists it into an attenuated rope form, known as *roving*, and winds this on to a bobbin. Twist impedes drafting so the amount of twist inserted in this machine is only that amount necessary to give sufficient strength to the roving to enable it to be further processed without breaking.

Further drafting and mixing of these rovings is then secured in the next machine, which is known as an *intermediate frame*. Again the roving emerging from the last pair of rollers is slightly twisted with the aid of a flyer and is wound on bobbins.

The next machine of the same type is known as a *roving frame*, and this delivers a roving of a degree of fineness and evenness such that it is ready for spinning most types of yarn. However, if very fine yarn is to be spun, it is better for this roving to pass through yet another of these drawing machines which is this time known as the *jack* or *fine roving frame*.

Spinning of cotton yarns   After passing through all the processes enumerated above, the cotton is now fully purified, thoroughly blended so as to have a uniform composition, and has its individual fibres aligned parallel to each other so that it is ready for spinning into yarn. The roving at this stage is about as thick as coarse string. But what a large amount of detailed processing is necessary to bring it to this point!

Cotton yarns are spun by one or other of two methods which are generally known as *ring* spinning and *mule* spinning. The latter is the older method and its discovery will always be associated with Crompton, of Bolton, the machine often being referred to as *Crompton's mule*. Ring spinning machines give a larger output than mule spinning machines and so there is a tendency today to develop ring spinning to the utmost. However, there are subtle differences in the qualities of the yarns which result, so that the mule is favoured for certain types of yarns. It is not intended to describe these spinning machines in any detail here but rather to indicate their essential functions.

The mule spinning machine occupies considerable floor space. At one end is a fixed frame on which are placed the bobbins of cotton roving ready for spinning. Also attached to this frame are numerous groups of three pairs of rollers in line and rotating at progressively

Figure 3.10 Ring spinning frame showing bobbing of cotton (top)
being twisted into yarn and wound on spindles (bottom)

increasing speeds just as they do in the draw-frames already described. Quite separate are carriages which run on the floor to and from the fixed frame carrying the bobbins, the length of travel being about 60–66 in (1·5–1·7 m). On the head of each carriage are numerous fine spindles. Such are the general features of the mule.

In spinning, the roving is drawn from the bobbins to pass between the drafting rollers and thence to be wound on the spindles of the carriage. This takes place in three separate stages. Firstly, the carriage moves away from the frame and draws the roving with it. The roving is drafted as it passes from the drafting rollers. Of course, if only this drafting took place the roving would break, but it is arranged that the spindle on the carriage to which the one end of the roving is attached steadily rotates. Thus the roving is twisted as it is drafted, the amount of twist being sufficient to give the necessary strength to the roving without preventing it from undergoing the draft. So, in this first operation, the roving is converted into fine twisted yarn extending between the carriage spindle tip and the drafting rollers of the frame. In the next operation the spindle is further rotated to insert additional twist, such as is required to make the yarn strong and have the characteristics required of it. Then follows the third operation in which the carriage moves back to the frame whilst the spindle again rotates in such a manner that the yarn just formed is wound on it without further twist. These three operations are then repeated over and over again with the formation of short lengths of yarn in each cycle until the spindles are full and need changing. This method of mule spinning is thus intermittent, and in this differs from ring spinning.

In a ring spinning machine the bobbins of roving are positioned on the upper part of a framework. The roving from each bobbin is led downward through drafting rollers of the usual type to be wound on a lower vertical spindle which rotates at the very high speed of 5000–10000 rev/min. Around the spindle is a light ring, and attached to this is a very small light ring-shaped traveller which can move easily around the ring. The ring, with its traveller, moves regularly up and down the spindle. Roving coming from the drafting rollers is threaded through a small yarn guide vertically over the spindle, through the traveller and thence to the spindle.

When this spinning machine is operating the roving is drafted by the rollers and is then wound on the spindle. The traveller imparts a slight drag on the attenuated roving passing through it and this has the effect of tensioning it whilst guiding it around the spindle. Thus the roving is drafted and suitably twisted into yarn as it passes from the drafting rollers to the spindle. The amount of twist is governed by the relative rates at which the drafted roving is delivered to the spindle and at which the spindle rotates. Ring spinning occupies but little floor space and is continuous, features which commend it in preference to mule spinning.

Linen

As we have seen previously, linen fibres are long and coarse and it is only within the last few years that it has been found possible to spin flax yarn satisfactorily on a ring spinning machine. The normal method of producing linen yarn, however, involves the use of a third method which is known as *flyer* spinning.

So far we have considered linen up to the point at which the retted flax stalks have been broken and scutched into coarse long fibres aligned fairly parallel to one another. From this point the fibres are *hackled* and this has the effect of making them firmer and separating the longer fibres from the shorter ones. The long fibres are known as *flax line* and the short ones are designated *flax tow*.

In hackling, a machine is used whose main function is combing. The fibres are moved in contact with a roving series of coarsely spaced pins which comb out the short fibres and at the same time straighten the longer remaining fibres.

Subsequently these two types of flax fibre, tow and line, are processed separately, since the former will be made into coarse yarns whilst the latter will be employed for spinning into the finest and highest quality linen yarns. In spite of this difference they will pass through the same kinds of machines which will now be used to mix, straighten and make uniform by drafting processes; they involve the use of carding and drawing machines. With suitable modifications these machines are based on the same principles of construction as those used for cotton.

The exception is that, in the draw-frames, use is made of pins for further straightening the long linen fibres. The linen fibres are drawn as a layer from one pair of rollers to another pair with a suitable distance between them. In this gap the fibres are combed by means of parallel rows of pins which rise up to penetrate the fibre layer and then move along through it to the point at which the fibres enter the second set of rollers. At this point the pins drop down and return to the point from which they started so that they can again rise into the oncoming fibre layer. It may be noted here that this method and type of machine for straightening the fibres is also much used for dealing with wool, when it is known as a *gilling machine*. The linen fibres finally pass through a frame to attenuate the roving sufficiently to enable it to be spun into yarn.

Spinning is carried out by wet or dry methods using a flyer spinning machine. In this the roving is drawn downwards from vertical bobbins through a thread guide, between two pairs of rollers which produce drafting, and finally through a flyer. This flyer is mounted over a vertical spindle around which is a bobbin on which the yarn is wound. The flyer is shaped like an inverted U and its function is to twist and guide the oncoming yarn around the bobbin which is regularly moved up and down during the spinning so as to build up a tight, regular package of yarn. In these spinning operations the roving is drafted to yarn thickness and is simultaneously twisted so that fully formed yarn is wound on the bobbin around the spindle.

In *wet spinning* the roving is led through a trough of hot water on its way to the flyer so that the fibres are softened and thereby brought into a better condition which is necessary for spinning fine yarns. Wet spinning is not required for Linron processed linen fibres (p. 249).

Silk

It has already been described in dealing with the production of silk how small groups of cocoons are unwound or reeled together to produce silk thread having a small amount of twist, just sufficient to hold the individual threads together. Silk thread produced in this way is used for producing the strongest and highest quality yarns and this will be done by twisting two or more of these threads together, it being recalled that up to a point increase of strength is obtained with increase of twist. This combination of the threads and their twisting together is carried out on machines similar in principle to those already described for the flyer spinning of linen yarns. If · these long-fibred threads are required to have the maximum lustre then the amount of twist is reduced to low limits, but if strength rather than lustre is desired then high twist is inserted.

Treatment of waste silk   A large proportion of all the silk produced is not so uniform and easy to convert into strong lustrous threads. In dealing with the cocoons a considerable amount of waste or entangled silk is obtained and this has to be worked up into yarns by methods and machinery quite different from that involved in *reeling* and *throwing* as indicated above. Silk yarns made from this waste silk are usually referred to as *spun silk yarns*.

In working up this waste silk it is first necessary to cut it up into fibres of about the same length as far as possible, 2 to 3 in (50 to 75 mm) being usual. This silk is then *degummed* by one of two methods. It is either boiled with a 1 per cent solution of soap or else subjected to a fermentation treatment. Practically all the silk-gum is removed by the first method, leaving the silk fibres soft and supple; the fermentation process removes only about half the gum, but this leaves the fibres sufficiently softened to pass through the succeeding operations.

Now follows a succession of processes including beating, filling and dressing, with the object of bringing this much-entangled silk material into the form of a sliver in which the fibres have about the same length and are arranged more or less parallel to each other. At this point the silk is amenable to the ordinary processes of combing and drawing and it is then finally twisted into yarn.

In converting waste silk into spun silk yarns there is a certain amount of a lower quality silk waste produced. This can be carded and worked up into comparatively coarse yarns which also find useful application in the silk industry.

Types of silk yarns The silk from the cocoons is ultimately made into two main types of silk yarn, *tram* and *organzine*, both obtained by reeling the cocoons to form long fibred threads which are lightly twisted for tram, but highly twisted for organzine. These two types of yarn are used for weft and warp threads, respectively, in making woven silk fabrics. Neither is degummed, so that the individual threads are really double. *Spun* or *schappe* yarns which are short fibred and free or partly free from silk-gum are made from silk waste. The tram and organzine yarns have the highest value and they show real silk at its best as regards fineness, lustre and strength.

Wool

The wool industry has to deal with many different types and qualities of wool. These vary widely in fibre length from, say, 2–14 in (50–350 mm) and also differ considerably as regards their thickness. On this account the manufacture of wool yarns has become divided into two main sections, *woollen* and *worsted*. Whilst some machines are used in both of these sections there are others which have been specially devised for use in one or the other.

Stages in the manufacture of woollen and worsted yarns

Raw greasy wool

Sorting, selection and blending to suit type of yarn required

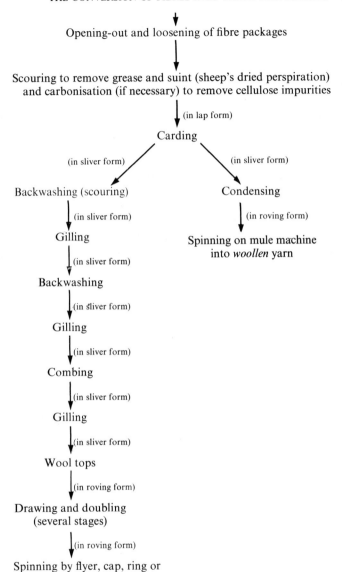

Opening-out and loosening of fibre packages

Scouring to remove grease and suint (sheep's dried perspiration) and carbonisation (if necessary) to remove cellulose impurities

(in lap form)

Carding

(in sliver form)   (in sliver form)

Backwashing (scouring)   Condensing

(in sliver form)   (in roving form)

Gilling   Spinning on mule machine
                                            into *woollen* yarn
(in sliver form)

Backwashing

(in sliver form)

Gilling

(in sliver form)

Combing

(in sliver form)

Gilling

(in sliver form)

Wool tops

(in roving form)

Drawing and doubling
(several stages)

(in roving form)

Spinning by flyer, cap, ring or
mule machine into *worsted* yarn

It has already been seen in dealing with the production of cotton yarns that the methods and machines used are very intricate. For the production of woollen and worsted yarns even greater ingenuity has been devoted to the design of machinery for carrying out the various processes in the most efficient manner. In this account it will therefore not be possible to deal in detail with the machinery involved.

Worsteds and woollens   For the manufacture of worsted yarns it is customary to use the finer qualities of wool, since in these yarns the fibres are arranged parallel to each other and it is generally desirable for these yarns to be strong and fine. The use of long fine wool fibres assists in achieving the production of a smooth even yarn. Woollen yarns are much less regular, since they are composed of shorter and often considerably coarser fibres which lie unevenly distributed in the yarn and with little semblance of parallelism.

Worsted yarns are employed for making fine clear-cut cloths, whilst the woollen yarns are useful for making rougher cloths in which a certain degree of milling or felting is invoked to give the fabric a more solid appearance. A worsted fabric is not milled except occasionally to give just a little surface cover. For the manufacture of worsted yarns the wool fibres have to be combed but this is not necessary for woollen yarns. More operations are involved in the production of worsted yarns.

Woollen yarn manufacture   Wool, which has been scoured in the loose form as described on p. 212, has to be prepared in the form of a roving which can then go forward for spinning into yarn. The wool is processed in a carding machine, considerably different in construction from that used for cotton but nevertheless based on the use of card clothing to straighten and attenuate the comparatively entangled mass of wool presented to it. The carding machine as used for wool usually consists of a group of two or three machines of the same type. If it is in three parts, the first is termed a *scribbler,* the second an *intermediate scribbler* and the third the *carder.* In passing through the scribblers and carder the wool lap is attenuated to the form of a film-like layer and is then finally brought into the form of roving in which the fibres are considerably more regularised than before they entered the machine.

Unlike the production of cotton yarn, in which the sliver or roving is drawn or drafted in a number of stages, no use is made of draw-frames in the manufacture of woollen yarn. Once the roving is formed it is taken straight to the spinning machine, usually a mule, so it is necessary that the roving itself should be even and uniform. To achieve this, the wool sliver in the form of a wide lap or web of non-parallelised fibres coming off the carding machine is split laterally into a number of sections across the web which then pass through an attached condenser in which each section is lightly rolled to form a separate roving. These rovings are led forward to be wound on separate bobbins. These bobbins are then taken to a spinning ring or mule machine and converted into yarn by drafting and twisting.

Worsted yarn manufacture   In this section the first stages of preparation of the scoured wool are for the purpose of making the wool suitable for passing through the combing treatment which is necessary for making worsted yarn. It is of considerable advantage to the comber if the wool fibres in the material presented have already been fairly well straightened.

Carding is, of course, an excellent method for straightening and attenuating fibres, but there is another method which is known as *gilling* and which has had some brief mention in connection with the manufacture of linen yarns. Gilling is not satisfactory if the fibres are short, so in the preparation of wool fibres for combing it is generally preferred to straighten them out by carding if the fibres are less than about 9 in (230 mm) in length and to gill if the fibres are longer, say up to 15 in (380 mm) in length.

The general principles of carding and the machinery used have already been indicated, so only gilling machinery will here be described. In the gilling machine the wool fibres in the form of a lap are run between two pairs of rollers spaced up to 1 yd (1 m) apart. At a lower level, between these two pairs of rollers and below the moving layer of wool fibres, is situated a mechanism which operates continuously so that parallel rows of vertical pins rise and penetrate the fibre layer. These rows of pins protruding through the layer of wool fibres move forward with the wool, but at a somewhat quicker rate. It is easy to understand that in this way the pins have a

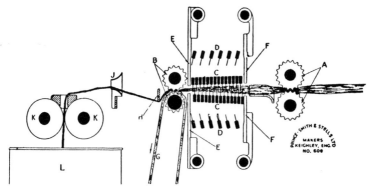

Figure 3.11 Sketch showing the path of a wool sliver through a gilling machine. As the sliver is drawn from the back rollers A to the front rollers B it is combed by the pins on the fallers C. These pins penetrate the wool downwards and then move in its forward direction but at a higher rate, thus combing it. After such a treatment the sliver with its fibres parallel drops into the can L. (Courtesy Prince-Smith & Stells Ltd)

combing or straightening effect. Then, just before reaching the second pair of rollers, the rows of pins drop down out of the wool and return to the starting point, there to rise and act on fresh wool coming forward. Not only does the gilling machine straighten the fibres, but it is also able to effect a certain amount of mixing. A number of slivers of wool are fed in at the same time so that these become combined and then drafted. The wool thus produced may then be washed in its sliver or roving form and again be gilled. It is then ready for combing.

Combing machines are exceptionally ingenious and it took several years for them to be perfected. The different types are named after their inventors—the Noble, Heilman, Holden and Lister combs. Of these the Noble is probably the most widely used. When the gilled wool is put through one of these machines it emerges with the wool fibres aligned to a high degree of parallelism.

The next stages are now mainly drawing operations in which the wool in the form of roving is progressively drafted in several stages until it is obtained in the form of a thinner roving ready for spinning. Spinning can be carried out on any of the different forms of spinning machines—mule, flyer, cap or ring, the last named becoming increasingly popular.

It is to be noted that in producing wool rovings for spinning into woollen yarns no effort is made to parallelise the fibres while the reverse is true for rovings to be spun into worsted yarns in which it is required that the fibres be parallel to each other so as to give a smooth non-hairy yarn.

It will now be understood how loose textile fibres are brought into the form of yarn. No matter what the type of fibre, the general principles of yarn production are much the same. The fibres have to be freed from impurities such as sand, dirt and stalk. They then have to be mixed and blended. After this comes attenuation and further mixing by combination of laps or slivers followed by still more attenuation. The object of this is to ensure that the fibre material is made uniform. Processes are then employed to eliminate the very short fibres and to a large degree ensure that the fibrous material in the form of sliver is composed of fibres all within certain extreme limits of length. Thereafter further parallelisation of the fibres is attained by combing and by repeated drawing or drafting. In drawing (drafting) slivers of fibres a degree of attenuation will be reached in which the fibres will not hold together unless at least a small degree of twist is introduced; the sliver then becomes known as *roving*. As soon as twist is used gilling becomes impossible. So long as drawing is effected the least possible twist is used since twist impedes drafting (attenuation). Finally the uniform fibre in the form of roving is further drafted and twisted into yarn.

Figure 3.12 Noble wool combing machine. (Courtesy Platt International Ltd)

Figure 3.13 Rectilinear wool comb. (Courtesy Prince-Smith & Stells Ltd)

The autoleveller

Obviously all the various operations described above such as doubling and drafting, frequently repeated, with the object of ultimately producing a yarn having maximum uniformity, involve considerable cost, and so there is always an incentive to devise 'short cuts'. Two main methods of shortening the fibre-to-yarn process are possible—one is to reduce the number of doublings of slivers and rovings (such doublings secure a better mixture of the fibres) and the other is to increase the degree of sliver or roving extenuation at each drafting stage (each extenuation stage carries further the conversion of a thick loose bundle of fibres into a thin twisted yarn). An ingenious device known as the Raper Autoleveller (Prince-Smith & Stells Ltd), which allows a shortened procedure for worsted yarn production by reduction of the number of doublings, is now available, and by its use the necessary time and machinery can just about be halved. This Autoleveller is a self-contained unit which can most advantageously be added to the feed-end of a gilling or a drawing machine.

The function of the Autoleveller is to measure automatically the thickness of the ingoing wool material and then appropriately and automatically control the back rollers and fallers of the gilling machine, or the back rollers only of the drawing machine, so that a constant weight per unit length of wool sliver is delivered by the drafting rollers to pass forward to the next process. In the gilling machine the Autoleveller can reduce a ± 15 per cent variation in the ingoing sliver to ± 1 per cent variation, and in a drawing machine it can reduce a ± 25 per cent variation of the ingoing sliver also to a ± 1 per cent variation. The Autoleveller has thus a very powerful levelling effect and because of this it can allow a reduction of the processes normally employed for this purpose. The Autoleveller is not concerned with the drafting conditions.

Doubling of yarns

All the operations so far described produce plain, uniform, single yarns. As is well known, use is made in the textile industry of two- and three-fold yarns and also of fancy yarns with thick and thin places. These are produced by twisting together the plain single yarns which have been made as described above. Thus, for the purpose of manufacturing fabrics and garments, the spinning and doubling section of the textile industry produces hundreds of different kinds of yarns, thick and thin, plain and fancy, single, double and treble or more, and of different types of fibres. All these yarns are made with a degree of control such that a yarn consignment ordered this year will be expected to be exactly the same as a similar consignment obtained, perhaps, two or three years ago.

Textured yarns

Recently advantage has been taken of the thermoplastic and other special properties of synthetic fibres to produce what are now generally termed *textured* yarns which are characterised in having either a high bulk or a very high extensibility coupled with almost perfect elasticity, so that when highly stretched they return to their original length with the release of the stretching force. Such yarns are usually made of nylon, Orlon, Acrilan, etc. and they are being used in the manufacture of so-called 'stretch' ladies' hose and shaped underwear, and also for 'stretch' men's half hose where their high extensibility allows them to shape themselves naturally in wear to meet any movement of the human body. The high-bulk yarns also have a high elasticity but they are mainly employed in textile materials where softness and warmth of handle are required.

Stretch yarns have proved exceptionally useful for the manufacture of men's socks and ladies' stockings. Such yarns are usually made of nylon and thus can be worn for very long periods (say months or even years) before a hole is worn in them. However, the yarn used in ladies' stockings is liable to ladder long before a hole is caused by wear (abrasion).

Stretch yarns can be made by highly twisting, say, a nylon yarn and heat-setting the twist, following this by controlled untwisting and then doubling the resulting yarn with another similar yarn but of opposite twist. The doubling of the single untwisted yarn with another yarn is for the purpose of giving it the stability required in handling.

Agilon stretch yarns are made not by the insertion of false twist in the yarn as described above but by drawing yarn of ordinary twist over a hot knife-edge so that the individual fibres are simultaneously stretched and irregularly distorted (compression on the one side on contact with the knife-edge and by extension on the other side remote from the knife-edge) thus causing them during subsequent shrinkage to acquire a saw-tooth crimp and thus a high degree of elasticity.

High- (often contracted to Hi-) bulk yarns can be produced in various ways including that of twisting and then untwisting, but much success has attended the method used for the production of Taslan yarns in which nylon or other yarn is run between two pairs of rollers with the first pair running faster than the second pair to allow the yarn to be suitably slack in the space between the pairs of rollers and also to allow an air blast to be directed into the yarn near to the second pair of rollers and thus entangle the fibres in it and expand its structure. It is in this way that the yarn can be made more voluminous.

A type of yarn having both stretch and high-bulk characteristics and thus somewhat desirably softer than a true stretch yarn is sold under the trade name of Banlon. It is made by an American process (J. Bancroft & Sons) in which the yarn is steamed while held in a tightly compressed and distorted state within a closed 'stuffing box' of a milling machine. The steam fixes the fibres permanently in their bent and distorted forms thus giving the yarn crimp associated with increased bulk and elasticity.

The 'stuffing box' resembles the 'spout' of the milling machine shown in Figure 4.19.

Other methods of producing high-bulk yarns have been developed

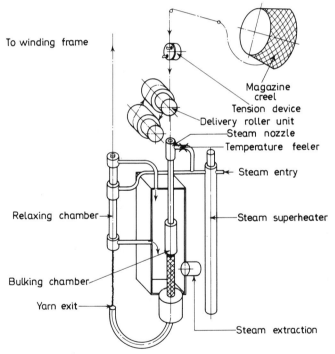

To winding frame

Magazine creel
Tension device
Delivery roller unit
Steam nozzle
Temperature feeler
Steam entry
Relaxing chamber
Steam superheater
Bulking chamber
Yarn exit
Steam extraction

Figure 3.14 Arrangement of the main parts of one unit in a Hi-bulk machine for shrinking yarn (that has been stretched first) by steaming it while relaxed, thus giving it high bulk. The degree of shrinkage is controlled by the slackness and 'dwell' period conditions as the yarn passes through the bulking chamber

which are based on the expedient of arranging that a yarn is composed of two types of cut-up man-made fibres (they can be of the same composition, say, polyamide, acrylic or polyester, etc.) which have been made so that they differ as regards the degree to which they shrink when dry-heated or treated in boiling water. The usual method for preparing such different types of fibre is to draw them to different degrees during manufacture—the greater the degree of drawing the more they shrink when subsequently heated. When these special yarns are heated free from tension the two types of fibre present shrink unequally and this causes a yarn-length contraction which by its irregularity makes the yarn correspondingly bulky and soft. This yarn relaxation in which hi-bulk is created by fibre shrinkage can be effected by heat treatment of the yarn as such or while in fabric. A unit for treating yarn with steam is shown in Figure 3.14. With yarn running at 1000 yd/min (900 m/min) and given a steaming for 10–80s at 100–110°C (for acrylic fibres) and 160°C (for polyester fibres), a length shrinkage of 20 per cent can be obtained.

Crimplene ester yarn (ICI) is a very important type of textured yarn very much used for all kinds of fabrics (men's and ladies' clothes). It is produced by the 'false twist' method largely developed by ICI Ltd (Figure 3.15). In this unit a single yarn, say of polyester, passes at a high rate through the tubular electric heater A, and then in its plasticised state is false-twisted (right hand and then left hand) in the device B which may rotate at up to 400 000 or more revolutions per minute. This yarn, left with a 'confused memory' to retain both types of twist, then passes with cooling to be cold stretched at C between pairs of delivery rollers C and E (rotating at different rates). It then passes through another tubular heater F with controlled relaxation to stabilise the yarn in respect of its bulkiness and firmness of handle. The degree of relaxation is controlled by the difference in rotation rates of draw rollers E and G. Thereafter the yarn passes over a small oiling device H to lubricate it before it is wound in a suitable form such as a cheese to be ready for knitting or weaving.

The resulting texturised yarn has a structure governed by the opposing right- and left-handed twists imparted and fixed in it so that it has a bulkiness, handle and resilience that make it very useful for woven and knitted fabrics.

Instead of using yarns made with cut-up fibres it is now possible to make them of composite fibres (p. 133) in which each fibre is composed of two different components (the difference which is required is that one will shrink more than the other under high temperature conditions) aligned side-by-side or in a core-sheath relationship. When such composite fibres are suitably heated (dry or wet) they

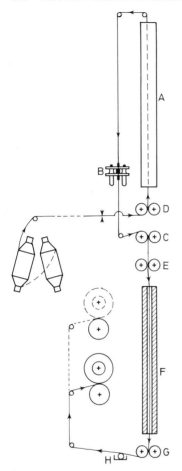

Figure 3.15 One of the many units spaced side-by-side in a horizontal row within a false-twisting machine for forming texturised synthetic fibre yarns such as I.C.I.'s Crimplene. (Courtesy Davide Giudici and Figli)

become crimped and bulky as a result of the uneven component shrinkage (without separation of the two types of component present in each fibre) and the yarn thereby acquires a high bulk.

It is possible to make use of fibres which differ, as regards their potential shrinkage, by as much as 50 per cent so that a very considerable bulking of the yarn can be obtained. Instead of bulking the yarns at an early stage this operation can be delayed until after weaving or knitting into fabric so as to increase the softness and bulk of the fabric, but usually under these conditions the fibre shrinkage is somewhat restricted.

Yet another method for producing hi-bulk yarn utilises the thermoplasticity of synthetic fibres. It involves running say nylon yarn at a

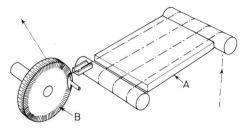

Figure 3.16 Thermoplastic yarn being crimped by running it round the heating plate A several times to become sufficiently thermoplastic. It then passes to the crimping head B where it is held at a moderate temperature and pressed into slots by a sinker. It is then carried forward and cools on leaving so that the deformation (crinkle) produced in the slot becomes permanent. Typically, nylon 6 yarn running at 540 yd/min (500 m/min) over the heating plate at 165°C is crimped in the crimping head at 120°C

high rate to be pressed against a hot slotted rotating circular plate, thus giving the yarn a reasonably permanent wrinkle (Figure 3.16).

For polypropylene textured yarns see p. 127.

Crêpe yarns

This type of yarn is widely used for the weaving of crêpe fabrics which when suitably wet-processed (as, for example, by holding the fabric fully immersed in the soap liquor in open width free from creases) gives a fabric having a surface uniformly and minutely all-over cockled. A crêpe yarn is usually of viscose rayon (acetate yarns are not satisfactory), made by giving it a high twist and then doubling it with another similar yarn but twisted in the opposite direction and then giving the doubled yarn a temporary stability sufficient to allow it to be woven into fabric by a steaming treatment while held taut in wound package form. Such yarn with its ends free rapidly untwists when placed in boiling water, but if its ends are firmly held (as in a fabric) so that it cannot untwist, then the same release of latent energy is achieved by 'snarling' and it is this change which produces a crêpe appearance when the fabric is treated with a boiling soap liquor as mentioned above.

The methods for spinning yarns, as described above, are mainly concerned with fibres which are relatively short, such as the natural fibres, cut-up waste silk and linen tow, and cut-up (staple) man-made fibres which have been cut to lengths similar to those of natural fibres so that mixtures of these man-made and natural fibres can be

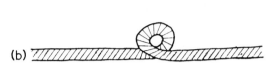

Figure 3.17 How a highly twisted crêpe yarn (a) snarls (b) when immersed in boiling water with its ends firmly held so it cannot untwist

more conveniently processed on the usual types of yarn-spinning machines. It is a valuable feature of man-made fibres that they can be made having those characteristics of thickness and length to make them most suitable for blending with natural fibres and for processing such mixtures on existing types of machinery.

The types of machines now in use for spinning yarn are based on principles which were developed in the very early days, but it must be noted that in recent years many improvements have been introduced. It is obvious that in converting natural fibres into yarn many operations are required and the recent improvements have been designed to reduce these operations so as to achieve more rapid production and reduce labour and machinery costs. These improvements have been more widely obtained in the conversion of man-made cut-up (staple) fibres into yarn for the reason that these fibres can be produced having a uniform thickness and length thus allowing a simplification or avoidance of those operations which are necessary in dealing with irregular natural fibres and which are designed to ensure that a uniform mixture of the fibres is obtained in order to produce yarn which is uniform throughout its length.

Of course, in the case of non-cut man-made fibres, yarns are obtained direct from the fibre-spinning (extrusion) stage and it merely remains to give these the necessary degree of twist or to combine them to give doubled and other forms as, for example, by texturing them (p. 163).

New and simplified methods for producing yarn

Although the older established methods for producing yarns have been largely perfected there remains a desire to change them yet further to accelerate and cheapen them by eliminating some of the machinery and processing stages originally considered necessary. Considerable success has already been obtained with indications that further progress can yet be made.

One new such self-twist spinning method of yarn production is now known as *self-twist* (ST and STT). By it two identically similar rovings (p. 151) are specially twisted in a simple manner so that along its length short lengths are produced in continuous succession, say of about 8 in (200 mm) each of righthand twist and then lefthand twist (Figure 3.18). At this stage the yarn produced has little strength, for on stretching it looses its twist, allowing the fibres to slide over each other and the yarn to break. But when the two yarns are closely placed side by side and allowed to untwist this causes them to self-twist together to produce an ST yarn of reasonable strength. However, this strength is not sufficiently robust for general use in weaving fabric and moreover it would give undesirable patterning in the fabric. These faults can be easily corrected by giving the self-twist yarn ordinary twist; then as so-called STT yarn (somewhat comparable to a doubled yarn) it has proved useful for use as a warp or weft yarn in fabric weaving.

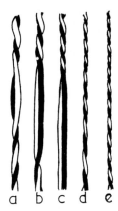

a   b   c   d   e

Figure 3.18 ST yarn in various stages of production. (a) S and Z twisted yarns united and allowed to intwist (self-twist). Yarns (b), (c), (d) and (e) result by progressively S-twisting yarn (a) through stages (b), (c), (d) to give yarn (e) that has acceptable uniform twist and strength. To give yarn (e) increased stability and thus make it easier to knit and weave, two such yarns can be twisted together to give a two-fold yarn

Of course various difficulties have been encountered in perfecting such a process and in fact it has proved more especially useful in making wool yarn as developed by CSIRO in Australia and by Platt International Ltd in Britain whose Repco Spinner machine is now available for large-scale self-twist spinning (Figure 3.19).

Open-end or break spinning   Another novel method for producing cotton, synthetic and other fibre yarns is that now known as *open-end* or *break* spinning. It was developed in Czechoslovakia and is now receiving attention in all textile manufacturing countries. The Czech BD200 machine is in commercial use and the comparable Rotospin Types 884 and 885 (for long and short staple cotton and synthetic fibres and their mixtures respectively) of Platt International Ltd is

Figure 3.19 Repco spinning machine as used for the production of self-twist (ST and STT) yarns in which eight wool rovings are given along their length, alternating right- and left-handed twist sections each of about 8 inches and then brought together in pairs with a suitable degree of drafting to give 2-fold yarns each of which, while free from restraint, untwist into each other to give an ST yarn so that this can then be further uni-directionally twisted to give an STT yarn having uniform twist and an acceptable degree of strength. The production rate of such yarn is exceptionally high. (Courtesy Platt International Ltd)

now available (Figure 3.20). It is proving more difficult to make this type of yarn spinning machine suitable for wool yarn.

In open-end spinning a sliver of, say, cotton is fed into a relatively small device (Figure 3.21) in which the fibres under air pressure are impelled by centrifugal action and an existing air vortex on to the internal surface of a high speed (up to 40 000 rev/min) rotating cylinder and are therefore carried forward with their twisting together to an exit point from which they are drawn in the form of yarn having an appropriate degree of twist. The yarn production rate on such

Figure 3.20 A close-up of the spinning positions of the Platt Roto-spin machine for open-end spinning cotton yarn. (Courtesy Platt International Ltd)

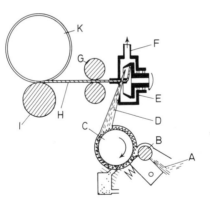

Figure 3.21 Principle of the Rieter Rotondo open-end spinner for taking away solid impurities from the cotton A, feed system; B, feed roller; C, opening roller; D, fibres in transit to rotor; E, rotor or turbine; F, suction; G, delivery rollers; H, yarn; I, friction take-up roller; and K, yarn package. (Courtesy Rieter Machine Co)

a machine can be up to four times greater than on a conventional ring frame spinning machine. Such a method can give acceptable yarn more free from trash and nep particles (natural impurities in cotton) and having desirable greater bulk than ring-spun yarn, but its strength can be lower. It would seem that open-end spun yarn can be very useful for the manufacture of carpets while also allowing the production of high quality finer yarns. It is an important advantage that sliver in open-end spinning is used instead of roving, for the latter requires additional pre-drawing of sliver.

It is important to notice that with the various new yarn spinning methods now being developed, each process produces a yarn which differs in its own peculiar way from conventionally spun yarns as produced, for example, by ring and mule spinning. In the case of cotton yarns resulting from open-end spinning they are usually somewhat weaker than ring-spun yarn but by contrast they can be produced to be more uniform, easily dyeable and to have good wear resistance with the further advantage of having a high bulk so as to produce very satisfactory woven and knitted fabrics. Such differences can be traced to the peculiar structure of the open-end spun yarn as shown in Figure 3.22. It is believed that the higher wear resistance of this yarn (open-end spun on a well-known type of BD 200 machine) is due to its unlevel surface as opposed to the level, smooth surface usually found with ring-spun yarn.

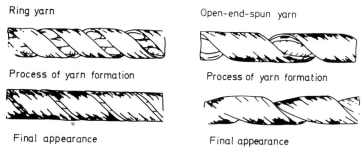

Figure 3.22 *Differences in appearance and structure of ring-spun and open-end-spun cotton yarns*

Bobtex yarn spinning

The commercial production of this type of yarn (it can be made from almost any type of fibre in its cut-up staple form having a fibre length of 3–6 in (75–150 mm) is now being developed in Canada by the Bobtex Corporation Ltd in Montreal and is yet another attempt to simplify yarn manufacture by higher production rates than hitherto possible by conventional systems as for example by ring spinning. As already mentioned, open-end spinning can achieve this same end but the claim is made that Bobtex spinning by comparison can give a stable fibre yarn at lower cost and several times faster, yet have properties fully amenable to normal methods for conversion into acceptable fabrics by both knitting and weaving procedures.

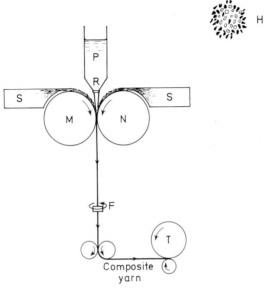

Figure 3.23 Bobtex-yarn spinning apparatus. Molten fibre-forming polymer P is extruded through spinneret R to give a multi-filament thread which passes downwards between a pair of rollers M and N. Staple fibres are fed from sources S and are made to pass around these rollers and to be pressed into the thread (suitably soft and adhesive) so as to adhere together. They then pass through the false twisting device F as a fibre-coated thread to be wound on roller T as a strong thread of cross-section as shown at H. The staple fibre adheres strongly to the outside of the core of polymer fibres P. The ratio of polymer core to the adhering staple fibres can be variable to the extent that the resulting fibre-coated thread may contain only 33 per cent of staple fibre. Bobtex yarn of this type can be made at the very high rate of 2000 ft/min (610 m/min)

The Bobtex spinning technique (Figure 3.23) consists of extruding a molten fibre-forming polymer as a multi-fibre yarn from an orifice in much the same manner as for nylon but with the modification that a sliver of cotton or other fibre in staple form is continuously fed to the issuing side of the orifice where the freshly extruded yarn fibres are commencing to solidify. The conditions maintained there are such that the fibres in the sliver are suitably simultaneously drafted and slightly twisted as they heat-bond to the yarn at their points of contact so that a sheath-core type of yarn is formed. The texture of the fibre sheath can be controlled to be smooth or hairy as desired and in fact it can have the same character as a textured (p. 163) yarn now so widely used. Under commercial conditions the yarn can be produced at the high rate of 600 ft/min (180 m/min) but it is anticipated that it can reach 2000 ft/min (600 m/min) since synthetic fibres can be melt extruded at this rate.

Present uses for such a low price yarn appear to be in carpets, upholstery and curtain materials and for industrial fabrics and coverings.

Fasciated yarn spinning

In producing fasciated yarn a sliver of say, substantially non-twisted parallel Orlon fibres, is led through a pair of delivery rollers to emerge in the form of a flat ribbon of fibres and pass through a so-called torque jet able to impart false twist. Thus as the ribbon approaches the torque jet it receives twist in one direction while on leaving the jet it receives equal twist in the opposite direction (with loss of twist) and then passes through a pair of draw rollers to be wound into package form.

Under these conditions and because the fibres are of the cut-up type those fibres on the edge of the ribbon will (in the first twisting) be less twisted than the inner ones while (in the second reverse twisting) they will be left more untwisted than the inner fibres and thus will be left spirally wrapping the inner fibres which thereby form a core of substantially parallel untwisted fibres. The conditions of such a procedure can be arranged so that a suitable proportion of the fibres in the original sliver form the wrapping fibres and give the resulting fasciated yarn strength and other properties which are acceptable and allow their satisfactory use in producing woven and knitted goods. One interesting feature of such yarn is that it has a higher lustre and cleaner appearance than comparable yarn ordinarily spun from staple fibre and this peculiarity arises from the way in which light is reflected from the inner bundle of parallel fibres to pass through the wrapping fibres.

Twist-free yarn spinning This type of twist-free yarn (also known as *TNO* because of its Dutch origin) is made from a moderately twisted wet cotton sliver by the equipment shown in Figure 3.24. Initially the sliver is scoured and bleached (and optionally also dyed) and then in its wet state and impregnated with a 10 per cent aqueous suspension

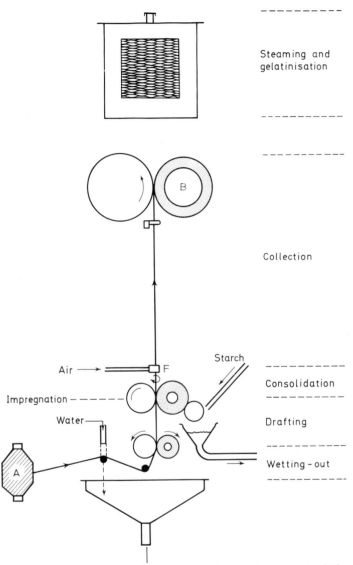

Steaming and
gelatinisation

Collection

Starch

Air →   F

Consolidation

Impregnation – – – –

Drafting

Water

A

Wetting-out

*Figure 3.24 Dutch Textile Research Institute's twistless spinning process, in which wet cotton sliver of low twist is drawn from package **A** to be wetted and then drafted in the usual manner to about yarn thickness by passage between the two pairs of drafting rollers while being impregnated with a 10 per cent aqueous suspension of starch. Thereafter the attenuated sliver, whose fibres adhere to each other sufficiently aided by the starch present, is led through the false-twisting device **F**. Such a device is now frequently used in producing texturised yarn, and imparts, in rapid succession, right-hand and then left-hand twist so that it can then move forward to be collected, substantially twist free, on the perforated bobbin **B**. In this form it is stabilised by steaming it for about 1 h and is thus left having a tape-like form ready for use in weaving fabric. The starch can be removed by washing, and the fabric is left with a useful strength*

of starch in granular form (not paste) it is drafted in the usual manner by passage between two pairs of rollers, the second pair rotating at a higher peripheral speed, to about $\frac{1}{10}$th of its original thickness. The attenuated sliver with its fibres adhering together aided by the starch present is then led through a false-twisting device assisted by compressed air, in which it is rapidly twisted right-hand and then left-hand thus rendering it substantially free from twist. The twist-free sliver is then drawn forward free from tension to be collected cross-wound on a perforated bobbin. Afterwards it is steamed on the bobbin for about 1 h to stabilise the yarn in a flat tape-like form.

The advantages of such a yarn are that in weaving it passes more easily through the healds and reed of a loom and gives a fabric having a more compact and lustrous nature than fabric woven from comparable but ordinarily twisted yarn. These characteristics are obtained following a washing of the fabric to remove the starch. In spite of the absence of twist the fabric has acceptable durability since the fibres hold together as yarn because of the compressive forces which normally exist in a woven fabric where the warp and weft yarns are closely interlaced.

The twist-free yarn produced in this way can be used normally as weft or warp in a fabric but its resistance to wear and tear can be increased by doubling it with a woollen yarn. It is to be noted that this new method for preparing yarn is at present in a stage of development and commercial exploitation, and it represents yet another attempt to avoid the complicated methods and spinning machinery normally employed. It is attractive since it can be 2·5 times as rapid as ring spinning.

FABRIC MANUFACTURE

When the yarns have been manufactured from textile fibres by the methods just described, they are used for the production of fabrics by weaving or knitting. These two methods of fabric manufacture will now be considered.

Weaving

All woven fabrics are made up of two sets of threads (the warp and the weft). Usually fabric is woven in long lengths, say of 40 to 100 yd (35 to 90 m) or more and from, say, 20 to 60 in (0·5 to 1·5 m) wide. The threads which extend throughout the length of the fabric are termed *warp* threads, while those which go across are termed *weft* threads. More technical names for these same threads are *ends* and *picks,* respectively. Since it is generally necessary for the warp threads to be strong in order to withstand the considerable strains to which they are subject in weaving, they are the more important and the inferior weft threads are frequently referred to as 'filling'. If the warp threads are strong, then quite inferior threads can be used as weft, for in the fabric they will be held together by the warp threads. In cutting fabric to make a garment or other article it is advisable to ensure that the cut pieces are arranged in the garment so that their warp threads lie in the direction where most strain is likely to occur during the use or wear of the garment.

In order that the fabric may have strength and compactness combined with a fair degree of elasticity, it is necessary that the warp and weft threads be interlaced in the fabric. If the interlacing is such that the weft threads pass alternately under and over the warp threads, then the fabric will be a plain one and will represent the simplest type of weave. These weft threads may also pass under two, then over one, then under two and over one, and so on. A twill fabric is built up in this way. A sateen fabric is made with the weft threads passing over and under different numbers of warp threads so as to avoid producing any pronounced patterned effect.

It is obvious that it can be arranged that any pattern produced by weaving may leave the fabric with a larger proportion of warp threads or a larger proportion of weft threads showing on the fabric surface. Thus we may make a warp twill or a weft twill fabric.

If either the warp or the weft threads are considerably thicker, then a rib type fabric will be produced.

There are a very large number of variations of the methods for interlacing the warp and weft threads and so it becomes possible to weave a wide variety of fabrics all of which have their special properties and uses.

Preliminary processes   Previous to weaving, the warp threads will have been brought together side by side, just as they will be in the fabric, and wound on a beam (roller). The length of these threads must, of course, be approximately that of the fabric which it is intended to weave after making allowance for the undulatory character of threads in a fabric. The weft threads will be wound on small tubes to form cops. Since the weft threads have to be moved between the warp threads in weaving, it is not possible to make these cops large, so a great many of these have to be used one after the other in making a long length of fabric.

In order that the warp yarns may not break during weaving, they are previously strengthened by coating them with a thin film of

Figure 3.25 Winding warp yarn from a yarn creel C on a Warper's beam B. (Courtesy Wilson and Longbottom)

size, which is usually a starch paste containing softening and other ingredients, and then drying. Starch sizes are usually not satisfactory for synthetic fibre threads and have to be replaced by special sizes (often containing synthetic polymers) which will adhere better to the threads.

The plain loom   The weaving or interlacing of warp and weft threads is accomplished with a machine which is known the world over as a *loom* and the essential details of this will now be considered.

Referring to Figure 3.26, the beam of warp threads is placed at W and the warp threads are then drawn from it across the loom from back to front to be wound on the cloth roller A; the warp threads are supported by the breast beam O and the back bearer B. Each warp thread passes through a heddle H and there are two sets of these heddles held between the top roller Q and the treadles T. These heddles are made of strong cotton twine and each has an eye N about its middle point through which a warp thread is passed.

As we have already mentioned, in making plain fabric the weft threads pass over and under alternate warp threads, and it is for this reason that the warp threads are led through either of these two sets of heddles. This arrangement allows half of the warp threads to be slightly lifted by raising the corresponding heddles,

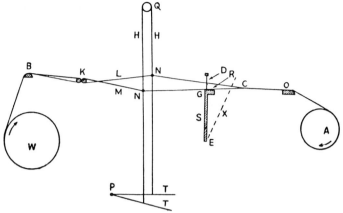

Figure 3.26 Diagram showing the essential parts of an ordinary loom. W is the warper's beam on which are wound the threads which will form the warp of the fabric. Two of these threads, represented by L and M, pass over the back bearer B. They diverge from the lease rods K to pass through the two sets of harness H (with a separate heddle eye N for each thread) until all the threads unite again in fabric at the point C where the last weft thread was inserted. The woven fabric is regularly wound on A. D is the reed or comb through which all the warp threads pass. The two sets of harness are attached to treadles T, which are pivoted at P, so that one set of warp threads may be raised above the other to form a shed at R. At this stage a weft thread is inserted in the shed R by a shuttle impelled along the sley G. The sley sword S swings forward about point E (to position X) and the reed D beats up the weft thread close to the preceding one. The sley sword then returns to its original position whilst the harnesses change their relative positions so that the shuttle can return through a different shed

and the remaining warp threads to be lowered by depressing the other heddles, this action being produced by operating the treadles. In this manner the two groups of warp threads are caused to diverge from the back to the front of the loom so that they form a *shed* at R. Also attached to the loom is a sley sword S hinged at E which carries the sley G and the reed D, the latter being in the nature of a comb through which the warp threads pass as they leave the heddles.

In weaving, a cop of weft thread contained in a shuttle is impelled along the sley G from one side of the loom to the other. Thus one weft thread is laid down to form part of the fabric. Then the heddles change position so that the warp threads which were higher now

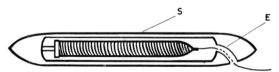

Figure 3.27 Sectional diagram of a shuttle containing a pirn or cop wound with yarn. E is the eye of the shuttle and S is the body

become lower; they thus form a fresh warp shed which encloses the weft thread just laid down. The sley sword now moves forward and the reed beats up or forces the weft thread close up against those previously laid down in this same manner. Then when the sley sword has moved back the shuttle is impelled back along the sley and so another weft thread is laid down. Again the heddles change position, the new warp shed enclosing this second weft thread, and again the sley sword moves forward for the reed to beat the weft thread up close to the preceding weft threads. These motions continue in rapid succession. The fabric is steadily formed and at the same time it is slowly wound up on the cloth roller A.

When a cop of weft thread is exhausted it has to be replaced. Formerly this was done by the operative attending the machine but, today, looms are provided with devices whereby this is done automatically. This same automatic device allows differently coloured cops of yarn to be used in turn and so produce fabric having a coloured pattern.

It will be understood that, in the plain loom just described, the arrangement of the interlacing of the warp and weft threads is settled at the outset of weaving by the manner in which the individual warp threads are threaded through the two sets of heddles. Once these arrangements have been made they must persist until the whole length of the warp threads is used up. For the production of some patterned fabrics, however, it is desirable to be able to change these arrangements frequently through the weaving. By such changes it becomes possible to make patterns based on variations in the way the warp and weft threads are interlaced. To be able to do this a special type of loom known as a dobby loom is necessary. A dobby loom allows only a number of simple pattern effects to be obtained. For elaborate patterns it is necessary to use a Jacquard loom.

Jacquard looms   In the Jacquard loom each heddle is attached to a wire or cord. These numerous cords pass upwards to a device by means of which, at any moment, any heddle can be raised or lowered independently of the others. This device operates somewhat after the manner in which a pianola plays. Cards have holes punched in them to correspond to the pattern which it is desired to weave. As these cards move, feelers sensitive to the holes react and raise or lower the heddles to which they are attached. Thus, to weave fabric automatically to a pattern, it is necessary to prepare the corresponding punched cards and then arrange for these to control the heddles.

Of course, there are all kinds of devices attached to looms to produce some particular effect or allow special fabrics to be made, or to correct or reveal automatically faults which arise during weaving.

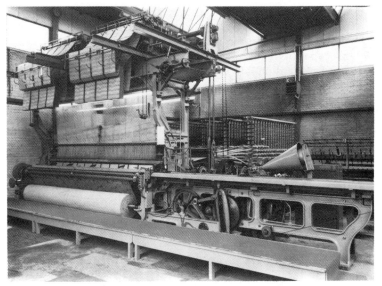

Figure 3.28 Coloured yarns being individually drawn from bobbins in a creel (rear of machine) to be woven into carpet. The yarns being woven are controlled by the Jacquard attachment so that the desired pattern is produced. The attachment consists of a series of punched cards which, via a cord from each hole, control the individual warp threads being used in the loom below. (Courtesy Wilson and Longbottom Ltd)

It is not possible to deal with them here, but it is useful to mention that looms are now available which do not use a shuttle to carry the weft thread across and between the warp threads.

Several types of shuttleless looms are now in use and they are favoured because the weft yarn can be in large package form and positioned conveniently by the side of the loom and not be required to be wound on pirns for insertion in a shuttle as a yarn carrier. These special looms differ essentially in the manner by which the weft yarn is propelled across the loom within the shed of warp yarn.

The three main types of shuttleless looms are known as *rapier, gripper-shuttle* and *jet*. In a rapier loom the weft yarn is propelled by two rigid or flexible rapiers one operating from each side of the loom. The weft yarn is carried half-way across the loom by one rapier and is there at once automatically transferred to the second rapier which pulls it across the remaining distance. In a gripper shuttle loom an element grips one end of the weft yarn and then carries it right across the loom. In a jet loom the weft thread is propelled across the loom by a jet of compressed air or by water.

The selvedges of fabrics woven by these special looms differ from the normal selvedges, and to give them satisfactory resistance to fraying, the loose weft yarn ends may need to be left tucked in the selvedge, or if thermoplastic synthetic fibres are present in them it may be most convenient to heat-seal them.

Other types of looms are circular. However, the great majority of looms are conventional as described previously.

In one form of rapier loom the rapiers are tubular so that the weft yarn can be blown from one selvedge through one rapier by compressed air to the middle of the fabric being woven, and then pass into the other rapier and be sucked through this to the other selvedge.

Ribbons are woven on special narrow looms or on wide looms suitably divided. Alternatively, acetate and synthetic fibre ribbons can be cut with hot knives from wide woven fabric—the hot knives fuse the ribbon edges so that these do not unravel in use.

Knitting

Now, instead of making fabric by interlacing warp and weft threads with the assistance of the loom as in weaving, there is an alternative method, i.e. knitting. Rows of stitches are formed so that each row hangs on the row behind and the row in front of it. Simple knitted fabrics are made in this way either by hand or by machine.

The modern knitting machine is a very complicated piece of equipment. One operative attending to several machines is able to produce fabric, socks, hose or garments shaped and patterned automatically at great speed. There is a wide variety of knitting machines to suit the different types of knitwear which are now produced, but all of them have a common basis. In the knitting machine there are a number of needles evenly spaced and with the spacing proportional to the size of stitch being knitted. Around each needle is a loop which is a stitch in the course of formation. Then thread is guided to each needle (or vice versa) and movements of both then take place such that a stitch is formed from the loop and each needle is left with a fresh loop of thread around it. Again the thread is guided to the needle, a fresh stitch being formed and a fresh loop left around the needle. The needles are side by side and the operations apply to each needle in turn. A row (usually termed a *course*) of stitches is therefore formed with each complete cycle of operations, and a continuous length of knitted fabric results.

In many machines the needles are arranged in a circle so that the resulting fabric is tubular. Obviously, circular knitting machines of this kind are required for knitting socks and various kinds of garments. Ladies' hose of the seamless type are knitted on circular machines. If the needles are arranged in a straight row then under ordinary circumstances a single width of fabric is produced. This is termed *flatknit* fabric.

Knitted fabrics produced by machines as described above have the defect of easily laddering on the breaking of perhaps only one stitch. This defect is associated particularly with rayon and silk materials since these fibres are smooth; laddering does not occur with the rougher and more hairy cotton and wool yarns because the projecting fibre ends get caught in the stitches and prevent their slipping through each other. However, by knitting the fabric so that a locking thread moves through the different courses this laddering can be prevented.

Figure 3.29 Machine for knitting single Jersey-type colour-patterned fabric for outer-wear. (Courtesy Paolo-Brizio)

All knitted fabrics easily suffer distortion when pulled but they return to their original shape when relaxed. For this reason knit-wear is exceptionally suitable for making underwear, and also many outerwear garments—it so readily accommodates movements of the body. Yet it is possible to knit in such a way that the fabric has special elasticity. This is achieved by rib knitting, which is much the same as purl knitting as practised in the home. In ordinary knitting the thread is presented to the loops on the knitting needles always on the same side, and the freshly formed stitches are drawn off the needles always in the same direction. It can be arranged that these movements are

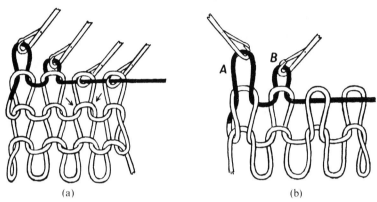

(a)   (b)

Figure 3.30 Stitch structures of circular knitted fabrics. (a) The formation of rows of stitches in plain fabric where all the new loops are pulled through the old loops from front to back. (b) Purl or rib knitting where new loops (or groups of loops) are pulled through the old loops alternately in opposite directions—back to front at A and front to back at B

changed in direction with every other stitch or with selected groups of stitches. The fabric which then results has a ribbed appearance and because of this rib it is very elastic. The tops of men's socks are almost always ribbed so as to secure a grip and in many cases the whole sock is ribbed in order that the fit may be better. Ladies' garments often have ribbed portions so that these fit the body more snugly.

At the present time there is a tendency for knitting to be so modified that it is able to produce fabrics of a more rigid construction so that they can take the place of woven fabrics. Such knitted fabrics are characterised by having straight yarns introduced to be part of the knitted structure and act as warp and weft threads passing through the stitches. They are termed *knit-weave* fabrics, and can be produced faster by knitting than by weaving and this is an incentive behind such developments.

The type of knit-weave fabric shown in Figure 3.31 (right) can be produced on an ordinary circular single jersey knitting machine provided with a modification devised by A. Marvin of Strathclyde University in Glasgow. Other modifications, devised with the basic idea of introducing warp and weft threads into knitted fabrics to stiffen them and make them more competitive with woven fabric,

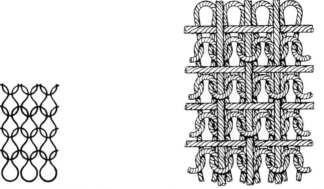

Figure 3.31 Knitted fabric (left) without warp or weft threads inserted, and (right) with warp and weft threads to produce a fibre more resistant to distortion and thus more like woven fabric

Figure 3.32 Warp fabric knitting machine. (Courtesy Platt International Ltd)

are available on Raschel and Tricot knitting machines. Warp knitted fabrics are now increasingly replacing woven fabrics in the production of wearing apparel and it now appears likely that in this particular field woven fabrics will become less important.

Seamless and fully fashioned hose   In connection with ladies' hose, distinction is often made between seamless and fully fashioned hose. The *seamless hose* are knitted on circular machines with the needles evenly spaced in a circle. The stocking therefore must have the same number of stitches throughout from top to toe, except at the heel where it is specially shaped. Any shaping so as to make the stocking conform more closely to the natural shape of the leg can only be obtained within small limits by tightening or loosening the thread during knitting. Actually the stocking appears to be well shaped when it is sold in the shop but this is only because it has been pressed to this shape. When wetted it will at once return to its true knitted tubular form.

Fully fashioned hose and garments are knitted on flat knitting machines which allow the fabric to be produced in a shaped form so that the flat fabric which results has its edges seamed together, and in this way it is brought into the form of a stocking or garment having a shape very close to that of a human leg or body. The knitted article will retain this shape when it is wetted and dried because it is its true shape. Obviously, fully fashioned hose and garments should wear better than seamless hose and garments because they are subject to less strain when worn.

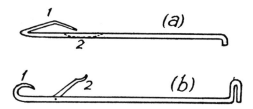

Figure 3.33 Two main types of needles are used in knitting machines: (a) the spring beard; (b) the latch needle

Mention should be made of so-called *plated* knitted fabric. It is possible to run two threads at once into the knitting machine and arrange that one of these predominates in the back of the fabric whilst the other is mostly seen on the front. Thus, if one is cotton and the other wool, the fabric can be made to look like a cotton fabric lined with wool or vice versa. This method of knitting is known as *plating*.

The knitting industry has a wide range of machines at its disposal for the knitting of many types of material, but it is not possible to deal with these in further detail here.

Lace making

Quite another method of making fabric out of yarn is seen in the lace making industry. This is perhaps more restricted in volume than weaving and knitting but nevertheless it is extremely interesting. The machinery used is ingenious and complicated.

The basic principle of lace making is that the weft threads are made to twist round the warp threads and move diagonally across the fabric. The weft threads do not run at right-angles to the warp threads or pass over and under them as in weaving.

Ornamental lace of the best quality is made with real silk threads as the groundwork and with effect threads of rayon or cotton. This material may then be dyed with contrasting colours. Many improved types of lace are now available which are less liable to shrink and change their patterned shape and appearance—this comes from the use of synthetic fibres instead of silk and other natural fibres.

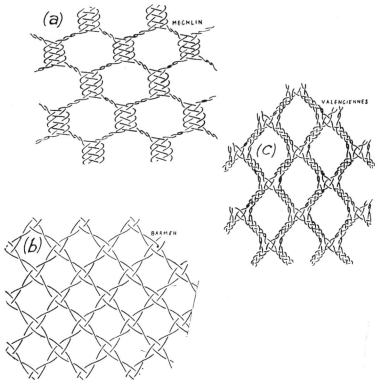

Figure 3.34 Sketch illustrating the manner in which the threads are interlaced in laces: (a) Mechlin (b) Barmen (c) Valenciennes. (Courtesy The Textile Institute)

Felt manufacture

Felt is a type of fabric in which there is no interlacing of yarns to give strength and character. In fact it is not made with yarns at all but with loose fibres. Its production depends on the fact that when a layer of wool or fur is rubbed whilst wet or in the presence of steam then the fibres become entangled and interlock as a result of felting. This closing up of the fibres, which is due to the action of the epithelial scales which cover each fibre, leads to formation of a compact fabric in which the fibres are so tightly interlocked that it has sufficient strength to withstand considerable pulling and distortion.

Rough felts are used for many domestic purposes but the best felts are made into men's and women's hats. For this last-named purpose rabbit fur is prized more highly than wool.

Wool and non-wool felts are now being made by a mechanical process in which a layer or web of loose fibres is pierced at innumerable points all over its surface by barbed needles. The barb on each needle penetrates and passes through the thickness of the web, carrying with it several fibres. On withdrawal of the barbed needle it leaves these fibres deeply embedded in the web. The web thus gradually becomes more dense and stronger as the fibres become more intermingled. Each needle can pierce the web at the extraordinarily high rate of up to 900 times per minute. The machine used is often called a *needle loom* but it is in no way related to a loom such as is used in weaving fabrics.

Production of other non-woven fabrics

The experience of some 3 000 to 4 000 years has taught that the present method of constructing fabrics by twisting fibres together to form yarns and then interlacing these yarns by weaving or knitting is a very satisfactory one. Such a construction is favourable to the production of fabrics which are flexible, soft and, above all, resistant to the various intermittent strains to which they are subject during use. This resistance to breakage is due to the fact that, when the fabric is suddenly strained, the fibres are able to move relatively to each other and so allow a useful degree of stretching or distortion. The strain thus expends itself without producing a breaking of the threads such as could lead to the formation of a rent or hole. However, a whole series of various expensive operations is required to prepare yarns and then weave or knit them into fabric, so in recent years very determined attempts have been made to devise an alternative cheaper method of making fabrics, *non-woven* fabrics as they are now generally termed.

The following basic method is now being employed to produce non-woven fabrics. First, a mixture of different fibres is made. One of the fibres which is evenly distributed within the mixture is a special type of fibre which can, at any suitable stage of processing, be brought into a tacky condition and thus become able to play the part of an adhesive or bonding substance. Then the fibre mixture is brought into the form of a comparatively thick layer or web of width corresponding to the desired width of the fabric which will ultimately be formed. Now comes the final stage where the fibre layer is hot pressed so that the special fibres within it partially melt and become sticky, thus causing all the fibres in the mixture to become securely bonded together. When the pressure is removed there remains a non-woven fabric in which the fibres are simply held together by the bonding fibres.

For the production of non-woven fabrics of the type just described, it is often very convenient to use synthetic fibres as bonding fibres, especially those such as Vinyon HH which have low softening temperatures. Of course, it is sometimes more convenient or desirable to use bonding fibres which do not ordinarily become tacky when suitably heated. These can be made to become so either by first coating the fibres with a substance which will become tacky or act as an adhesive when heated, or by treating them whilst in the fibre mixture with a gas or liquid which will induce in them a suitable degree of tackiness.

The essential idea underlying the manufacture of non-woven fabrics is to use fibres, temporarily made tacky, to bond together a higher proportion of fibres which remain unchanged under the processing conditions employed. There is one feature about these fabrics which should be noted. When the fibre mixture is being formed into a layer or web there is an opportunity to lay down the fibres parallel to each other and all pointing in the same direction or to lay them down at random pointing in all directions and criss-crossing each other. When these two types of web are finally formed into fabric it will be found that the first type of fabric will be especially strong in one direction (the direction in which all the fibres lie) whereas the fabric composed of fibres criss-crossing each other will be equally stong in all directions. A carding machine (Figure 3.8) is often used to prepare the fibre webs for conversion to pre-woven fabrics. Loose fibres are fed into this machine where they are combed free from useless very short fibres and specially delivered in the form of a web (layer) with the fibres all parallel to one another and aligned to the web length. A number of these webs can be superimposed so that the fibres in the composite web are all aligned in the web-length direction or cross at right angles in alternate webs. If the latter method is used then the resulting non-woven fabric will be equally strong in length and width—otherwise the strength will be greatest in the web length direction.

In another method for forming a web with its fibres randomly arranged, the loose fibre is air blown into a large chamber and allowed to settle in web form on a horizontal belt moving through the chamber at a suitable rate.

The production of so-called non-woven fabrics (that is those where textile fibres are not brought together by the conventional methods of weaving and knitting which involve an interlacing of threads to confer an acceptable degree of strength combined with flexibility, elasticity and resistance to fairly drastic handling and perhaps repeated washing while still being reasonably soft) has in recent years increased greatly. It involves the use of most types of fibres. The uses of non-wovens are numerous, ranging from disposable polishing cloths and towels to the more durable and washable bed-sheets and pillow covers, curtains, interlinings for wearing apparel, linings for footwear, and cloths for industrial filtration and for surgical use, in addition to their more recent uses for blouses, skirts and even outer wear.

For the production of these comparatively new non-wovens a whole range of textile machines has been brought into use that differ considerably from the machines currently employed in the manufacture of conventional textile goods.

Non-wovens have a strong appeal to both the manufacturer and the public for generally they can be produced rapidly, cheaply and give consumer satisfaction when used for the purposes for which they have been designed.

It is possible to mention only some of the more important aspects of non-woven fabrics and this with the reservation that they may be rapidly outdated.

The chief stages employed in bringing loose textile fibres together to form non-woven fabrics involve (1) the formation of a web (layer) of the fibres mixed at random or specially aligned relatively to each other; and (2) a compacting of the web, usually with fibre-to-fibre bonding obtained with the aid of an added adhesive applied as a liquid or powder or by needling the web so as to align a proportion of the horizontally laid fibres into a vertical position and thus increase the solidity of the web.

Many types of adhesive can be used including those allied to natural or synthetic rubbers and those which are synthetic polymers having thermoplastic or thermosetting properties. The selection of

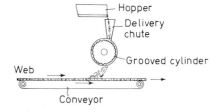

Figure 3.35 Method of applying dry adhesive to a web as a stage in the manufacture of non-woven fabric

an adhesive is governed partly by the type of fibre in the web and partly by the properties such as strength, resilience, softness, etc. required in the resulting non-woven material. Thus a viscose solution comparable to that employed in the manufacture of viscose rayon fibres and consisting essentially of cellulose xanthate can be employed more particularly for the fibre bonding in non-wovens made from cotton and rayon fibres. By contrast a butadiene-acrylonitrile copolymer can be used, and in the bonded fabric it is resistant to washing and dry cleaning with chlorinated hydrocarbon organic solvents. If this copolymer also contains methacrylic acid then these fastness properties are further improved. It is an important point that the presence of the fibre binder can improve the overall properties of fabric in addition to its function as a binder. The bonding strength is largely influenced by the surface roughness and the cross-sectional shape of the fibres and their porosity since these features in their turn influence the adherence of the binder to fibre.

In handling the fibre webs in the early stages of processing and often throughout it is necessary to support the web on a brattice or belt which moves with it to prevent the web being unduly stretched or damaged. This applies especially when the web is being impregnated with a liquor containing an adhesive and possibly also while it is being dried.

Polypropylene and polyethylene are among the thermoplastic staple fibres which may be used as binders and they have the advantage of giving a bond which is very durable although it may temporarily soften at high temperatures. Mention must also be made of plastic binders which can be used as a continuous film having an open lace or net structure which favours flexibility in the bonded fabric. Sheets of fibrillated polypropylene or other type of polymer (p. 125) may also be used. Such films can be sandwiched between fibre webs and the composite web, and then be hot calendered to promote softening of the film to allow its easy pressing in and between the fibres.

It has been mentioned that one advantage of making non-woven fabrics over weaving and knitting is the high rate at which loose fibres can be converted into them. The following data illustrate this by showing the relative production rates which modern textile machines allow:

Process	Type of processing machine	Relative production rate of fabric
Weaving	Automatic loom	1
,,	Shuttleless loom	2
Knitting	Circular	4
,,	Tricot	16
Non-woven	Stitch bonding	40
,,	Needle bonding	500
,,	Roto former	2 500
,,	Carding	300
,,	Spun bonding	1 500

Non-woven floor coverings

With the invention and manufacture of heterofil and bi-component fibres of the side-by-side and sheath-core types (p. 133) has come the development of a new method for making non-woven fabrics. It is based on one of the two components of a bi-component fibre having a lower melting point than the other so that a web of these fibres can be heated causing the one to soften and melt first and then act as a specially effective fibre binder. ICI Ltd have recently perfected such a method for making a floor covering brand-named Tultrim having several desirable properties, by the use of polyamide sheath-core fibres. A description here of the essential features of this method can well serve to indicate the general procedure and principles involved in making high quality non-woven fabrics (p. 188).

The bicomponent fibres used are of the sheath-core type, one component being of nylon 6·6 (m.p. 260°C) and the other of nylon 6 (m.p. 215°C), or a copolymer of these two different nylons; these component fibres are chosen because of their usefully different softening and melting temperatures and partly because woven, tufted and needle loom carpets made with them have proved to have good strength, abrasion resistance and dyeing properties.

In the first stage these staple fibres (optionally pre-dyed) are formed into a web using a carding or garnetting machine or by air-laying using a Rando-Webber machine. Then to give this web useful improved strength it is needle punched to cause a proportion of the fibres to be aligned vertically within the web and actually pass through it to form a tufted surface on the underside, and so make the web more compact. Then follows the combination of a suitable number of webs by super-position and further needle punching. The resulting thicker web, supported by a wire brattice, is then fed through a hot sintering oven to effect fibre bonding where the fibres cross each other. The heating is effected by hot air which is directed onto the web from above. It is beneficial to admit some steam into the oven since this can accelerate drying and at the same time prevent any excessive yellowing (by aerial oxidation) of the fibres. Fixation of the pile on the dried fabric by a high temperature heating (curing) stage contributes a desirable soft feel. It is an interesting point that this pile fixation and accompanying softness are not satisfactory if the bi-component fibre is of the side-by-side type.

Then follows dyeing of the reasonably strong web in a winch machine or by a pad-steam process and finally the coloured web is rendered dimensionally stable by attaching to it a hessian or other backing fabric. It is necessary to attach this backing because polyamide fibres lengthen or shorten with changes in the moisture content of the surrounding air. The adhesion of the backing to the web is obtained by use of powdered polythene and a short heating to melt it.

Such a floor covering can be modified by introducing into the first formed fibre web a proportion of ordinary nylon staple, say up to 30 per cent; it confers additional softness at the expense of a moderate reduction of wear value.

In the sheath-core polyamide fibres used, the nylon 6·6 should form the core and the nylon 6 (having the lower melting point) should form the sheath. This structure ensures that in the high temperature cure (bonding) stage, in which a kind of spot welding occurs at numerous points where the fibres cross one another in intimate contact, the core remains substantially intact without melting while the sheath melts and runs around the fibre crossing point and then solidifies to make a strong joint. It is due to the strong type of fibre bonding and the high porosity of the bonded web which confers desirable resilience, that this special floor covering has excellent durability in use.

Tultrim materials are not only used as carpets. They can be made into blankets, upholstery fabrics and articles of clothing. In these uses the ability to modify their stiffness, softness and draping properties, as indicated above, is valuable.

Spun bonded non-woven fabrics

Spun bonded non-woven fabrics are made by a one-run process in which molten polymer is extruded in the form of a bundle of fibres, which at once fall to give an evenly spread web on a belt which conveys this forward through a bonding unit, where it is optionally sprayed with a binder and then consolidated as a non-woven fabric. Such fabric is available from the American Du Pont Company in three types, with brand names of Reemay (polyester), Typar (polypropylene) and Tyvek (polyethylene).

The fibres in the web may be straight or crimped and randomly arranged, but the important feature is that as spun they are long continuous fibres. These fabrics are thus quite different from the usual non-woven type of fabric made with cut-up staple fibres of relatively short length. Partly because of this difference, spun bonded fabric requires only about 15 per cent of binder as compared with 40 per cent for a comparable fabric conventionally made with staple fibre, yet this structure gives a high tensile strength coupled with a more open fibre spacing.

Spun bonded fabrics are judged to have a uniquely different balance of physical and aesthetic properties from that of ordinary non-woven fabrics. The fabrics containing straight fibres are relatively stiff as compared with those fabrics consisting of the crimped fibres and they thus have greater flexibility.

These new non-woven fabrics are useful as backing fabrics for needle-felted carpets but for this use it has been found necessary to give the fibres increased toughness and pliability so as not to be excessively damaged by cutting by the barbed needles commonly used in making needled carpets. Such barbed needles cut the fibres more than the smoother needles (non-barbed) used in making tufted carpets.

Flocked pile and non-woven fabrics

This type of fabric is widely employed for a considerable variety of purposes including imitation fur, coverings for hat bodies, floor and case coverings, draperies, shoe fabrics, etc. Essentially it is made by producing a uniform scatter of loose textile fibres within an enclosed space through which an adhesive-coated fabric in open width is carried at a steady rate supported on a conveyor belt; under such conditions the fibres fall by gravity (usefully assisted by a downward air flow) or are impelled towards the fabric by electrostatic force. In the gravity fall method it is necessary for the fibres to be very short and they can be almost dust so that the maximum proportion of fibres will become embedded in the adhesive coating in an upright state. With the use of electrostatic force the fibres can be longer, for in passing through the applied electric field they can become oriented perpendicularly to the fabric before reaching it.

In the electrostatic method the fabric is maintained in contact with or adjacent to a positively charged plate while the fibres are scattered from an overhead container which is negatively charged. The density of the resulting flock or pile on the fabric is largely determined by the density of the fibre scatter, the rate of travel of the fabric and the strength of the electric field. In any case not all the fibres applied become fixed in the adhesive surface and any excess is pneumatically sucked away.

In some electrostatic systems for flocking, the fibres are drawn upwards to become fixed in the fabric passing overhead. This method has the advantage that the excess of unfixed fibres fall from the fabric by gravity assistance.

In practice the electrostatic conditions are not quite so simple as indicated above for it must be noted that the fibres passing through the electrostatic field become similarly electrically charged and in this state will repel each other to cause their irregular distribution on the adhesive fabric.

After the flocking, the fabric is led through a heating zone to complete fibre fixation in the adhesive coating. It can then be colour printed to any desired pattern or have its surface lightly brushed to remove loose material and simultaneously create a nap, or just regularise the pile. It will also be sheared by passage through a machine having cutting rollers similar to those of a lawn mower and generally be finished to give it a level pile height and more attractive appearance.

By having the adhesive coated fabric free from adhesive in selected parts and by other modifications of the processing, a patterned pile surface can be obtained.

Locstitch pile fabric

This new double pile fabric suitable for a wide variety of industrial and domestic uses including sportswear, beachwear, towels and upholstery fabrics, is based on a new stitching principle devised by Wray and Ward of Loughborough University and is now being made on machines exclusively made by Pickering Locstitch Ltd in England. Its essential feature is that its uncut loops are very securely locked into a base fabric. Thus a pull on either an upper or lower loop tightens the fabric structure a lot and a broken loop causes no unravelling. The fabric can be cut in any way without causing unravelling.

The yarns used for the upper and lower loops can be quite different from each other in colour and structure and the fabric surface can be afterwards raised or cropped if desired. The loops need not have

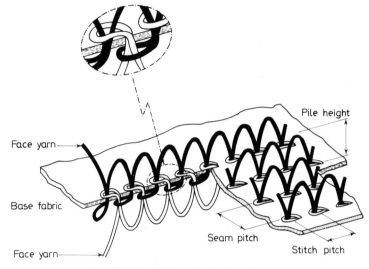

Figure 3.36 Locked loop pile fabric and locstitch

the same height on both sides of the fabric. In the stitching operation the fabric moves forward vertically through the machine instead of horizontally as in ordinary weaving.

It is obvious that non-woven fabrics can never entirely take the place of woven or knitted fabrics and that in general they will be of a lower grade but, nevertheless, non-woven fabrics appear capable of satisfying many varied demands for a spongy soft absorbent fabric of moderate strength. Their production is steadily increasing and their quality, strength, durability and wash fastness is being much improved. Non-woven fabric is used very much for pre-shaped garment linings.

Pile fabrics

The conventional method for producing material having a pile (hairy) surface involves weaving a fabric so that closely and uniformly spaced loops of yarn are left on one side which are simultaneously or subsequently cut to leave the fibre ends free to open-out (this can be assisted by a steaming or other treatment) and so better cover the fabric surface. In the case of some fabrics such as are used for upholstery purposes, for example moquette fabrics, the loops are left uncut, it being considered that the fabric is then more resistant to wear. Obviously the character of the fabric so produced depends on the closeness and length of the loops formed. When the loops are cut the material has the appearance of a base fabric in which tufts of fibres are anchored. To improve the anchorage the material can be sized on the under-side with an adhesive composition which may be based on a starch mucilage or a more expensive but superior adhesive such as rubber latex, or else a synthetic polymer may be used.

More recently quite different methods for producing pile fabrics have come into use which do not require the use of a loom. Such fabrics include tufted and needle-loom carpets (this name is a misnomer since no loom is used for their production) which are now competing very strongly with the traditional loom-woven carpets of the Wilton and Axminster types. With this change in the method of manufacturing carpets there has at the same time come a change in the types of fibres used. It is generally agreed that wool is the most useful fibre for carpet manufacture and has hitherto been exclusively employed for the best and most durable types of carpets. But now man-made fibres—more especially the regenerated cellulose fibres but also the more costly synthetic fibres such as the polyamide, acrylic and polyester fibres—are being used. Many wool carpets of the woven type now contain 20 per cent of nylon to increase their wear resistance. In the case of viscose rayon fibres these are attractive because they are so much cheaper than wool.

Tufted carpets

The tufting process   A few years ago nearly all carpets were produced by weaving using a special type of loom and as in the case of the more expensive types such as Wilton carpets it was possible by the use of attached Jacquard equipment (p. 180) and coloured yarns to produce very attractive multi-coloured patterns representing the utmost skill and perfection for this type of floor covering. But in the mid-fifties the method of tufting instead of weaving carpets was introduced from America and this method for producing floor coverings more cheaply and rapidly has now found world wide acceptance.

In the tufting process a bar carrying a row of closely spaced needles is positioned above a flat backing fabric (usually of hessian made from jute but now often a non-woven synthetic web, say of polypropylene scrim). Each needle is supplied with a yarn drawn from a separate yarn package (a cheese) forming one of a number in a creel in the background. The needle bar is lowered so that the needles pass through the backing fabric to a controlled distance where a corresponding number of loopers are positioned.

As the bar is then raised to bring the needles back through the fabric to their original positions each looper holds its yarn as a loop projecting from the underside of the backing fabric (Figure 3.37). At this stage the bar is again lowered and raised but with the fabric having first been advanced a controlled amount so that the loopers hold another row of loops similar to the previously formed

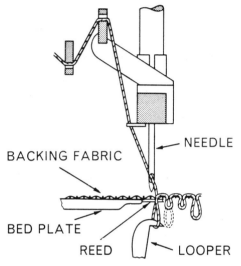

Figure 3.37 Action of a single needle which is one in a row of closely spaced similar needles fixed in a bar so that they can all act simultaneously. The needle fed with a thread of cotton, rayon, etc. drawn from a yarn package such as a cheese, is lowered to pass through a hessian backing fabric to a controlled distance so that on its upward return a so-called looper holds it in the form of a loop extending downwards from the underside of the backing fabric. After this lowering and upward return of the needle bar a single row containing many loops is thus formed. Then the fabric is advanced a controlled amount and another row of loops is formed as before. Thus with repeated needle movement the carpet is formed with as many rows of loops as is desired. The resulting carpet is termed a looped pile tufted carpet and it only remains for the back to be coated with rubber latex or other pliable permanent adhesive to secure a firm anchorage of loops in the backing fabric

row. This lowering and raising of the needles is continuously repeated and so rows of loops are formed in the backing fabric. All loops and rows are uniformly spaced to form an uncut loop-pile carpet which is then coated on its underside with rubber latex, foamed rubber or some other similar coating, to anchor the loops securely in it. In this manner a tufted uncut-loop pile carpet can be steadily produced in one machine say to a width of 15 ft (4·5 m) at a rate exceeding 500 rows of loops per minute with all the loops of the same height and uniformly spaced so as to produce carpet at the rate of, say, 10 yd^2/min (85 m^2/min).

Most tufted carpets have a cut-loop pile. The cutting of the loops to give tufts with free fibre ends can be achieved simultaneously with the formation of the loops in the tufting machine as shown in Figure 3.38. In this case there are small cutting knives which operate in conjunction with the loopers to cut the loops while still held on the loopers.

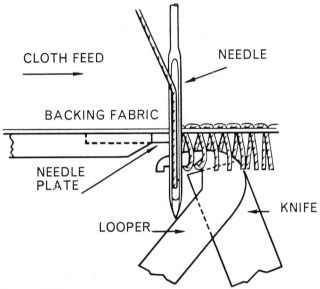

Figure 3.38 This shows how the loops can automatically be cut by means of a combination of looper + knife at the time of their formation to give a so-called 'cut pile' carpet. By use of a special type of looper + knife combination it is possible to form the carpet so that it has a texture pattern in which some areas have low loop pile and others have high cut pile

As thus described the carpet is produced having its loops or tufts of uniform predetermined height. It is now possible, by means of a special attachment to the tufting machine, to cause some areas of loops to be higher than others conforming to a pattern, thus giving a sculptured appearance which can then be enhanced by passing the carpet through a shearing machine (operating as a lawn mower) to cut off the tips of the non-cut higher loops. Thus a carpet is obtained that is formed to have un-cut loops at a lower level than cut-loops at a higher level. Further modification can be obtained by suitably pattern embossing the carpet surface.

In an alternative so-called 'Honesty' tufting machine, hollow needles instead of the usual type are employed. This allows a controlled length of yarn to be delivered to each needle whose tip is piercing the hessian or other backing fabric to be blown through it, and thus to form a loop of controlled height on the underside of this fabric, as shown in Figure 3.39. It is claimed that this method of tufting allows greater ease of loop formation with less wear on the yarn, and also allows more flexible control of loop height variation and thus the production of more attractively patterned carpets.

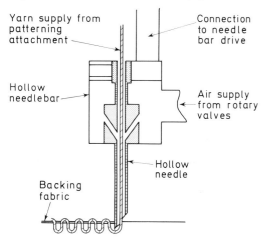

Figure 3.39 Manufacture of carpet by the Honesty process

The production of fabric having an especially dense high pile can be obtained by a knitting technique where a pocket size carding system feeds sliver into a circular knitting machine so that tufts of this fibre are picked up by the hooked needles just before they engage in knitting to give the base fabric. The sliver tuft, whose size is controllable, thus becomes anchored on one side of the resulting fabric to give a pile whose character and appearance can be usefully modified by subsequent finishing treatments (steaming, brushing, shearing, etc.). Such pile fabric has many uses in garments and as linings where warmth is required together with softness.

For the manufacture of needle-loom carpets a machine is used which comprises rows of vertical needles each having a barb. These are positioned over a hessian base fabric on which is laid a suitably thick layer or web of loose fibres, and as this moves forward the needles are caused to punch their way downwards through the fibres and the fabric underneath and then return. The barb on each needle moving through the web collects and carries a tuft of fibres downward to become anchored in the base fabric. Since the anchorage of the fibre tufts at this stage is less than in the case of woven carpets it is important that it should be further strengthened by sizing the underside of the base fabric with a rubber, polyurethene or other polymer composition and it is desirable for this to penetrate slightly through the fabric so as to give increased solidity and firmness to the resulting needle-loom carpet.

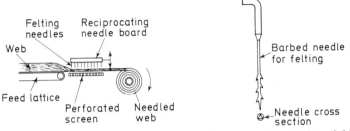

Figure 3.40 Production of non-woven fabric by needle-loom method. Needle felting (left) and barbed triangular-section needle (right) for needle felting

The hessian backing fabric is now being replaced by a polypropylene non-woven fabric (so-called *scrim*) in both tufted and needle-loom carpets.

It is possible to vary the carpet structure widely according to the thickness of the fibre web used and the closeness of the barbed needles. It is also possible to increase the thickness and density of the first formed carpet by similarly needling on to it a second fibre web. It will be understood from its method of manufacture that needle-loom carpet and fabrics can have a felt-like structure.

Tufted and needle-loom carpets are now being made backed with a layer of synthetic polymer foam. This layer may be caused to adhere to the carpet by use of an adhesive but success is now being obtained to form the foam layer directly on the back of the carpet itself.

Many difficulties have been encountered in further backing freshly made tufted carpets (already backed with a coarse hessian fabric) with a foamed rubber backing to improve the anchorage of the tufts over and above that provided by the primary hessian backing particularly when the carpet is above 9 ft (2·7 m) wide and when it is necessary to cure the additional backing at a high temperature to stabilise it permanently. These difficulties have now been overcome by applying, in a one-run process with specially adapted plant, a selected elastomeric resin (containing useful additives) in its molten state and at a usefully high rate.

Using this plant the carpet is led with its pile surface uppermost over a roller rotating partly submerged in the molten polymer, which is contained in a low-positioned tank provided with a 'doctor' blade to scrape off excess polymer from the back-coated carpet. Then follows the application of a secondary hessian backing fabric. The combination of carpet and secondary hessian backing fabric is then hot pressed by passage between a pair of rollers while the resin present is still in a partially cooled, soft, tacky condition. At this stage the fabric is further cooled while being led over a stenter to finalise the bonding action of the resin adhesive. Thereafter the carpet has its selvedges trimmed and its tufted pile brought to a uniform desired height by passage in succession over four shearing machines. Preceding this shearing the pile surface is brushed so that it can better respond to the shearing. Figure 3.41 shows needle-punched non-woven fabric production. Two fibre webs drawn from rolls (1) and (2) which sandwich a third web from roll (3) are led forward on a travelling conveyor belt (4) to be needle-punched [200 punches/in^2 (30/cm^2)] right through from one side at (5) and then from the reverse side at (6). It then passes at 10 ft/min (3 m/min) through a hot oven (7) at 230°C and is finally brought into roll form at (8).

The special features of such processing are: (a) while outside webs (1) and (2) are entirely conventional as regards the type of staple fibres present in them, the staple fibres in web (3) are of the bicomponent type, each consisting of equal proportions of one component fibre of nylon 6·6 and another component fibre of an 80/20 copolymer of nylon 6·6 plus nylon 6, these component fibres being in a side-by-side relationship within the bicomponent fibres (because the copolymer melts at a lower temperature than the nylon 6·6, the bicomponent fibres shrink and crimp when heated to a suitably high temperature, say 230°C) and (b) the needle punching is sufficiently deep for the barbed needles to carry some of the fibres right through the composite web and leave them protruding as small tufts. Thus, in the high temperature treatment to which the composite web is finally exposed, the web becomes compacted by shrinkage of the bicomponent fibres within it and at the same time it acquires a tufted surface by shrinkage and crimping of the protruding bicomponent fibres from both sides of the composite web to produce a fully stabilised attractive non-woven fabric.

Needle-punched mattresses and pads   Needling (needle punching) methods are also used for the production of mattress and pad materials. For these, cotton or wool rags and similar waste material obtained from hospitals and similar institutions, etc. are thoroughly purified by boiling scour liquors and are then freed (by carding, for example) from very short fibres and laid down on a steadily moving brattice in the form of a web. Thus the web is brought to a needle punching machine after having first underlaid it with a hessian fabric on to which it is then needle punched to form a non-woven resilient material that has many uses. The thickness of the web will of course be controlled to suit the end-use of the non-woven product.

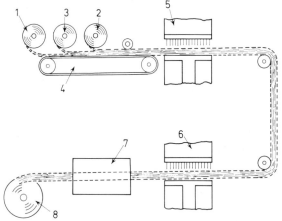

Figure 3.41 Production of needle-punched non-woven fabric using barbed needles

Figure 3.41 shows the production of mattress and pad materials by a continuous method in which the webs advance to be super-imposed on each other and then to be needle punched with barbed needles first from one side and then from the other.

Most of the processing, apart from the initial purification of the rags, is carried out in a continuous manner using Garnetting machines for pulling the rags into loose fibre form so that there is a steady flow of fibre to mattress or other pad form.

Two important features of these non-woven carpets must be mentioned. The first is that their rate of production can be much higher than that of the woven carpet so that it is possible to produce them more cheaply. The second is the question as to whether they wear as well as the conventional carpet normally made of wool.

It is believed that a tufted carpet made with wool can be equal to a conventional carpet as regards wear but most of the tufted carpets are made either with a mixture of wool and man-made rayon or synthetic fibres. Ordinary rayon fibres are inferior to wool in respect of their resistance to abrasion and resilience so that an all-rayon carpet is correspondingly less satisfactory although it may satisfy selected conditions of use. Because this inferiority is recognised rayon manufacturers have given much attention to the use of especially durable so-called polynosic fibres (p. 41) of the viscose rayon type so that they will have better wear value and in addition be more resilient and less liable to soil in use. Remarkable progress has been made in this direction. By the increasing use of synthetic fibres, particularly nylon, carpets can now be made at least equal to the conventional carpets, for it is well known that this fibre is much more resistant to abrasion than wool is. Wool has several properties which make it an ideal fibre for the manufacture of carpets but it has some defects (for instance it is attacked by moth whereas synthetic fibres are immune to attack by moth and most micro-organisms) and so it can be anticipated that man-made fibres will steadily become the equal of wool and perhaps excel it for the production of carpets. At present about equal amounts of wool and rayon fibres are used in Britain for carpet manufacture.

Two defects encountered, more especially in the use of carpets made with man-made fibres, arise from their hydrophobic properties; they become electrostatically charged following the rubbing of people's shoes against the fibres. This 'static' can cause sparking and can thus be a fire risk but there is the second disadvantage that they may become soiled by electrically attracting dust from the resulting air. Soil trodden into the carpet may also become more firmly fixed. To mitigate these defects carpet yarns are frequently treated with compounds (so-called anti-static agents) which confer a degree of electrical conductivity which enables the 'static' to dissipate to earth and so become ineffective (p. 123).

As an alternative to incorporating anti-static agents in the carpet yarns it has been found effective to arrange in its manufacture to distribute in it about 1 to 5 per cent of viscose rayon yarn. This yarn being hydrophile and electrically conducting readily dissipates to earth any static electricity formed on the carpet by people walking over it. Alternatively the carpet may be made to include a small proportion of uniformly spaced fibres pre-coated with about 0·5 per cent of finely divided silver and held there by a washfast adhesive, thus making the carpet electrically conductive (see p. 122 and Figure 2.7).

Now available for the manufacture of antistatic carpets is a Zefstat synthetic fibre made by the Dow Badische Co which contains an aluminium core having a degree of electrical conductivity to make it anti-static. It is also flameproof and it is substantially modacrylic so it can be satisfactorily dyed with cationic dyes.

The recent discovery of a satisfactory method for making extremely fine stainless steel fibres (somewhat finer than a cotton fibre and just as flexible) allows a carpet to be made anti-static in the same manner as with viscose rayon fibres. The effectiveness of such fibres (eight times as strong as comparable nylon fibres) is greater as they are distributed nearer to the underside of the carpet. However it is effective if 1 per cent of these are among the carpet pile fibres since the static electricity appears able to jump from one metal fibre to another reasonably distant metal fibre without difficulty.

Ordinary wool blankets are made by milling (p. 252) a woven wool fabric until it is sufficiently dense and then raising on it a pile surface by means of suitable raising machinery (p. 341). The yarns used for weaving such blanket fabric are soft spun so that the fibre ends can be easily brought out of the fabric surface during the raising operation. It has now been found that such blankets can be made much more quickly by first forming a tufted pile fabric instead of one which is woven. This fabric can then be raised to give it a blanket-like pile.

In one type of tufting machine on which the needles penetrate from the underside of the base fabric the projection of the yarn loops above this fabric is assisted by compressed air. The advantage obtained is that quite a high pile can be obtained without requiring correspondingly longer needles.

If the tufted fabric has to be raised on both sides then it is quite possible in the raising of the underside to weaken the anchorage of the fibre tufts. To overcome this, special types of tufting machines have been devised to produce fabrics having loops on both sides of a plastic film so that it is only the loops which are affected by the raising.

Blankets are also being made by the use of needle looms. One type of blanket being made 'needles' a fibre web on to a nylon base fabric which is then given a pile surface by raising.

These new methods for making blankets give manufacturers wide scope in the choice of fibres alternative to wool. Nylon, Terylene, acrylic fibres and rayons are all being used either singly or in admixture. Now polypropylene fibre blankets are being made; they have the advantage of being very light. Blankets made from synthetic fibres can be stain-resistant and can be very easily washed and rapidly dried. Although the synthetic fibres are usually considered to have a cold handle the blankets made from them can be very warm because of the large amount of non-heat-conducting air occluded between the fibres.

Sliver-knitted pile fabric   A special type of knitted pile fabric can be made by steadily feeding to the needles of a knitting machine a sliver of, for example, Terylene fibres, preferably prepared on woollen carding machinery so as to have a uniform open texture free from entanglement of the fibres and which can advantageously be pretreated with an anti-static agent such as Cirrasol TCS (ICI Ltd). The slivers are blown by compressed air into the stitches as they are being formed and so become firmly anchored in the knitted fabric being made. Afterwards this fabric is stabilised dimensionally to resemble a woven pile fabric more closely by causing it to adhere to a backing fabric with the aid of a suitable acrylate latex adhesive whose tenacity is then improved by stentering under conditions such that it is heated with hot air at 160°C for 3 min while being held and thus set to a uniform width.

It is usual to finish the fabric by shearing it down to about the desired pile height, calendering it by contact with a hot metal surface to polish its surface, and then again shearing it to secure the desired pile finish. The fabric may be embossed or given a plain hot press such as may be required when the fabric is to be used as a lining for rain-coats. Sliver knitted pile fabrics can have a very attractive appearance (subtle lustre) when made from Terylene bright fibres having a trilobal cross-section (p. 133).

It is evident that there are many new and interesting possibilities attending the production of textile materials by these new methods of tufting, needle-punching, stitch bonding, etc. instead of by the conventional weaving and knitting processes.

In a recently proposed method the punching needles are replaced by powerful fine jets of water to carry fibres through the web of loose fibres.

CHAPTER FOUR

Bleaching, dyeing, printing and finishing—methods and machinery

In the previous chapters we have considered the properties of the various textile fibres and the methods by which they can be made into yarn and then into fabrics and garments. But as yarn and fabric come straight from the spinner, weaver or knitter they are in a rough condition. Frequently the material is harsh in handle and nearly always contains impurities, either added to facilitate the processes of manufacture or which are natural to the fibres. In many cases it is soiled and may have oil stains. Altogether, freshly manufactured textile material is unattractive. Owing to their grey or dirty brownish colour such materials are often referred to as being in the 'grey' or 'brown' state. It then becomes the task of the 'finishing section' of the industry to scour, bleach, dye, print and finish or otherwise process these yarns, fabrics and garments so that they acquire an attractive appearance. At the same time it is also possible to improve their serviceability in many ways.

SCOURING

The principles underlying the scouring and bleaching of all textile materials are simple. Methods and purifying substances are used which, under the conditions employed, have a minimum harmful effect on the textile material, yet are able, by destruction, solubilisation or emulsification to remove the impurities present. Since the different fibres have different susceptibilities to attack by the various purifying agents available, so must the different classes of textile material such as wool, cotton, silk, rayon and synthetic fibres be treated accordingly.

The machinery used for carrying out these purifying processes, as indeed also for all other kinds of wet processing, must be devised to suit the textile material as it is in the form of yarn, fabric or garments.

Different machinery may also be required according to whether the fabric is woven or knitted, or non-woven.

Impurities to be removed

The main impurities in textile materials which have to be removed during scouring and bleaching comprise starches, fatty and oily substances, natural nitrogenous bodies, gums and mineral impurities. In addition, the natural colouring matters present in fibres must be destroyed when the finished material is to be left a pure white or to be coloured in bright clear shades. Most of the impurities are removed by scouring; the residual organic impurities and colouring matters being easily removed afterwards by chemical bleaching.

Of the various fibres the rayons and synthetic fibres are usually the cleanest and most pure; their purification is relatively easy. Cotton is a fairly pure clean fibre, but wool is very impure. It contains a large proportion of wool fat and perspiration residues from the sheep. (Purified wool fat is the basis of the useful skin salve Lanolin.) Flax contains a large proportion of woody impurities whilst raw silk has a 20 to 30 per cent content of silk-gum. Starch and oil substances are introduced into almost all textiles to strengthen and lubricate the yarns as they pass through weaving and knitting machines. Synthetic fibres are usually lubricated immediately following spinning to facilitate their further treatments where inter-fibre friction must be avoided or at least reduced. Antistatic treatment may be used.

Looking at this purification problem broadly, it may be stated that starchy, nitrogenous and gum products can be largely removed by treating the textile material with alkaline liquors or with hot or boiling water. Fats, waxes and oils respond to treatment with hot alkaline soap solutions when they become emulsified and wash out. The natural colouring matters are usually resistant to these treatments and must be destroyed by oxidising substances, especially hydrogen peroxide and sodium hypochlorite or other substances containing active chlorine. Acid liquors assist to remove mineral substances such as calcium and magnesium compounds.

In any of these operations it is necessary to move the textile material in a stationary liquor or circulate the liquor through the stationary textile material.

Treatments for various types of fibres

Natural fibres   The cellulose fibres are resistant to alkalis and so it is permissible to give these a good and thorough boil with solutions of such substances as sodium carbonate and even caustic soda. On the other hand, wool and silk are sensitive to alkalis and so it is better to use the weaker alkali, ammonia. However, where the full purification demands it, just a small proportion of sodium carbonate may be used with wool provided that the temperature of treatment is kept low, say, not exceeding 40°C. Boiling of the animal fibres, wool and silk, under neutral or alkaline conditions is deprecated, especially so far as wool is concerned. Wool is least reactive, in purifying or other liquors, when these are weakly acid, say of pH 4·2 (its iso-electropoint), and in such liquors it is least likely to suffer damage. It can thus be beneficial to scour wool under these acid conditions using scouring assistants which are stable and effective. Such assistants are those of a non-ionic type. Silk is more resistant to alkalis than is wool (more so the wild as opposed to the cultivated type), and so it is common practice to boil raw silk in a slightly alkaline soap solution to remove the silk-gum. In this degumming process real silk is not appreciably harmed, but the treatment should not be continued longer than is necessary.

The fats, waxes and oils to be removed from textile materials are affected in one or other of two ways during any purification process which uses alkali. In the first these impurities are simply emulsified in the alkaline soap liquor. Secondly, they may be saponified, that is, converted into water-soluble soaps. It does not matter which change takes place, so long as the oily impurity can ultimately be washed out.

Acid purification treatments must be cautiously applied to cellulose fibres for these may be tendered in the process. Generally, cotton and linen should not be boiled with even dilute acid; such treatment should be cold, or at most warm, and the concentration of acid kept low. After the treatment the acid should be thoroughly washed out or neutralised by use of an alkaline liquor. If even a small amount of acid is dried into cotton or other cellulose material the strength and durability will be very much lowered. In contrast, there is no objection to boiling wool or silk in dilute acid liquors of reasonably low concentration. After such treatment it is sufficient to wash well with water before drying as a small residue of acid will do no harm in the drying.

Rayon and synthetic fibres   Viscose and acetate fibres are all more sensitive to acids and alkalis than cotton or wool and so their treatment should be made as mild as possible. Viscose can be boiled in dilute alkaline liquors, but if this treatment is made too drastic a loss of weight and harshening of handle may result. Polynosic regenerated cellulose fibres are less affected by chemical treatments than the ordinary types of viscose rayon and this applies especially to alkali treatments for in these they behave more like cotton. Acetate fibres should not be treated with hot solutions of sodium carbonate or particularly with solutions of caustic soda since this saponifies the cellulose acetate and converts it to regenerated cellulose; the acetate fibres are thus brought back to a type of regenerated cellulose rayon and at the same time lose their unique properties with an undesirable loss in weight.

However, for some special purposes it may be desirable to effect saponification, such as to give the acetate fibres an affinity for cotton dyes.

The synthetic fibres such as nylon, Vinyon, Terylene, Orlon, Courtelle and Saran are generally highly resistant to purification treatments. However, it is not desirable to treat nylon materials with liquors containing active chlorine. Boiling dilute mineral acids such as hydrochloric acid can split the synthetic fibre molecules into the constituents from which they have been built up. With the possible exception of Terylene they are immune to alkalis. In hot alkaline solutions Terylene and other polyester fibres can lose weight by dissolution. Generally it is sufficient to purify synthetic fibre materials by scouring them in a warm liquor containing a synthetic detergent and tri-sodium phosphate.

It may be recalled that the rayons and synthetic fibres are usually brought to a good white colour during their manufacture. Their bleaching when in the form of fabric is thus simply to remove dirt which has been picked up by these fibres in the loom or knitting machine. In weaving and knitting and associated operations synthetic fibres can become electrostatically charged (arising from fibre friction) and this enhances their attraction of soil from the surrounding air. Of course, when used with other fibres such as cotton and wool for the production of mixture materials, these rayon and synthetic fibres may have to undergo the purification treatments usually applied to these other fibres.

Purification of loose fibres

As previously mentioned, the wet processing of textile fibres is left as much as possible to the later stages of manufacture. It is easy to understand the reasons for this. Firstly, there is always a certain amount of waste in converting loose fibre into yarn and then this yarn into fabric, so that treatment of loose fibre or yarn would mean the processing of a proportion to no good purpose. Secondly, the owners of textile material like to leave their decision as to how it shall be coloured and finished right to the last moment and so be able to keep abreast of fashion demands. On this account it is better to process textile material at the fabric or garment stage.

However, it is necessary to scour and bleach a limited amount of loose fibre, especially in the case of wool since this fibre contains such a large proportion (often 40 to 50 per cent) of impurities.

Kiering of cotton   Cotton and wool are mainly involved in scouring in the loose fibre form. The usual practice with raw loose cotton is to pack it within large cylindrical vessels termed *kiers* (Figure 4.5) and circulate through it a boiling liquor containing sodium carbonate or

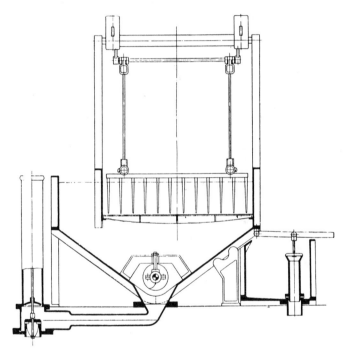

Figure 4.1 Cross-section of a self-cleaning wool scouring bowl showing the swinging forks which move the wool along the length of the tank. (Courtesy Petrie & McNaught Ltd)

caustic soda, or both, with perhaps a small addition of a soap or similar detergent. The boiling liquor is sprayed over the top of the cotton material so that it percolates downwards through the cotton until it passes through a perforated false bottom. From here it is pumped through a heater to bring it up to boiling temperature and thence it goes again to be sprayed at top of the kier. In this way the boiling alkaline liquor is continuously circulated through the cotton for some 3 to 4 h, and frequently during the whole night. At the end of the operation the dirty liquor containing most of the impurities solubilised or emulsified originally present in the cotton is discharged to the drain and the cotton is then thoroughly washed with fresh water. Sometimes the scouring treatment in the kier is repeated so as to effect a more complete purification.

If the scoured cotton is to be used as absorbent cotton wool then this purification must be very thorough for it is only by substantial removal of the natural waxy impurities that satisfactory absorbency is secured. Scouring will then almost certainly be followed by a bleaching treatment with active chlorine, and reference will be made to this later.

Scouring of loose wool   Loose wool is not usually scoured in the way described for cotton. The general practice is to move the wool through a series of long shallow bowls (tanks) containing warm and slightly alkaline soap solution. This is an instance of moving the textile material relative to the treating liquor. It is considered that the wool becomes less matted by this method for it must be recalled that if wool is handled too much in aqueous liquor it felts and the fibres become so entangled that it is scarcely possible to pull them apart without breakage when later working them up into yarn.

A conventional wool scouring machine (Figure 4.2) comprises from three to five of these shallow iron tanks which may be up to 6 ft (1·8 m) wide and 20 ft (6·1 m) long. They are arranged end-to-end and between each adjacent pair is a mangle. Each bowl is filled with the scouring liquor whose composition may vary from bowl to bowl in order to secure special advantages. Thus the last bowl may contain pure running water so as to complete washing out of the dirty detergent liquor brought forward by the wool from the previous bowls. The first bowl may contain a relatively strong soap liquor or it may be fed with a proportion of the spent liquor drawn from later bowls, for the wool will be most dirty as it enters the first bowl.

The wool is fed into the first bowl in the form of a thin layer or lap and a device at once submerges it in the soap liquor. Then an overhead system of swinging harrows or forks impels the wool slowly through the bowl to the remote end. There the wool is moved out of the liquor, supported on a moving inclined brattice, to pass through the mangle and drop into the next bowl, where it is at once submerged in the fresh liquor. It is then impelled through this bowl as in the first

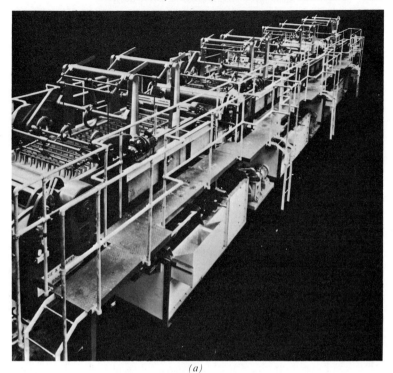

(a)

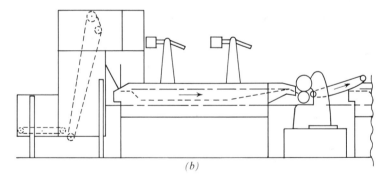

(b)

Figure 4.2 (a) Petrie/Wira improved machine for washing loose wool having the advantages of being shorter in length than earlier types and causing less entanglement of the wool fibres so that less fibre breakage is incurred in subsequent processing. Wool enters the first bowl to pass forward through 3 or 4 more bowls with mangles between each pair. (b) Passage of wool through the last bowl where it is washed with running water so that it leaves the mangle free from all scouring liquor. (Courtesy Petrie & McNaught Ltd)

bowl and then passes through the second mangle and into the third bowl. The wool is thus moved continuously through the series of bowls to become cleaner as it passes into each succeeding bowl. Finally it emerges from the last bowl and is then ready for drying. Loose solid impurities on the wool fall to the bottom of each bowl to facilitate either periodic or continuous removal as a sludge.

The important idea behind this scouring is to keep the wool in a soft, free and lofty condition completely free from fibre entanglement and matting.

In an alternative jet type of scouring machine the wool is carried through the tanks on a moving brattice while being sprayed overhead with the scouring liquor which is continuously withdrawn by pumps from the bottom of each bowl and filtered before its re-use. The object of this non-use of harrows is to avoid more positively any entanglement of the fibres. However, more recently Petrie & McNaught Ltd have, in collaboration with the Wool Industries Research Association (Wira), devised a much improved type of wool scouring machine which returns to the use of harrows but has several important advantages. It is likely to replace previous types of wool scouring machines following large scale successful experience with it (Figure 4.2).

Improvements to be noted are that owing to a closer packing of the loose wool and a slower movement of the harrows with a shorter stroke there is less movement of the fibres relative to each other and thus less fibre entanglement (this is revealed in subsequent combing of the scoured wool by a useful reduced amount of entangled and broken fibres). A further advantage is that the bowls can be shorter, so that a 4-bowl machine will occupy a length of only 71 ft (22 m) as against the usual 106 ft (32 m).

By having spare liquor-containing tanks one on each side of a scouring bowl provided with pump and liquor filter (to remove impurities from the wool) it is possible to run the machine continuously and without stoppage when changing the scouring liquor as may be desirable after running for 8 h.

This Petrie/Wira machine uses only about half the usual amount of water, say about $\frac{1}{2}$ lb (0·23 kg) of water per 1 lb (0·45 kg) of scoured wool.

In all such wool scouring machines it is customary and advantageous for the detergent liquor to move countercurrent to the movement of the wool.

Hydroextraction Whatever the type of the loose fibre being scoured it will eventually have to be hydroextracted or 'whizzed' to remove excess water before drying. It is obvious that if this removal of water is incompletely done then the drying will be the more difficult and expensive. Modern hydroextractors are very efficient

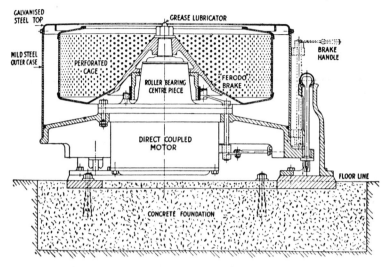

Figure 4.3 Cross-section of a hydroextractor. (Courtesy Thomas Broadbent & Sons Ltd)

and seldom leave in the wet material more than 80 per cent of moisture; frequently, depending on the nature of the fibrous material, they leave in only 40 per cent. The 'whizzer' is simply a circular cage or container for the loose wool or cotton which can be made to rotate at about 1000 rev/min and so throw out the excess water through the perforations in the wall of the cage. The cage is within a fixed outer casing which has an outlet for the water.

Carbonisation   When the wool contains a high proportion of vegetable impurities such as plant stalks, seeds, motes and leaves often designated as 'burrs', it is better to subject it to a process of carbonisation in order to remove these. This process depends on making the vegetable impurities so friable that when the wool is led through a machine which crushes and shakes it the disintegrated impurities easily separate from it.

For carbonisation the wool is saturated with a weak solution of sulphuric acid or sometimes a solution of aluminium chloride, and is then hydroextracted to remove all excess liquor. The hydroextractor used for this purpose has a vulcanised cage and other parts are similarly protected to resist corrosion by the acid liquor. The wool is then led through a drying or baking machine. The acid dried in the wool does not harm this fibre, but by contrast the acid dried into the vegetable (cellulose) impurities disintegrates them. The dyeing properties of wool can be modified by carbonisation so that the wool can have its affinity for some dyes reduced.

Drying loose fibres Drying is effected by passing the loose fibres as a thin layer supported on a perforated brattice through a large chamber within which hot air is vigorously circulated. The wet material goes in at one end and after traversing backwards and for-wards at different levels within the chamber it emerges at the other end quite dry.

Purification of yarn in skein form

Quite considerable amounts of cotton, wool, silk and linen yarns are scoured preparatory to dyeing. These yarns are either sold later for home knitting or are used by weavers and knitters who wish to produce colour patterned fabrics which could not so easily be obtained by dyeing or printing methods. Much of the yarn used for coloured carpets is pre-dyed.

It is possible to process yarn in different forms and the decision on this point is usually made with regard to the fineness of the yarn. It may therefore be treated in the form of skeins or it may be treated whilst wound in the form of cops or cheeses, or whilst wound many layers thick on perforated beams (rollers). If the yarn is very fine it is best processed as a wound package since there is then less chance of entanglement ensuing. Coarse yarns can be handled com-paratively roughly in skein form. Wool yarns require special care to avoid felting.

Skeins of cotton and linen yarn can be most conveniently scoured in kiers by practically the same methods as previously described for loose cotton.

Wool yarns in skein form are often scoured in large shallow tanks somewhat similar to those described for the scouring of loose wool, but in this case the skeins are carried forward through the warm soap liquor in each tank held between two endless, moving, upper and lower wide belts or brattices. These brattices consist of parallel spaced laths so that the scouring liquor can pass freely through them whilst the yarn is held loosely between the upper and lower brattices. In passing from one tank to the next the skeins are mangled so as to squeeze out the dirty liquor and prevent it from being carried forward into the next tank.

Such machines are used where it is desired to process expeditiously large amounts of cotton, linen or wool yarns. But there is always the opportunity to operate on a smaller scale for smaller or odd lots of yarn. For this, use is made of long rectangular wood tanks which are filled with a detergent liquor. The skeins are suspended from smooth wood or stainless metal rods (about 2 lb (0·9 kg) of yarn from each rod). The ends of the rods are allowed to rest on the sides of the tank or vat so that practically the whole of each skein hangs completely submerged in the liquor. From time to time men move the rods from one end of the vat to the other, thereby deftly drawing the skeins through the liquor. At the same time they invert the skeins or at least turn them partly around so that the portion formerly over the rod and therefore not immersed now becomes submerged to receive its fair share of scouring treatment. The scouring obtained in this manner is quite satisfactory but it is obviously not a mass-production method.

Skein scouring machine   Scouring of skeins can also be effected in a Hussong type machine which is generally employed for dyeing (Figure 4.26). This machine consists of a rectangular vat filled with the soap or other detergent liquor. An upper framework serves to carry a suitable number of rods from which the skeins of yarn can be hung. This framework filled with skeins is lowered so that it just fits over the top of the vat to allow the skeins to hang completely submerged in the liquor.

A small pump is attached to one end of the vat so that the liquor can be circulated continuously and evenly through the skeins. When scouring is complete, the detergent liquor can be run off and replaced by fresh water for washing. Finally the framework of skeins is lifted from the vat. The skeins can then be removed, hydroextracted and dried.

Drying of skeins   The drying of skeins of yarn is most conveniently carried out by passing them through a large chamber in which hot air is continuously circulated. The skeins are carried through this chamber either on a brattice or apron or, better still, they are carried forward whilst hanging from horizontal rods. As the hot air blows through the skeins it not only dries them but at the same time it opens out the separate threads and finally leaves them in a soft, lofty condition.

Purification of yarn in package form

It is quite a different matter scouring yarns in 'package' form, that is, as wound on cones, bobbins or beams, or as cheeses and the like. Under such circumstances it is, of course, necessary to pump the scouring liquor through the yarn, and it is important to ensure that the flow of liquor is even and that all parts of the yarn receive adequate treatment. With this aim it is usual to have the yarn wound on skewers or spindles which are really perforated tubes. It is then possible to circulate the liquor through the tubes and yarn alternately in both directions, that is, from the inside, outwards, as well as from the outside, inwards.

With small yarn packages a number of these are stacked on perforated tubes within a large outer container. Then the scouring liquor is forced by means of a pump up through the inner tubes and outwards through the packages of yarn. From time to time the direction of liquor flow is reversed so that the liquor in the containing vessel is sucked through the cops or cheeses of yarn from the outside. The liquor then escapes through the central perforated tubes and so back via the pump to the containing vessel. Quite good results can be obtained by these continuously circulating liquor machines. Finally the yarn can be washed with fresh water in the same machine. The scouring of yarn wound on perforated beams follows much the same plan. These beams are placed within liquor containers and then the liquor is pumped in either direction through the yarn.

As we shall see later, these machines for treating yarn in package or wound form have been mainly devised for dyeing yarn. But, since the scouring and dyeing processes follow each other, there is often no necessity to remove the yarn packages from the machine between the two treatments. If necessary the yarn can be dried after scouring whilst still in its package form, but this is a somewhat slow process.

Purification of fabrics

We can now deal with the scouring of woven and knitted fabrics and this is the most favoured form for this processing. Both methods are employed—movement of liquor in relation to fabric and vice versa.

Singeing and kiering of cotton fabric Cotton fabrics are generally scoured in kiers similar to those described for loose cotton fibre. Kiers capable of holding 3–5 ton (3–5 t) of fabric are in use. The fabrics are first sewn end to end so as to make a continuous length, but the sewing is by means of a chain-stitch so that after the processing the thread can be easily withdrawn and the fabrics separated.

Actually, before kiering, it is often necessary to pass the fabric through a singeing machine with the object of burning off fibre ends projecting through the fabric surface and thus to remove hairiness. Two types of singeing machine are in use.

The hot plate machine comprises two half-circular copper plates which are heated to redness by means of coal, gas or oil, or even electrically. The two plates are spaced side by side and parallel to

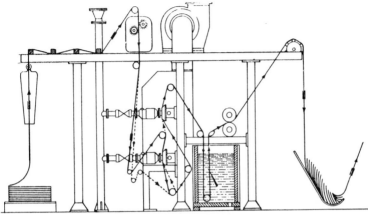

Figure 4.4 Two-burner gas singeing machine for singeing off protruding fibre ends from the surface of cotton or other fabric. The fabric is led through horizontal gas flames so that first one side and then the other is singed. After singeing the fabric is led through the water trough to quench sparks which may have fallen on it. (Courtesy Mather & Platt Ltd)

each other with the curved surfaces uppermost. The fabric in full open width is led over these plates at high speed and so as just not to touch them. If the fabric travelled slowly then it would simply be scorched. But, by suitably adjusting the rate of travel and the extent to which the fabric presses towards the plate surface, this scorching is completely avoided and it is only the surface hairiness which is burnt off.

In the other more used type of machine the fabric is led over a horizontal row of gas flames or a long gas flame emerging from a slotted pipe. Here the flames impinge on the surface of the fabric and burn off the loose hairs without scorching it.

The action of the hot plate is not quite the same as that of the gas flame and so sometimes it is preferred to pass the fabric both over a plate and through the gas flames. Further, it is frequently desired to produce a sheer surface on both sides of the fabric. In this case the fabric can be run over suitable guiding rollers so that this is accomplished in the 'one run'.

Although singeing appears at first sight to be a very risky process, it is only occasionally that damage to the fabric is caused, usually by running the fabric too slowly. Moreover, cotton fabrics containing viscose and acetate threads can be singed without damage. When the fabric leaves the singeing plate or gas flames it may possibly carry some smouldering fibre ends, or sparks may have fallen upon it. These can be immediately extinguished by leading the fabric through a small mangle provided with a trough of water. The fabric passes through the water and then through the mangle rollers to ensure that the water is spread uniformly through it.

Fabrics are singed not only to give them an improved appearance but in the case of synthetic fibres and some wool fabrics, to remove any tendency of the fabric to pill during use—a hairy surface promotes pilling (p. 410).

Starch liquefying enzymes At this point it is convenient to refer to a special treatment which is mainly applicable to cotton and rayon fabrics and which has for its object the removal of starch impurities without the use of a kiering process. Some cotton and rayon fabrics are fragile and can be damaged by the somewhat rough handling involved in kiering. It may also be necessary to avoid kiering if the fabric contains coloured portions which are not sufficiently fast and which would bleed or run during an alkaline boil.

There are available to the scourer and bleacher certain enzyme preparations which have the power of liquefying starch so that it becomes water-soluble. Thus it is very convenient if the mangle trough used in conjunction with the singeing machine is provided with a liquor containing such an enzyme product. Then the fabric leaving the singeing machine can be saturated, not with water, but with the enzyme liquor, and immediately afterwards piled down into a heap on the floor or within a tiled cistern, or otherwise stored to allow the enzyme to react fully. Rapid-acting enzymes are now available. After lying for about 2 h all the starch in the fabric will be soluble, so that it just remains to give the fabric a thorough wash. There is then no necessity for it to be purified by kiering although this latter method is more thorough.

Normal kiering method Returning now to a well-established procedure, the singed fabric, quenched by wetting, is run into a kier. It is not possible or necessary to keep the fabric in open width but it is brought into rope form. Thus it is evenly and closely packed in the kier. Formerly, small boys used to stand within the kier and guide the fabric, but nowadays it is more usual to make use of mechanical plaiting devices which ensure that the fabric is uniformly distributed within the kier. It is most important that the fabric in the kier should be evenly packed for, if it is not, the boiling alkaline liquor circulating through it from the top downwards to the bottom will find easy channels and so some parts of the fabric will be better scoured than others.

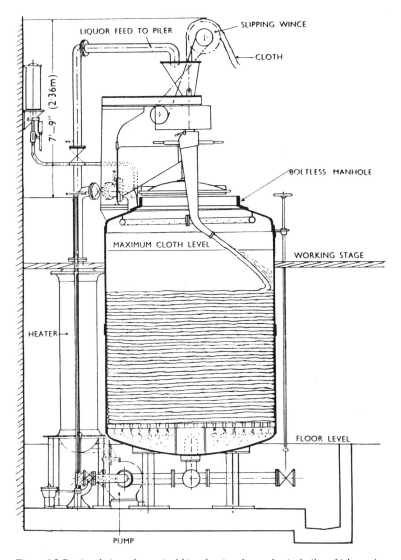

Figure 4.5 Sectional view of a vertical kier showing the mechanical piler which evenly distributes the fabric prior to treatment with the detergent liquor. The detergent liquor is continuously withdrawn from the bottom of the Kier by the pump, and is led to the top, via a heating unit, where it is again sprayed over the fabric and percolates through it.
(Courtesy Mather & Platt Ltd)

222

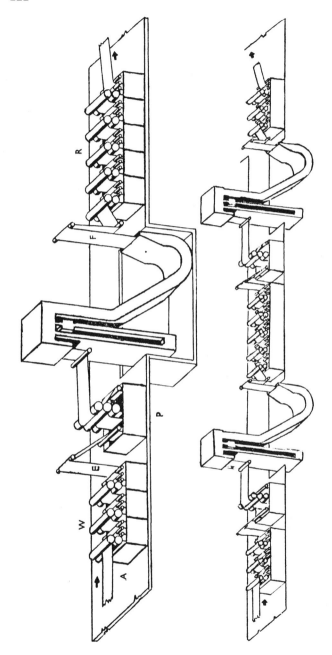

Figure 4.6 Plant for scouring (top) and for scouring and bleaching in succession (bottom) cotton fabric in rope or open width form. The scouring section W is followed by the J-box and the washing-off section R. The section between the two J-boxes (bottom) is for washing-out the alkaline scouring liquor and impregnating the fabric with the hydrogen peroxide bleaching liquor. (Courtesy E. I. Du Pont de Namours & Co)

With cotton fabrics it is usual to maintain the alkaline liquor slightly above the normal boiling point of water and so make it more effective. This means that the kier must be sealed with a large screw-down plate and the treatment within carried out under pressure, say up to 10 or 20 lb f/in^2 (70 or 140 kN/m^2) steam pressure. Apart from this, the operations and kiering plant are as described for the kiering of yarn in skein form.

Generally the fabric is kiered twice, the fabric being withdrawn temporarily from the kier and washed in a roller washing machine between the two boils. Each boil may extend overnight, and seldom for less than 4 h.

After the second boil the fabric will be drawn out and run through a roller washing machine to remove all traces of the dirty alkaline liquor and the dissolved impurities which it contains. Then it is drawn over a scutcher device which very simply opens it out to full width again. Of course, if the fabric is to be bleached, then it can be kept in rope form for this operation.

Continuous method   The method of purifying cotton fabric by use of separate scouring, kiering and washing operations, as described above has hitherto been widely practised in Britain and, in general, it is efficient having in mind the conditions which apply. However, it is desirable, whenever possible, to effect the purification of cotton fabrics more especially by the use of methods by which the fabric, either in rope form or in open width, is led through a range of machines so that in one run it is scoured and bleached so as to be satisfactorily prepared either for subsequent dyeing and finishing or to be finished as white fabric. Such methods of processing were first developed in America rather than in Europe, but now it has been adopted world wide and is even used in the remote parts of countries such as Korea and Malaya. Of course the incentive behind this progress has been the possibility of achieving a higher rate of production at lower cost.

The Du Pont type of plant shown in Figure 4.6 initiated progress along these lines.

In operating this type of Du Pont plant the separate pieces of cotton fabric are first sewn end to end to form, in either rope or open width, a continuous length of several thousands of yards or metres.

The fabric, in rope form, is run through a 2 to 4 per cent solution of caustic soda contained in the trough of a padding machine and excess of this liquor is squeezed out by mangling rollers. The fabric then passes forward through a steaming device which raises the temperature of the fabric to about 100°C as it is being piled or plaited downwards into the long arm of a *J*-box. The fabric may be

224

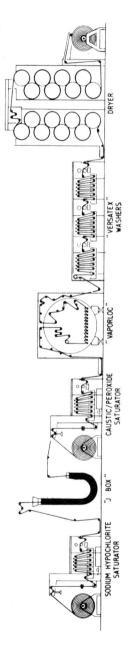

Figure 4.7 Plan of the one-run (continuous) bleaching of cotton fabric. The fabric, impregnated with a cold dilute bleaching liquor containing sodium hypochlorite, undergoes a substantial degree of bleaching while passing through the J-box with a suitable dwell period. It then passes through the Vaporloc section (see Figure 4.13) after having been impregnated with a solution of caustic soda and hydrogen peroxide (the latter destroys residual sodium hypochlorite in the fabric), where it is steamed to complete the bleaching. Thereafter, the fabric is washed and dried.
(Courtesy Mather & Platt Ltd)

led into this *J*-box at a rate of, say, 100 yd/min (90 m/min), but the capacity of the box is so large that the folds of fabric take 1 or 2 h to work downwards into the short arm so that the fabric can be withdrawn at the same high rate to pass through a roller washing machine where the caustic soda and loosened and solubilised impurities are completely washed out of it. If it is now desired to bleach this purified cotton fabric it is either led through the same or a modified but similar plant where it is first impregnated with an alkaline solution of hydrogen peroxide instead of caustic soda, suitably heated in the steaming device which precedes the *J*-box, and is then allowed sufficient time within the *J*-box to become fully whitened. Finally the fabric is thoroughly washed in a roller washing machine. This bleaching treatment can be carried out at the same high rate as the kiering treatment. Optionally, both scouring and bleaching can be effected in one run using a combined plant.

British makers such as Mather & Platt Ltd, Sir James Farmer Norton & Sons Ltd, etc. (possibly first stimulated by the success and wide use of the Du Pont plant) now provide comparable plant with modifications and improvements to the Du Pont plant, but its high cost can only be justified when long runs of fabric are to be regularly processed.

Figures 4.7 and 4.8 show plants which allow one-run bleaching of cotton fabric using sequences of treatments with hydrogen peroxide and with sodium hypochlorite with intermediate washing. Figure 4.9 shows alternative related plant for scouring cotton fabric followed by a final bleaching in which sodium chlorite is the bleaching agent instead of sodium hypochlorite.

These continuous one-run processing plants can be adapted for the treatment of the cotton fabric in open width instead of in rope form.

The essential part of such a continuous kiering and bleaching plant is the *J*-box, for it is this with its high fabric storage capacity which allows the fabric to pass through the plant continuously at a high rate. Although it is usual to treat cotton fabric in rope form, similar plant is available for treating fabric in open width. This form is used when there is a possibility that the fabric will become permanently creased or frayed by handling it in a folded state. For such continuous kiering and bleaching plant to justify its installation there should be about 1 000 pieces of fabric for processing in each 8-h day.

Roller washing machines   The roller washing machine referred to above is one of the most useful of all the scouring machines to be found in a textile works. Essentially it comprises two large-diameter

226

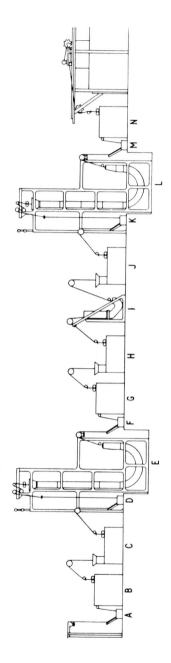

Figure 4.8 The continuous bleaching range for cotton fabric: A–scray. B–5-stage washer. C–caustic soda+peroxide saturator. D–scray. E–hot caustic soda+peroxide J-box. F–scray. G–5-stage washer. H–sodium hypochlorite saturator. I–cold sodium hypochlorite J-box. J–caustic soda saturator. K–scray. L–hot peroxide caustic soda J-box. M–scray. N–5-stage washer. (Courtesy Mather & Platt Ltd)

Figure 4.9 Plant for the continuous bleaching of cotton fabric in open width, comprising units for carrying out in succession the necessary treatments associated with the use of sodium hypochlorite. (Courtesy Sir James Farmer Norton & Sons Ltd)

horizontal rollers mounted over a large shallow trough. Frequently the rollers are of wood but they may also be of hard rubber. In the washing operation the fabric enters at one end of the pair of rollers and, leaving these, it is guided down at the back of the machine to the rear of the trough which is filled with running water. Here it passes over and under a long roller which extends the length of the trough and then along the bottom of the trough to the front where it passes upwards and over another similar roller and thence through the mangle rollers again, slightly to one side of the place where it went through the first time. The fabric is then guided through the trough from back to front, again parallel to its first path of travel, and again through the mangle rollers. Guiding pegs keep the strands of fabric parallel and so that they do not become entangled. In this way the fabric gradually works its way across the pair of mangle rollers to emerge at the far side having been through the trough of running water and the mangle rollers some 10–20 times, according to the width of the machine. It is obvious that a very thorough washing is obtained in this way.

Of course, such a machine can be used not only with cold running water in the trough but also with detergent liquors to secure better and more penetrating washing treatments.

In washing fabrics and removing excess of water by passing them through mangles they become distorted; usually the weft threads become askew and the fabric narrows in proportion to the degree to which it becomes extended in length. This is a disadvantage since fabric straightening and widening have to be carried out in later finishing operations. It has also been found that the efficiency of washing is reduced as the fabric is held taut while passing through a washing liquor. Thus in one type of washing machine the fabric in rope form is maintained in a condition of minimum tension. In such a machine there are upper and lower series of rollers over which the fabric is led in its path of travel which includes partial immersion in a lower trough of detergent liquor. The upper driven rotating rollers are tapered and there are guiders which direct the fabric to that part of the tapered roller over which it should be drawn. By means of control handwheels the fabric can be directed over a tapered roller where it has a large diameter, so as to leave where the diameter is smaller, thus suitably reducing the fabric's lengthwise tension.

Washing machines can consume very large amounts of water and today it can be difficult to obtain an adequate supply. Also the disposal of waste liquors can be a problem now that pollution has aroused much public attention. Modern washing machines are therefore specially designed to be very efficient and so use the minimum amount of water.

The consumption of water may be reduced by scouring with organic solvents such as trichloroethylene or perchloroethylene which have the advantage of being non-flammable. An ICI apparatus is available and is very satisfactory for treating cotton fabric in a one-run through process while a Bentley Rapide machine is available for the treatment of knitted goods batchwise with the same solvents. This latter apparatus allows finishing agents such as fluorescent whitening agents to be applied simultaneously.

While it can be advantageous to wash a fabric in a substantially tensionless state from the viewpoint that superior purification can be obtained arising from the greater ease of fabric penetration and through-flow by the detergent liquor, there is the further advantage that the washed fabric is left in a less stretched state. Knitted fabrics are especially liable to excessive length stretching.

In Figure 4.10 is shown an Italian (Mezzera Spa) washing machine especially designed to allow cotton knitted fabric to be scoured in any appropriate detergent liquor while free from width and length tension.

The tubular knitted fabric is first slit to form a single thickness open-width fabric, which is led into the machine guided by photoelectric selvedge guiders and passed through selvedge uncurlers.

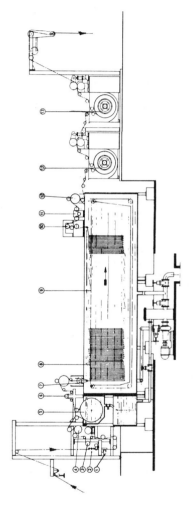

Figure 4.10 Machine for scouring knitted (especially polyester jersey) fabric in open width and with its full relaxation: such fabric may normally shrink in length to 80–85 per cent of what it was originally when it left the knitting machine, and may contain 7 per cent of oil which should be reduced to 0·2 per cent at the most. If, in the subsequent dyeing with disperse dyes, risk of colour staining is to be avoided when dye-carriers or antifoaming agents are used. 1, photo-electric selvedge guides; 2, selvedge uncurlers; 3, expanding rollers; 4, overfeed device; 5, immersion roller; 6, expanding rollers; 7, main tank; 8, fabric supporting rods; 9, saw toothed guides; 10, driven expanding rollers; 11, suction unit; 12, mangle rollers; and 13, final washing tanks.

(Courtesy Mezzera Spa)

After passing over two groups of fabric-expanding rollers, the fabric, in its fully extended flat state, is led around a roller immersed in a small tank containing detergent liquor to wet it before entering the main long washing tank. It moves through this while carried in vertical loop form on parallel horizontal rods. The loops of fabric are fairly closely spaced but the hot detergent liquor circulating in a direction counter to that of the fabric is able to move sufficiently freely between the fabric loops to effect efficient scouring.

The fabric leaving the washing tank passes over expanding rollers, a suction slot and through a small mangle before passing through two rinsing tanks provided with mangles so that it can then be plaited into trucks ready for drying.

Since at all stages, and especially while in the washing tank, the fabric is not subject to stretching, it is finally left dimensionally stable and subject only to a small amount of shrinkage in subsequent washing.

Another but different type of machine for washing knitted fabric with minimum stretching and distortion is shown in Figure 4.11.

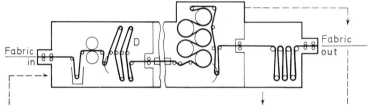

Figure 4.11 Machine for washing knitted fabric with minimum stretching and distortion; the passage of the knitted fabric through this machine is assisted throughout in order to eliminate tension

It may be recalled here as mentioned earlier, that, fabrics made of synthetic fibres have a tendency to shrink during their first hot wash, and if they are in a creased or folded state during this washing stage the creases are liable to become permanently fixed so as to spoil the finished appearance of the fabric. To avoid such creasing a simple expedient is adopted. It consists of subjecting the fabric to a high temperature while it is dry and free from creases and which is high enough to confer on it a stability so that it has no tendency to shrink (with crease formation) when afterwards entered in the hot washing liquor. The temperature required to produce this stability must exceed any temperature which it is likely to encounter in washing. In this high-temperature heating, latent strains within the fibres (left from the stretching to which they are subjected during manufacture) are relieved and rendered ineffective. If the fabric is subsequently exposed to a higher temperature then residual latent strains may come into play. It is thus necessary to use a high temperature higher than any to which the fabric is likely to be exposed to later.

The heating of the fabric may be carried out by running it over a stenter while hot air is directed upon it. This is a very convenient method since during the heating the fabric is held in open width taut and free from creases since its selvedges are gripped by pin plates or clips as the fabric is travelling from one end of the stenter to the other. Another method is to run the fabric in a crease-free form over a number of hot cylinders. The fabric may also be wound in roll form on perforated tubes through which steam can be blown. In general a high temperature treatment to give synthetic fibre fabrics increased dimensional stability is increased by having moisture present.

Winch machines In the case of lighter, more fragile fabrics, such as those made mainly of one or other of the different rayons or of wool, there is another type of machine which is very useful for scouring and which makes unnecessary any kiering treatment. This is the winch machine. It is also very suitable for knitted fabrics, especially those which are distorted too much when pulled lengthwise—a defect which can scarcely be avoided in drawing long lengths of fabric into and out of a kier or in running them through roller washing machines.

The winch machine is essentially a vat, usually having the back curved and the front straight, and above which is mounted a horizontal winch or roller. There is a row of guiding pegs in front of the winch just above the level of the detergent liquor in the vat. In threading up a winch machine with fabric the pieces are kept separate and are not joined end to end since they are not required to pass continuously through the machine from one side to the other, as is the case with a roller washing machine.

Thus, taking a piece of fabric, which may be from 40 to 80 yd (37–73 m) in length and, say, 20 to 60 in (0·5–1·5 m) in width, one end is drawn over the winch and then sewn to the other end. Thus the piece is made into an endless band with the winch inside it. Except for the short length of fabric which is rising out of the liquor in the trough to pass over the winch and then fall down into the liquor at the back of the machine, practically all of the fabric is fully submerged in the liquor. A number of pieces of fabric are threaded up in this manner and arranged side by side with the aid of the guiding pegs.

The liquor is prepared with soap and alkali and is heated by blowing live steam into a partitioned portion in the front of the machine. Through the same perforated partition can be added detergent or other substances which will not come into direct contact with the fabric until they have become diluted with the main bulk of the liquor.

Rotation of the winch draws fabric out of the liquor, between the guiding pegs, over a guiding roller, and then over the winch to fall back into the liquor. This movement takes place continuously, and ensures that the fabric receives a uniform treatment in this scouring operation. The rate of rotation of the winch is such that it takes about 10 to 15 min for each piece of fabric to pass over the winch. When scouring is complete then the detergent liquor can be replaced by running water to remove all traces of the dirty liquor. The fabrics can then be removed and hydroextracted preparatory to drying, or they may pass forward for bleaching or dyeing. The winch machine is very mild in its action and so is very suitable for fragile fabrics of all kinds.

Wool fabrics are frequently scoured in machines similar to the winch machine, but instead of having a winch they are provided with a pair of rollers which press upon each other lightly. Thus the fabric receives a light squeeze as it passes between the nip of the mangle rollers which squeezes out the emulsified, oily and fatty matter impurities commonly found in wool fabrics. The scouring liquor will in most instances contain soap together with a small proportion of ammonia or sodium carbonate.

Winch machines have in recent years been much improved in various ways to make them more effective and less likely to damage the fabric by excessive friction and tension and also more economical in respect of heat consumption (for example, they are now often hooded).

Another type of machine suitable for washing fragile fabrics and those which should not be stretched is known as the 'ripple' washing machine. In this the fabric is led in open width down a glass slide and then falls on an endless moving apron which carries it forward whilst being sprayed from overhead sprays of hot liquor.

Scouring of lace Cotton and silk lace fabric is often made in long lengths so that it can be easily cut up into small portions after the material has passed through all the wet processing. Such lace can be conveniently scoured in winch machines. For silk lace it is, of course, necessary to employ small and easy running machines so that there may be no possibility of causing damage. Some cotton laces, and most cotton curtain materials, are sufficiently robust to withstand the normal kiering treatment. The machinery used for making lace goods is lubricated with graphite instead of oil and so it frequently happens that black stains are present in the freshly manufactured material. It requires a strong soaping and rubbing by hand to remove such stains. Special stain-removing substances are now available.

Scouring of smallwares   It is evident that the scouring of small-wares and garments cannot be effected with large or medium-sized winch machines and kiers, for these require that each piece of the textile material should be at least 40 to 60 yd (37 to 55 m) long. It is not satisfactory to string together large numbers of small garments so that they can be kiered. So with these goods it is necessary to scour in alternative machines.

One very convenient type of scouring machine is known as the *rotary* washing machine. It consists of two cylinders, one inside the other, and both mounted on the same horizontal axis. There is a distance of 2–3 in (50–75 mm) between the two cylinders and it is in this space that the detergent liquor is placed.

The outer cylinder is watertight except when the doors are open to permit access to the inner cylinder. The inner cylinder is usually divided into three compartments extending the whole length of the cylinder but, unlike the outer cylinder, this is perforated so that the liquor initially placed between the cylinders can flow through the perforations into the inner cylinder.

In operating such a machine, the garments are packed fairly loosely within the radial compartments and the covering flaps

Figure 4.12 Rotary dyeing machine for stockings. The inner drum, shown separately, is divided and revolves alternately in opposite directions. (Courtesy Samuel Pegg & Son Ltd)

fastened down. Then the watertight doors on the outer cylinder are securely fastened. The inner cylinder is now rotated slowly. Usually it is automatically controlled so that it reverses periodically. Arrangements are provided so that steam can be blown into the liquor within the outer cylinder to maintain it at a suitable temperature. Thus the garments are moved in relation to the hot detergent liquor which flows through the perforations and so among the garments. Altogether this is a very satisfactory method for scouring loose articles.

The Pegg Toroid machine (Figure 4.32) can be used for the scouring of garments. It is cylindrical having a false (perforated) bottom and a lower closed steam coil. The garments are in the upper part of the cylinder. The scouring or dye liquor is continuously circulated so that it passes upward past the coil to maintain its high temperature, through the false bottom, and is then split into two streams to maintain a uniform tumbling of the garments so that those on the outside tumble inwards continuously.

Yet another popular machine is for scouring small articles and in particular it is very convenient for the treatment of men's socks. This is known as a *dolly*. Such a machine may be either round or rectangular but its action is the same in both cases.

Briefly, the dolly comprises a vat to hold the detergent liquor within which are placed the loose goods to be scoured. Over the vat are mounted a number of heavy wooden 'fallers', which are positioned in a single row to extend across the width of the vat. They are arranged side by side and almost touching each other. Each faller is a long length of wood somewhat like a plank standing upright. A mechanical device lifts each faller in turn to a height of about 8 in (200 mm) and then allows it to fall by its own weight on the garments below it in the vat. Meanwhile the round vat is slowly rotated or the rectangular vat is steadily moved backwards and forwards. Thus the garments submerged in the soap or other detergent liquor are continuously pounded by the fallers and thus thoroughly washed. The wooden fallers are exceptionally smooth, and although quite heavy it is found that no damage is caused to the garments provided that there is a sufficient quantity of them in the vat to 'cushion' each blow of the faller. Scouring in this machine can be very efficient.

It will be realised that dyeing machines can often be used for scouring purposes. For example, the paddle machines referred to later (p. 275) as being suitable for the dyeing of men's socks are equally satisfactory when employed for scouring them. Often scouring and dyeing operations are carried out successively in the same machine.

BLEACHING TREATMENTS

So far we have dealt with scouring. In most cases the textile material will be whiter after this operation than before, owing to the removal of impurities, but this is not necessarily the case. Much depends on the type of material. For example, after kiering with alkaline liquors, cotton goods may be much browner or yellower than before the treatment. This is because some residual impurities have become more highly coloured by partial decomposition. The essential feature of the scouring is that it has removed most of the impurities, and those that are left, including the natural colouring matters, are such that they are easily destroyed by the bleaching treatment which follows. Bleaching is a final treatment to complete the purification, which at the same time ensures the production of a good white colour.

Bleaching silk and wool

Wool and silk materials may be bleached white by one of two processes or in some cases a combination of both. One method is that of exposing the textile goods to fumes from burning sulphur within a large closed chamber, whilst the other method is that of steeping the goods for some hours in a warm dilute solution of hydrogen peroxide.

Stoving   The first method is known as *stoving* and depends on the reducing action of the sulphur dioxide gas which arises from the burning sulphur. Stoving does not give a permanent white. Gradually, as the wool or silk material is exposed to the air, it reverts to its original yellow colour. This is because the natural colouring matters have not been destroyed by the sulphur dioxide; they have simply been reduced to a colourless form which reverts to the coloured form as a result of aerial oxidation. In contrast, the colouring matters are destroyed once and for all by the hydrogen peroxide treatment and this method is to be preferred although it is somewhat more expensive.

The bleaching of silk, and particularly of wool, by stoving is commonly applied both to skeins of yarn and to fabric. It most conveniently follows a scouring treatment, for it is an advantage that the textile material shall be fairly damp and preferably slightly alkaline, such as is the case when the wool contains a little soap liquor. Skeins are hung from horizontal poles or rods and fabrics similarly, but in long folds. Many pieces of fabric can be hung in the 'stove' which is generally a wood chamber about twice the size of an ordinary room at home. Ventilation is secured by the presence of a small chimney.

To bleach 100 lb (45 kg) of wool requires about 8 lb (3·5 kg) of sulphur. This is placed in an iron tray in one corner of the room. The sulphur is lighted by touching it with a red hot iron poker and the doors are closed. This operation is usually started in the evening. During the night the fumes from the burnt sulphur are absorbed by the wool or silk and gradually reduce the colouring matters present in the fibres. The next morning the yarn or fabric is well washed with water and is then found to have a good white colour. Bleaching by stoving is used only to a small extent.

The burning of lumps of sulphur to provide sulphur dioxide is somewhat crude but it is simple and cheap. More controllable is sulphur dioxide obtained from the liquefied gas but hydrogen peroxide is now universally used.

Hydrogen peroxide bleaching   For bleaching with hydrogen peroxide a vat is filled with 2-volume strength hydrogen peroxide solution, which is obtained by diluting the strongest solution commercially available about fifty times. A small amount of sodium silicate or sodium triphosphate is added. This addition gives the liquor the necessary slight alkalinity and at the same time stabilises the liquor so that all its oxidising power is spent on the wool or silk and none is wasted by liberation of oxygen gas.

When the liquor is at 50 to 55°C the wool or silk is immersed in it and is packed fairly tightly. About 6 h is allowed and during this time the liquor cools gradually. At the end of the bleaching period the goods are withdrawn and thoroughly washed. They have thus been given a permanent white colour.

It is not permissible to bleach wool and silk with solutions containing active chlorine as for cotton, since under such circumstances the wool becomes weakened and moreover it acquires a yellowish colour and a certain degree of harshness. However, as we shall see later, wool is treated with a small proportion of chlorine for the purpose of reducing its felting power.

Bleaching cotton and linen

All cellulose fibres, such as cotton, linen and the various rayons, can be bleached with hydrogen peroxide after the manner used for wool and silk, but just as good a white can be produced more cheaply by using a solution of active chlorine. This solution is conveniently prepared by acidifying a solution of sodium hypochlorite.

Loose cotton and other cellulose fibre materials and yarn are usually bleached by circulating the bleach liquor through them in any simple form of vessel. Fabrics are more conveniently treated by a process in which they are saturated with the bleach liquor and allowed to lie in a cistern for 1 or 2 h to give the bleach liquor time to produce a good white. They are then thoroughly washed, slightly acidified and again washed to remove all traces of the bleaching liquor.

The fabrics are sewn end to end as for kiering and then run through a small mangle provided with a trough for holding the bleach liquor. The fabric thus saturated with this liquor is piled down into tiled cisterns which must be free from metal. Wherever the impregnated fabric lies in contact with iron or copper a special rapid bleaching action takes place and the cotton or other fibre is very much weakened. After about 2 h the bleached fabric is run out of the cistern through a roller washing machine provided with running water, then through another provided with weak acid, and finally through yet another washing machine to give the fabric a final wash.

Attention was drawn on p. 223 to modern practice and equipment for scouring and generally purifying cotton fabric, by which the fabric can be led continuously through a plant so as to enter in the impure state and emerge at the far end fully purified. This continuous method can combine both scouring and bleaching. Thus a first section can effect scouring using a caustic alkaline liquor while the second section can use hydrogen peroxide to effect the whitening or bleaching. Hydrogen peroxide is being increasingly employed for the bleaching of textile materials—it has the advantage that it leaves in the bleached goods no harmful residues which cannot be removed by simple rinsing. Alternatively, sodium chlorite can be used instead of hydrogen peroxide, with the advantage that it is less harmful to the cotton.

While plant which will allow all the operations to be carried out in one run will always be economically attractive, it will be appreciated that careful consideration must be given in making a decision on this point when it is mentioned that the cost of erecting such a plant is very high.

The layout of such a plant is shown in Figure 4.6 and it would be operated for woven cotton fabric in three stages: (1) impregnation with a 4 per cent caustic soda solution, followed by heating to 98·9 C and a passage through a J-box occupying 2 h; (2) thorough washing, impregnation with a cold solution of sodium hypochlorite, which may optionally be weakly acid or neutral according to the character of the fabric, and then passage through a smaller *J*-box occupying 1 h; and (3) impregnation with a weak alkaline solution of hydrogen peroxide, followed by passage through the third *J*-box occupying 1 h at 98·9°C, followed by a final thorough washing and drying to give the desired fully bleached white.

In the third stage, indicated above, the hydrogen peroxide destroys any residual sodium hypochlorite carried forward by the fabric and at the same time further whitens the fabric.

Scouring and bleaching can be more effective, or at least can be carried out more rapidly, if the treatment with alkali or alkaline hydrogen peroxide liquor is conducted under pressure in suitable apparatus so that the operating temperature is above 100°C. Such a treatment can be carried out within a kier but the difficulty encountered if a continuous processing is required is that of maintaining a satisfactory seal in the top of the kier where the fabric would have to pass into and emerge from it—if leakage of steam occurred the high temperature and pressure within the kier could not be maintained.

The so-called Vaporloc reaction chamber, devised by ICI Ltd and made by Mather and Platt Ltd and used in the scouring and bleaching shown in Figure 4.7 overcomes this difficulty. The fabric to be treated passes (in open width) downwards through a slot in the top of the horizontal chamber to pass over guiders in a circuitous path within the chamber and then emerges upwards through the slot. All the guiding rollers are positioned on a frame which can be easily run into and out of the chamber; some of these rollers are positively rotated so that the fabric can be free from tension and thus promote a more efficient treatment. The slot is provided with a static air-pressurised special seal. As the fabric passes through the chamber (it is pre-impregnated with the scouring or bleaching liquor) it is subjected to a temperature which may be as high as 134°C arising from a steam pressure of 30 lbf/in² (200 kN/m²). Under these high-temperature high-pressure conditions cotton fabric can be scoured or bleached within 45s to a degree equal to that obtained by treatment in a *J*-box at 100°C and requiring 90 min. Thus by using a Vaporloc reaction chamber a fabric scouring or bleaching treatment can be effected at the high rate of up to 500 ft/min (150 m/min).

To effect a continuous scouring and bleaching process it is most convenient to use two Vaporloc reaction chambers in series thus taking the place of the first and second J-boxes in the scouring and bleaching plant previously described and shown in Figure 4.8. A Vaporloc reaction chamber is shown in Figure 4.13 and the

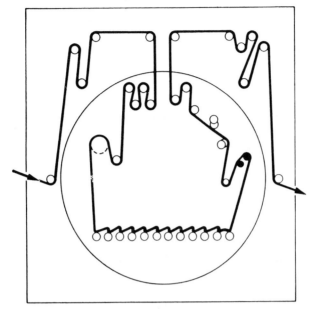

Figure 4.13 Fabric in open width passing over guiding rollers within a Vaporloc reaction chamber. The lower horizontal row of rotating rollers assists the forward movement of the fibre. (Courtesy Mather & Platt Ltd)

passage of fabric through it is shown.

The Vaporloc reaction chamber can also be used for the steaming of printed fabrics to fix their colours (p. 288) and also for their flash ageing (p. 290).

Sodium chlorite is another bleaching agent which can be used for the bleaching of cellulose fibre goods instead of sodium hypochlorite. It has disadvantages, for example, the bleaching liquor can liberate obnoxious vapours harmful to the health of the operatives and it can be corrosive towards the materials ordinarily employed in the construction of bleaching equipment, but by contrast it can give a very good white without appreciable deterioration or weakening of the cotton. It has been found that a number of substances can be added to activate the sodium chlorite bleaching and also to prevent undue formation of the obnoxious chlorine dioxide. Titanium metal resists corrosion and is now being used for bleaching plant. Sodium chlorite finds but limited use as a bleaching agent for fibres which can be bleached equally satisfactorily with sodium hypochlorite.

Bleaching of flax (linen) fibres

The raw material flax straw has a much lower content of pure cellulose than has raw cotton and so it can be anticipated that in removing these impurities by the preliminary processes of retting and scutching to produce so-called 'grey' flax, and in the subsequent chemical treatments of this to secure a fully bleached and stabilised linen fibre of high and uniform quality suitable for conversion into yarn and fabric, considerable objectionable degradation of the fibres could result unless suitable processing conditions are employed. Essentially these conditions must leave the fibre with two of its most important properties and in which it exceeds all other cellulose fibres. These are its high absorbency of water and its high resistance to stretching coupled with great flexibility.

The preservation of these properties are more the concern of those who carry out the final stages of fibre purification, including a boiling with caustic soda solution to remove wax and pectic impurities and a bleaching with active chlorine to destroy natural and other colouring matters, for under inappropriate conditions the alkali and the chlorine can impoverish cellulose.

Research by Dr Sloan of Kirkpatrick Bros at Ballyclare in Ireland has resulted in establishing combined processing conditions which are now employed to produce a standardised type of Linron (brand name) flax fibre having optimum properties which allow it to be spun into yarn by any commercially used method (including worsted, woollen, linen and cotton systems) by itself or when blended with most other types of fibre. Especially useful are blends with wool (Linron adds crisp durability to wool warmth) and polyester fibres (it adds coolness and comfort to the easy-care of polyester fibres).

An important feature of Linron fibre is that it can be spun into fine yarn by dry rather than the less convenient wet method of spinning which it has hitherto been found necessary to use when spinning any but a coarse yarn. An interesting feature is that the method of dew-retting is employed to assist the separation of the woody matters from the fibres in the freshly gathered flax.

Bleaching of synthetic fibres

It is generally more difficult to obtain pure white synthetic fibres since they are less responsive to the bleaching treatments ordinarily applied to such fibres as cotton, wool, silk, linen, etc. Further, some synthetic fibres, especially those of the polyacrylonitrile type, are slightly yellowish or brownish or otherwise off-white as they come from the synthetic-fibre manufacturer. In the high-temperature heat-setting treatments to which most synthetic fibre fabrics are subjected to make them more stable in subsequent hot wet processing, a yellowish discoloration can be produced in the fibres.

On the whole sodium hypochlorite and hydrogen peroxide are only moderately effective and when applied to nylon these bleaching agents may be harmful. Sodium chlorite and peracetic acid under acid conditions have been found most satisfactory, but the conditions of treatment are usually more drastic than those used in the bleaching of hydrophile fibres such as cotton, wool, etc.

In devising a bleaching treatment for, say, nylon materials, consideration in the selection of a bleaching agent must be given to the possibility that the nylon may contain a metal compound of say copper or manganese to give it improved stability to heat and light. Such metal could accelerate the bleaching sufficiently to damage the fibres.

Blueing

Textile materials which have not responded satisfactorily to bleaching and which have a slight off-white appearance (usually slightly yellow) can be much improved by blueing them. This well-known housewife's method consists of tinting them with a blue dye or pigment such as ultramarine which takes away the yellowness by converting it into a small amount of grey which is not noticeable.

Fluorescent whitening agents (FWA)

As described above, it is customary to bleach protein fibres such as silk, wool and regenerated protein rayons with hydrogen peroxide and sometimes with sulphur dioxide gas. While it is more usual to bleach cellulose fibres such as cotton, viscose and acetate rayons and linen with active chlorine, in recent years there has been a tendency to use hydrogen peroxide also for these fibres. It is unfortunate that all these bleaching agents attack the textile fibres at the same time as they destroy the colouring matters and other impurities. Thus, if the textile material is exceptionally dirty so that the bleaching treatment is unduly prolonged or is otherwise made too severe, it is likely that the fibres will be attacked so much as to be impoverished. It is to avoid this impoverishment that increasing use is today being made of a new method for whitening textile materials without the use of dangerous bleaching chemicals.

It has been mentioned that perfectly bleached goods which are for this reason slightly off-white can be made to appear whiter by the simple process of blueing them. This expedient is used by those who wash clothes at home. The blueing simply counteracts the yellowish tone of the off-white fabric or garment and obviously the simple drying into the fibres of a reddish-blue pigment or dye can have no harmful effect. However, this method of correcting an off-white cannot be carried too far, for it depends on adding red and blue to the yellow of the fabric to make a grey colour. If only a small amount of grey is formed the greyer colour of the fabric is not noticed and the fabric appears to be whiter, but if much blue is required to correct much yellow, then the large amount of grey produced is sufficient to make the fabric appear dull or dingy. Thus, blueing is only satisfactory for goods which are slightly off-white.

Just about 45 years ago a German chemist named Krais discovered that the blueing could be effected not only with a blue dye but with a colourless substance which gave a blue fluorescence in ultra-violet light. Ordinary daylight, but not all kinds of artificial light, contains sufficient ultra-violet or invisible light for this purpose. In using a colourless substance of this type there is no formation of grey, so that a much superior whiteness is obtained. Some fluorescent whitening agents give a reddish white while others may give a greenish or pure blue white. It is thus possible, by selection of the agent, to correct any off-white and secure a whiteness having any desired hue. It is possible, by the use of these fluorescent whitening agents (FWAs), to obtain a whiteness having a brilliance not to be attained by bleaching alone, however well this may be carried out.

During the intervening years, this discovery of Krais has been well exploited, first by manufacturers of soap powders, who found it very convenient to add fluorescent whitening agents to their products and thus allow housewives to produce whiter garments, and then by the makers of textile auxiliaries who now provide them for large-scale bleachers. Today, by means of these special products, it is possible to enhance the white appearance of bleached goods by simple treatment with one or other of these FWAs which have no harmful effect whatever on textile fibres. It is not possible to do without chemical bleaching altogether, but it is possible by the use of these products to avoid bleaching to a degree where damage to the textile material occurs.

FWAs have been found very useful for the whitening of synthetic fibre goods since these are usually difficult to bleach to a good white without involving some deterioration of their properties. They are often introduced into synthetic fibres during their manufacture, during the formation of the fibre-forming polymer or by addition to the spinning melt.

So many FWAs have now become available that the manufacturers are able to provide selected agents to suit best the different fibres to which they may be applied. At the same time many of the early types have been discarded in favour of improved types.

At least 1000 different FWAs are now being marketed and they are of many different types including those made from derivatives of coumarin, stilbene, pyrazolin, naphthimide and benzoazole. They all have to be tested in respect of their toxic and carcinogenic (cancer-forming) effects.

There is one feature about these new products which should be noted Their blueing is dependent on there being sufficient ultra-violet light illuminating them. Some kinds of artificial light can be deficient in ultra-violet light and so fail to create a sufficient intensity of blue fluorescence in the bleached textile material to ensure that it appears white. Most of the FWAs appear slightly yellow in bulk so that if too much is applied to a fabric it will appear duller and yellower instead of whiter.

From the viewpoint of the large scale textile finisher it is convenient for him to use FWAs which have an affinity for the fibres to which they are applied, usually in aqueous solution. By contrast those FWAs added to proprietary detergents for home use should have only limited affinity so as to avoid a build up of the FWA in garments being repeatedly washed, for then the resulting high concentration of FWA in the fibres can dull whiteness.

MERCERISATION

Mercerisation has already been mentioned as a very useful method for giving cotton yarns and fabrics a higher lustre which has the special advantage of being permanent. It is a process in which the cotton material is treated with strong caustic soda solution for about 1 min or less, followed by thorough washing out of this alkali under conditions such that the cotton is stretched to prevent it shrinking. If the cotton is allowed to shrink then there is no increase of lustre, but only a thickening and rounding of the individual fibres with an accompanying shortening of yarn, or a closing-up of the fabric, according to which type of material is being treated. We can now deal with the essential facts concerning the large-scale application of this treatment.

Mercerisation of yarn

The mercerisation of cotton yarn is invariably carried out with the yarn in the form of skeins. A machine is used in which there are pairs of strong parallel steel rollers over which can be placed the skeins of yarn [usually about 2 lb (0·9 kg)]. Sometimes these pairs of rollers are arranged radially as the spokes of a wheel. Underneath each pair of rollers is a tank, whilst over the rollers is a perforated pipe to convey the strong caustic soda mercerising liquor. The rollers rotate so that the skeins of yarn are continuously moving around them. Also it is arranged that the rollers can be moved to and from each other thus relaxing or stretching the skeins around them.

In the mercerising process the caustic soda solution is sprayed over the skeins from the perforated pipes and at the same time the rollers are brought slightly closer together so that the yarn is loosened and thus better able to absorb the caustic soda solution. Then the liquor spray is stopped and the rollers are gradually moved further apart thus stretching the alkali-impregnated skeins. At the same time water is sprayed over the skeins so as to wash out the caustic soda solution. Washing and stretching are continued until most of the alkali is washed out and the skeins are slightly longer than they were originally. All this time the excess caustic soda mercerising liquor and the washing liquors are collected underneath the rollers separately in tanks so that they can be used again after suitable purification and concentration. The skeins of yarn are removed from the rollers and are treated in a separate machine with weak acid to neutralise the last traces of caustic soda before a final thorough washing with water. When they are then dried it is seen that the yarn has a much higher lustre.

The skeins show a very strong tendency to shrink in length during the treatment with the mercerising liquor so that very strong hydraulic force is required to move the rollers apart to bring these skeins to their original length.

Mercerisation of fabric

With cotton fabric an equally powerful type of machine is required. It uses a stenter of the type described for the finishing of fabrics. This machine treats the fabric in open width (Figure 4.14).

On entering the machine the fabric is led through a heavy mangle which impregnates it with the caustic soda solution. The fabric then passes on to the stenter which comprises two parallel chains

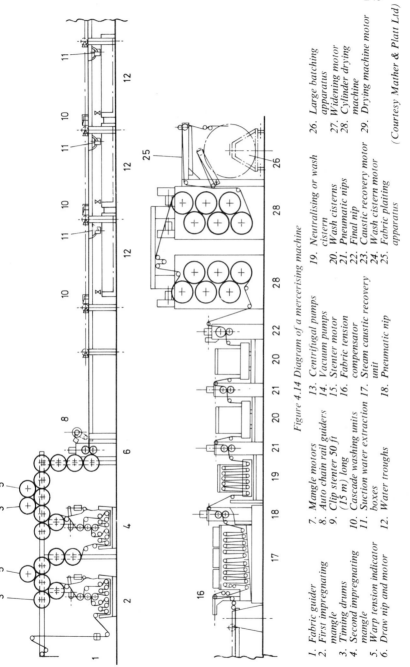

Figure 4.14 Diagram of a mercerising machine

1. Fabric guider
2. First impregnating mangle
3. Timing drums
4. Second impregnating mangle
5. Warp tension indicator
6. Draw nip and motor
7. Mangle motors
8. Auto chain rail guiders
9. Clip stenter 50 ft (15 m) long
10. Cascade washing units
11. Suction water extraction boxes
12. Water troughs
13. Centrifugal pumps
14. Vacuum pumps
15. Stenter motor
16. Fabric tension compensator
17. Steam caustic recovery unit
18. Pneumatic nip
19. Neutralising or wash cistern
20. Wash cisterns
21. Pneumatic nips
22. Final nip
23. Caustic recovery motor
24. Wash cistern motor
25. Fabric plating apparatus
26. Large batching apparatus
27. Widening motor
28. Cylinder drying machine
29. Drying machine motor

(Courtesy Mather & Platt Ltd)

of clips which are moving forward continuously throughout the length of the machine, which is about 50–60 ft (15–18 m). The clips grip the selvedges of the fabric and it is arranged that for the first 10 ft (3 m) these chains of clips diverge sufficiently to stretch the fabric in width so that any contraction in width already brought about in the mangle is counteracted. Thereafter the fabric is carried forward at the original or required width of the fabric while the clip chains run parallel to each other spaced at this width. Whilst the fabric is travelling forward it is washed nearly free of caustic soda by sprays of water directed downwards upon it, the water on the fabric being sucked through the fabric by slotted or perforated boxes placed immediately underneath it. By the time the fabric has reached the far end of the stenter frame it is practically free from alkali and no longer shows any tendency to contract. It is therefore led out of the clip chains and led through a washing machine where it is neutralised with dilute acid and then thoroughly washed before being dried. As with the mercerisation of cotton yarn, the spent mercerising liquor and also the washing liquors are collected so that the alkali which they contain can be utilised for the mercerisation of more fabric.

Provided that the fabric has not been allowed to shrink materially in length or width during this mercerisation treatment it will have acquired a high lustre thereby making it much more silky in handle and appearance.

Cotton fabrics containing ordinary viscose rayon threads can be mercerised under fairly simple suitably modified conditions to protect the rayon, but more latitude in these conditions is permissible when polynosic viscose rayon is present, since this is more resistant to the mercerising liquor (p. 41). When acetate rayon is present great care must be taken to avoid saponification of this type of rayon by the alkali. Such saponification changes the dyeing properties of the acetate rayon whilst at the same time making it harsher and of less weight. Special precautions have to be taken also if wool or silk are present in the cotton fabric since these textile fibres are easily deteriorated by the alkali used. Cotton fabric containing polynosic rayon threads can be mercerised under the usual conditions with little damage to the rayon threads.

It has been found by experience that in mercerising cotton fabric by this method the fabric must be in contact with the strong caustic soda solution for about 20 to 40 s. If the stenter frame is long it is possible to pass the fabric along it at a high rate, but a shorter stenter frame requires that the fabric should travel more slowly. To avoid having to use an expensive long stenter the expedient sometimes adopted is that of passing the fabric impregnated with caustic

Figure 4.15 Entry of cotton fabric to the mercerising machine. A–fabric guiders; B–fabric tensioning device; C–Alstonip chainless mercerising machine containing caustic soda solution. (Courtesy Mather & Platt Ltd)

soda over a number of unheated rotating drums immediately before it enters the stenter frame. Sometimes the arrangement is to have these cylinders positioned between two impregnating mangles and this is quite a good one.

If mercerised and non-mercerised cotton fabrics are dyed together the mercerised fabric usually dyes to a much deeper shade and it also appears to be dyed more evenly and solidly. This mercerisation treatment is much favoured for cotton fabrics which are intended for dyeing with aniline black.

In an alternative type of machine (see Figure 4.16) for mercerising cotton fabric the impregnating mangles and stenter, described above, are replaced by two long shallow liquor-holding tanks within each of which are upper and lower series of freely rotating hollow drums which are closely staggered relatively to each other. The lower series are buoyant in the liquor so that they press against the upper series to a controllable degree. The fabric being mercerised moves forward at a steady rate over and under the drums and is pressed firmly to their surface (according to the buoyancy force).

The first tank contains the mercerising liquor for impregnating the fabric and the second tank contains dilute mercerising liquor for washing the fabric. Because the fabric is so tightly pressed against the surface of the drums it is prevented from shrinking in both length and width as is required for satisfactory mercerisation.

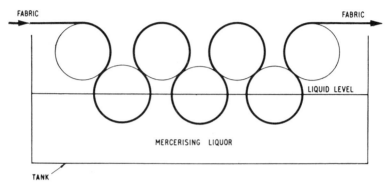

Figure 4.16 Alignment of the hollow drums within the Alstonip type of mercerising machine (see Figure 4.15) and the fabric passing over them during mercerisation

Following the second tank is a short washing machine which allows neutralisation of residual alkali in the fabric and a final thorough wash. Instead of leading the fabric over a series of hollow drums or along a stenter to counteract shrinkage while washing mercerising liquor out of it, there is an alternative machine in which the fabric is led, for a similar purpose, over a series of parallel, closely spaced curved rollers. A much simpler machine (Figure 4.17) is available for mercerising small lengths of cotton fabric batchwise, under the required conditions that while in contact with a mercerising liquor it is not allowed to shrink in length or width.

Recently mercerisation has been found useful for producing cotton fabrics which have much more elasticity than ordinary non-mercerised or mercerised (with stretching) fabrics and which are used for making 'stretch' garments. For this purpose the cotton fabric is allowed to shrink freely during mercerisation.

Recent discoveries in Sweden and Britain have shown that the caustic soda used in mercerising can, with advantage, be replaced by liquid ammonia to obtain several advantages related to the processing and the product.

The new processes now available for mercerising cotton yarn and also fabric use liquid ammonia at about −33 °C to effect changes in the cotton fibres similar to those produced by mercerisation with

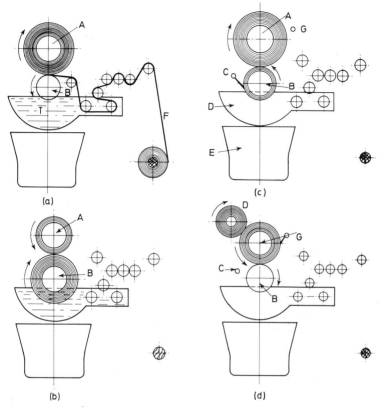

Figure 4.17 Machine for the batchwise mercerisation of cotton fabric. In operation (a) the fabric F is first led through the trough T which contains the mercerising liquor to be wound on top roller A which is hydraulically pressed hard against lower roller B. The fabric travel is then reversed (b) and it is wound on roller B to ensure complete impregnation of the fabric with the liquor and its mercerising action. In the next stage (c) the tank T is emptied of liquor (which is used again after strengthening), and the fabric is then again rolled on the top roller A, while at the same time being well sprayed with hot water to remove the mercerising liquor. The fabric, now substantially freed from mercerising liquor, is in a fully-mercerised stable lustrous state so that it can be wound off roller A on to another roller D and taken away from the machine (d). A built-in refrigerator unit allows the mercerising liquor to be maintained cold, for it is found that as the mercerising liquor becomes warmer, say up to 35°C, it becomes less effective.
(Courtesy Sir James Farmer Norton & Sons Ltd)

caustic soda. In particular the liquid ammonia swells the fibres and causes them to shrink in length (such changes are greater than with caustic soda) so that on stretching the yarn to somewhat more than its original length while removing the ammonia (this is re-covered for re-use and is necessary to make the process economical) the yarn acquires a much increased strength which can be about

90 per cent according to the type of yarn. Thus cotton yarn can be given a strength equal to that of a normal nylon or polyester yarn. In such processing the yarn is run at a speed of up to 150 yd/min (140 m/min) through a tube containing the liquid ammonia and has a dwell period therein of about 1s. Removal of the ammonia retained by the yarn is effected by washing in water and it is in this stage that the stretching is carried out. Then follows drying so that the yarn is left stronger, more lustrous and dimensionally stable during laundering, and it has the advantage of dyeing to deeper shades and being more resistant to heat.

In the treatment of cotton fabric this is padded with liquid ammonia, stretched widthwise and the ammonia evaporated from the fabric is recovered. It is claimed that the fabric can be treated up to 50 yd/min (45 m/min) and apart from giving it wash-and-wear properties it gives increased resistance to wet abrasion and to soil pick-up. To some degree the liquid ammonia treatment of cotton fabric can have the effect of resin finishing and confer crease resistance.

Recent research has indicated that the bean-shaped cross section of a cotton fibre which signifies an unevenness of fibre structure is a source of weakness just as is the reversal of fibre twist. At the point of twist reversal and immediately adjacent to this the fibre is weaker than in other parts. Permanent fibre swelling to give a cylindrical fibre and also to eliminate twist reversal strengthens it at these points and thus makes the fibre mercerised with caustic soda or liquid ammonia better prepared for subsequent wash-and-wear and durable-press resin finishing, in which a severe loss of fibre strength can occur (p. 395).

A recent improvement in the liquid ammonia process now branded Prograde (J. & P. Coats Ltd) allows considerable simplification such that smaller plant can be used and ammonia recovery is unnecessary in this stage. Whereas formerly the consumption of ammonia was 2·9 lb (1·3 kg) per 1 lb (0·45 kg) of cotton as compared with a theoretical consumption of only 0·2 lb (0·09 kg) it is now possible to reduce the ammonia consumption to about 0·8 lb (0·36 kg).

It is to be noted here that while the swelling action of liquid ammonia on cotton was known earlier it was not until about 1963 that real interest in the possibility of using this action to replace or supplement the action of caustic soda solution was taken. First Swedish, and later British, research has revealed that in many ways it can be advantageous to use liquid ammonia instead of caustic soda to modify cotton. Intensive research now being carried out has for example shown mercerisation with liquid ammonia to be

(a)

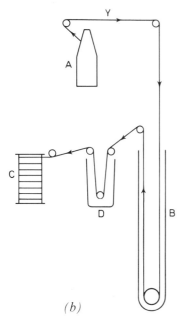

(b)

Figure 4.18 (a) Machine for mercerising (Prograde process) cotton yarns (separately and running side-by-side through liquid ammonia). (b) Running of a single yarn taken from bobbin A through liquid ammonia in tube B followed by washing in hot water in D to be wound on bobbin C substantially free from the ammonia, which is recovered. This machine has recently been improved to reduce ammonia loss. (Courtesy J. & P. Coats Ltd)

cheaper and mechanically simpler to carry out. It is being progressively discovered how in various directions such modification of cotton's normal properties can be more widely utilised not only in the manufacture of sewing cotton but also in dyeing and finishing, and in blending with other types of textile fibres.

Solutions of dyes in liquid ammonia can now be used for the instantaneous fast dyeing of synthetic fibres.

MILLING OF WOOL MATERIALS

It has already been mentioned that wool fibres readily close up on each other by reason of their unique felting power. This property is frequently used to give wool materials a thicker, more compact character and to partly obscure their thready appearance. To achieve this object the fabric or other material is subjected, usually

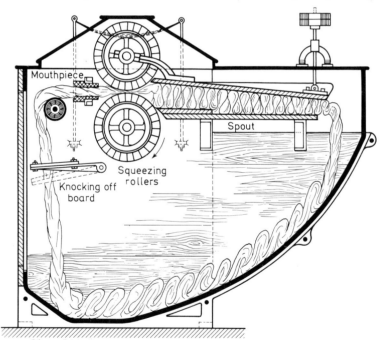

Figure 4.19 Milling machine for woven or knitted wool fabrics. The fabric is in the form of an endless rope. It lies in the soapy liquor, passes upwards and then in succession through the mouthpiece, between the squeezing rollers and into the tapering spout where it is squeezed. It then passes out of the mouthpiece and falls back into the liquor and the cycle of operations is repeated. The alternate squeezing and relaxing causes the fabric to felt and so acquire a denser structure. (Courtesy The Textile Institute)

while wet with a soapy or an acid liquor, to a process of milling which is essentially compression and relaxation. One method is to pound the fabric, garments or socks with wooden 'feet' as in a dolly washing machine. Woven and knitted fabrics made into endless strands are usually milled in a machine where they repeatedly pass through rollers and a compressing device. All milling methods involve intermittently squeezing the fibres together and with each squeeze the fibres remain a little closer together and more inter-locked than they were before.

DYEING

Dyeing is distinguished from printing in that it only allows one shade to be produced all over the yarn or fabric unless two differently dyeing fibres are present. Dyeing is employed to give solid all-over shades, whilst printing is used to produce patterns which may con-tain up to twelve or fourteen different colours.

Dyed shades are obtained by treating the textile material with a solution of a dye which in most cases has an affinity for the material, either with or without previous special preparation. There are exceptions to this as will be described later. From this general statement it becomes evident that, if uniform well-penetrated dyeings are to be produced, the conditions of the dyeing must be such that the dye liquor is applied evenly and that it penetrates through the innermost parts of the yarn or fabric. So far as is possible, every individual fibre in the textile material must receive its fair share of exposure to the dye liquor. In turn, this implies that in dyeing the textile material must be moved continuously in relation to the dye liquor or vice versa. Thorough and even treat-ment of the yarn or fabric by the dye liquor is of the utmost im-portance, and is the basis of all successful dyeing.

The development of dyes

Present-day methods of dyeing are based on a supply of pure standardised dyes which are made synthetically from coal-tar and other products. In earlier days the dyes were extracted from plants and vegetable roots or they were made from mineral substances. They were generally difficult to apply because their purity could never be relied upon and they were not of constant strength. Only a limited number of these natural dyes are now used.

It was in 1856 that a young chemist named William Perkin was endeavóuring to make quinine synthetically by a process of oxidising aniline with bichromate of potash. He failed in this but produced a tarry product. He did not just throw this away to start again, but examined it closely and eventually extracted from it a mauve colouring matter which was found to be a dye for silk yielding shades fast to light. Messrs Pullar, of Perth, encouraged Perkin in his discovery and soon he perfected a method of making the mauve dye and, with a limited amount of capital, set up in business as the first aniline dye manufacturer.

Perkin was very successful. Not only did he succeed with his large-scale manufacturing operations, in spite of several disturbing explosions, but he persevered with his researches and discovered several other dyes all of which were added to his range of products.

Other chemists sought to travel in the same footsteps and in a few years they also discovered dyes—all made from aniline and similar products derived from coal-tar, a by-product of coal gas manufacture. Thus commenced a new era in dye manufacture and between 1856 and 1900 there was great activity both in the U.K. and in Europe in manufacturing new and better dyes.

One interesting development was that chemists analysed the chemical constitution of such useful natural dyes as indigo and alizarin and eventually discovered methods for manufacturing these from coal-tar derivatives with a higher degree of purity than the natural products. Moreover, it then became possible to manufacture numerous other dyes similar in structure to the natural dye but differing from it either in shade or fastness properties.

The result of all this effort (the full story is very interesting) is that the dyer has at his disposal today at least some 2 000 or 3 000 synthetically made dyes. Actually many more thousands of dyes are known and have been made, but it is only those which are easy to apply and which give shades of satisfactory fastness which are regularly provided to large-scale dyers.

Classification of dyes

Now these dyes can be classified according to their chemical structure or according to their affinities for the different textile fibres. Actually a compromise classification is adopted. The result of this is that dyers of textile materials acquire from experience, and also from the technical information service provided by the large dye-makers, a good knowledge of the dyes and classes of dyes which can be applied to any particular fibre, say, cotton or wool. Generally there are special methods of application for each of these different classes of dyes.

The main classes of dyes are as follows:
(1) Direct cotton dyes; (2) basic dyes; (3) acid wool dyes; (4) mordant wool dyes; (5) direct cotton dyes capable of being further developed in the dyed textile material by diazotisation and coupling with a naphthol; (6) sulphur dyes; (7) vat dyes; (8) indigosol and Soledon dyes (these are specially solubilised forms of vat dyes); (9) naphthol or insoluble azoic dyes; (10) disperse dyes; (11) reactive dyes; (12) metallisable-after-dyeing and pre-metallised dyes; (13) dyes that are self-polymerisable (to become insoluble within the fibres) and for copolymerisation with the fibre substance.

The last-named type of water-soluble dye claimed to give highly washfast colours on textile materials is distinguished by its molecules each consisting of two permanently linked components one of which is responsible for the colour and the other because it is polymerisable for ultimately causing the dye to become insoluble within the dyed fibres and possibly chemically linked to them. Such a dye is applied to the textile material in an aqueous dyebath containing a catalyst so that on raising the dyeing temperature to the boil polymerisation is induced by the catalyst of the dye absorbed in the fibres.

The general reader need not trouble to remember these classes, which are given here for reference.

Principles of dyeing

Simple dyeing The simplest form which dyeing can take is that in which the textile material is moved about in a solution of the dye. The dye is attracted to the material and it is seen that the dye liquor gradually loses its colour whilst the fabric becomes more deeply dyed. To hasten such dyeing or to ensure better penetration of the dye into the innermost parts of the fabric it may be necessary to warm or perhaps boil the dye liquor. In other cases exhaustion of the dye liquor (this implies that more dye goes to the fabric) may be assisted by adding an acid to it. In other cases common salt can assist increased absorption of the dye by the fabric. If the dye is only sparingly soluble and therefore likely to be absorbed unevenly, then the addition of soap or some ammonia or sodium carbonate may aid even dyeing. Thus with this simple dyeing process there are various methods for modifying it beneficially as for example the addition to the dyebath of one or more so-called textile auxiliaries which are usually complex organic compounds able to influence dye absorption (p. 359). Such dyeing assistants can be anionic, cationic or non-ionic to suit the type of dye being applied.

Other dyeing methods Now it may happen that the fabric is made of a fibre which has no direct affinity for the dye and so will not absorb it from the dye solution. Yet there are colourless substances which have an affinity both for the dye and for the fabric. Then arises the possibility of first treating the fabric with one of these substances (usually termed *mordants*) and afterwards dyeing the mordanted fabric by immersing it in the dye liquor. Here the mordant acts as a link between dye and textile material. Many useful dyes are available for application to most of the textile fibres in this special way. They are termed *mordant* dyes.

There is a further interesting way of colouring a textile material less simple than the method first described. It has been indicated that it is essential for the dye to be soluble in water in order that it may be applied evenly to a fabric or other material. But some dyes, the vat and sulphur dyes, are not soluble except under special conditions. A vat dye in its normal state is completely insoluble in water. However, if it reacts with a reducing agent, such as sodium hydrosulphite, then it forms a so-called leuco compound which is soluble in dilute solutions of an alkali such as caustic soda. When cotton or other material is immersed in this solution of the leuco vat dye there is a marked affinity between them and so the cotton absorbs the dye. The cotton has not at this stage acquired its final colour for the leuco form of the dye may be either colourless or a different colour from the dye in its normal state. However, the essential feature is that the cotton absorbs the dye in its leuco form. Thereafter the cotton can be withdrawn from the dye liquor and either oxidised by exposure to air or by chemical means so that the leuco dye within each fibre is changed back to the original dye in its true colour. Since the dye in this final form is insoluble in water it is easy to understand that colourings thus produced with vat and sulphur dyes have exceptionally good fastness to washing. In such a dyeing method the expedient of utilising a temporarily modified water-soluble form of the dye solves the difficulty of applying an otherwise insoluble dye.

Yet another somewhat more complicated method of dyeing is used for the insoluble naphthol dyes which are also of good washing fastness. These dyes are made by combining together two and sometimes three soluble components which are not, themselves, dyes. The dye made in this way is insoluble in water and therefore is not suitable for the dyer. He therefore adopts the following expedient. Firstly, a solution of one component is applied to the fabric. Generally the fabric has little or no affinity for this component so the solution is just dried into it. Then the fabric is immersed in a solution of the second component. At once this component enters into the fabric before the other component has time to move out of it and then combination takes place between them so that the resulting insoluble naphthol dye is formed within the fibres. It is scarcely possible to wash out this dye, even by prolonged hot soaping.

Reference has already been made (p. 102) to the special reactive dyes which chemically combine with cellulose fibres such as cotton and viscose rayon during their application and thus lead to the production of shades having a very high degree of fastness to washing.

These reactive dyes have now become very important for the dyeing of cotton, linen and rayon fibres; they are also applicable to silk, wool and polyamide synthetic fibres by special methods. Reactive dyes are valuable partly because they can be easily applied and partly because they give very fast and often very bright colourings. It is to be noted, however, that it does not necessarily follow that because a reactive dye gives a fast-to-washing colouring this will also be fast to light and other adverse influences. In fact one of the problems facing manufacturers of reactive dyes is to combine in the one dye reactive properties and a high fastness to all adverse influences.

The discovery of reactive dyes almost simultaneously as a result of British and Swiss research may be described as epoch-making in dye manufacture since they are now being used throughout the world and especially whenever very fast colourings are desired and which can in most instances be obtained by relatively simple methods. Such dyes thus justify particular mention here.

Simply described, a reactive dye consists of an ordinary dye whose molecules each carry a special group of atoms which under appropriate conditions can react with a molecule of a textile fibre and thus cause dye and fibre to combine chemically. This combination is very stable so that the dyed fibre has excellent fastness to most adverse conditions such as light and washing.

Many reactive atomic groups have in the past five years been discovered as being suitable for attachment to dye molecules and a number of different ranges of reactive dyes are now available. They are marketed under such brand names as Procion, Cibacron, Levafix, Drimarene, Remazol, etc.

At first the dyes could only be used to dye cotton and other cellulose fibres but their use has been steadily extended to wool, silk and nylon; their use is still being steadily extended.

In the dyeing process the reactive dye combines with the fibre either by addition to it (left) or by splitting of the reactive group X (right) thus:

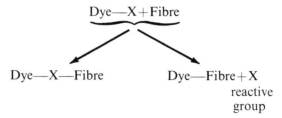

There are other instances where special dyes, such as aniline black, are formed within the textile material. For this particular fast black shade it is aniline which is oxidised in the yarn or fabric.

Another dyers' expedient is that of dyeing fabric with a dye in the ordinary way and then changing its fastness or its colour by combining it with another component. Many dyed materials can have the fastness of their colour improved either to washing or to light by after-treatment with a metal salt. For example, many cotton colourings can be made faster to light by after-treatment with a warm solution of copper sulphate. Treatment with chromium salts can be employed to make certain wool dyeings much faster to washing and to light. A very useful variation in this method for after-treating dyed materials with metal salts is used a lot today. It involves the use of pre-metallised dyes of a type which are water-soluble and thus can be easily applied to cellulose fibres as are direct cotton dyes. Once such dyes are within the fibres they become highly washfast and usually additionally fast to light.

Polypropylene fibres are very difficult to dye using the dyes commonly in use for dyeing the natural fibres. Their absorption of dyes can be much assisted by introducing special dye-attractive substances into the fibres during their manufacture—such substances may be nickel and other salts.

From the foregoing it is seen that the art of dyeing involves considerable skill, and this is especially the case where shades of the highest degree of fastness must be produced. The problem is, of course, made more difficult when the yarn or fabric to be dyed consists of a mixture of different textile fibres having different affinities for dyes.

Perhaps, before dealing in more detail with the methods and machinery used in dyeing, mention should be made of early difficulties associated with the colouring of acetate fibres. Incidently there is now a link between the dyeing of this type of fibre and the dyeing of synthetic fibres such as nylon.

Difficulties of dyeing acetate and synthetic fibres   When acetate fibres were first produced very few of the dyes then available could be used for dyeing them. Even those dyes which showed some affinity for this new fibre failed to yield shades of satisfactory fastness. So new dyes had to be discovered. But, curiously enough, the new dyes were quite different from any previously used in that they were insoluble in water. Even today most of the dyes used for acetate are of this so-called 'disperse' type as it seems that acetate has little or no affinity for water-soluble dyes.

Study of the manner in which acetate fibres absorbed various water-soluble substances (not dyes) applied to it from aqueous suspension showed that the absorption was really a case of the substance being dissolved by the fibres. Cross sections of the fibres revealed that the absorbed substance was uniformly distributed throughout the fibre, whereas when other textile fibres are dyed it is often the case that the dye is present mainly on the outside of each fibre. Thus it seemed that to dye acetate fibres it would be necessary to discover coloured substances which were soluble in cellulose acetate.

Such special dyes were soon found. To apply them to the acetate fibres it was found most convenient to grind them very finely and then paste them with soap or turkey red oil and so make a dye liquor in which the dye was in the form of highly dispersed insoluble particles. In the dyeing process these dye particles quickly come into contact with the surface of each fibre, there to form a thin layer; afterwards they pass slowly by solution from this into it. This is an instance of the solution of one solid in another.

In recent years it has been found that nylon and other synthetic fibres behave similarly to acetate fibres in that they dissolve the dyes applied to them and in fact the special dyes found suitable for acetate fibres, and now termed *disperse* dyes, are now proving very useful for colouring these synthetic fibres. Mention has been made earlier (p. 118) of certain modifications of synthetic fibres made by introducing acid or basic groups into their molecules and thus making them receptive to commonly used dyes for cotton and wool.

Today the disperse dyes have become very much more important than formerly, and increasing numbers of them are being manufactured. This importance arises from the fact that they are now being used for the colouring of practically all the synthetic fibres now available.

Materials used in dyeing machinery

So far nothing has been said about the materials used in the construction of dyeing machinery. In earlier days most of it was made of wood and cast iron. Both of these materials had disadvantages. Iron was liable to rust and wood readily absorbed the dyes so that it was liable to contaminate the next dyeing. For example, if a wood dyeing machine was used for producing a black shade, then it would be impossible to dye immediately afterwards a pale pink shade on fresh fabric as the absorbed black dye would bleed out and soil the pink.

Thus dye machine makers now use solely steel or other non-corrosive metal. The use of these metals not only assists the dyer to produce cleaner shades but it also speeds up dyeing, for between different dyeings it is only necessary to wash the machine down with water to make it absolutely clean and free from dye. In earlier days it was necessary to boil out the machine with soap and even bleach it to strip the wood of absorbed dye.

There is a further advantage that metal does not produce splinters that can catch in the moving fabric and so produce chafe marks.

Fabric dyeing methods

Dyeing fabric in a jigger   It will probably be most satisfactory to consider first the dyeing of a plain woven cotton fabric. This will be dyed in a jigger machine, and it is usual to dye four to six pieces at the same time to the same shade.

The jigger in its simplest form (Figure 4.20) is a V-shaped vessel provided with two upper draw-rollers, lower side guiding rollers, and a pair of rollers or a single guiding roller right in the bottom.

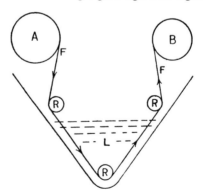

Figure 4.20 Sectional view of a dyeing jig in which fabric F is drawn backwards from draw-rollers A and B via the guiding rollers R. The dye liquor L is contained in a V-shaped trough

All the guiding rollers are free-running and only the draw-rollers are power driven. The dye liquor is contained in the V-shaped vessel. A closed steam pipe in the bottom of the jigger allows the dye liquor to be heated to any desired temperature.

Firstly, the fabrics will be prepared for dyeing by scouring and perhaps also by bleaching. If the shade is black, deep brown or navy blue, then obviously it would be a waste of effort to bleach it to a pure white before dyeing. About five of these pieces will be sewn end to end and wound in full open width on a roller. This is generally termed a *batching* roller. The fabric is first wound from the batching roller on to one of the draw rollers. Then one end of the fabric is threaded downwards around a side guiding roller, under the bottom roller and up around the other side guiding roller to be secured to the other draw-roller. At the commencement of dyeing the fabric is thus all wound on one of the draw-rollers.

Now the other draw-roller is rotated by a motor drive and it continuously draws the fabric from the first draw-roller through the dye liquor. When all the fabric has been drawn and wound on to the driven roller then the drive is transferred to the empty draw-roller. The fabric then passes in the reverse direction through the dye liquor and back to be wound again on the first draw-roller. As the fabric thus passes backwards and forwards it absorbs dye liquor, and whilst lying on the draw-roller it extracts from this absorbed liquor the dye which it contains. The rate at which the fabric is drawn through the dye liquor is usually too fast to allow the fabric to absorb much dye; dye absorption into the fibres mainly takes place on the draw-roller.

From time to time the dyer can cut a small test pattern out of the fabric and ascertain whether or not it has acquired the shade required. He can add further quantities of dye to the dye liquor and if necessary make additions of other substances such as common salt or Glauber's salt to assist its exhaustion. Meanwhile the fabric is run backwards and forwards all the time so as to ensure that it becomes coloured uniformly.

Finally, when the required shade has been attained, the dye liquor is run off and replaced by cold water so that the fabric can be lightly rinsed. Then the fabric is wound on to the batching roller and taken away to be mangled and dried. The jigger is then ready for dyeing another batch of fabric.

This method of dyeing in a jigger is very simple and straightforward so that it is preferred whenever the nature of the fabric allows it. There is, however, one feature which is at times objectionable. It is to be noted that the fabric is pulled from one draw-roller to the other and this involves a considerable lengthwise

262

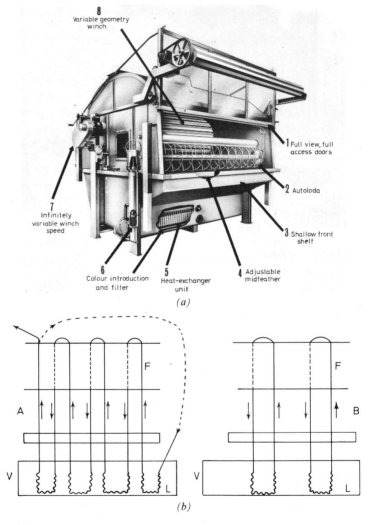

Figure 4.21. (a) A totally enclosed (to conserve heat) winch dyeing machine provided with a rising-and-falling glass front window. Such a machine allows lengths of fabric to be dyed by one or other of two different methods, in both of which the fabric lies most of the time loosely immersed in the dye liquor contained in the dye vat below, but is continuously drawn up in single strand form to pass over a guiding roller and then over the winch to fall back in to the dye liquor at the back of the machine. As shown in (b) in one method all the fabric as a single strand continuously advances from one side of the machine to the other while in the other method each length of fabric, as a closed loop, keeps its initial position in the machine as determined by passing between stationary spaced pegs in a guide-rail on its way to pass over the winch. F fabric passing over the overhead rotating winch, V vat containing dye liquor in which the fabric lies slack awaiting its upward pull to pass over the guiding roller and then over the winch. (Courtesy Leemetals of Macclesfield)

tension on the fabric. For strong fabrics this does not matter much, but there are other fabrics which would be damaged by this tension. Again, there are crêpe fabrics whose special pebble appearance, due to shrinkage in length, would be quite spoilt by stretching.

Length tension also hinders dye liquor penetration and dye absorption of yarns and fabrics. In recent years much has been done to improve jigger dyeing machines so that they cause only the minimum of length stretching. However, there is a limit to this, since if there is no length pull at all the fabric tends to crease and does not keep out to its full width. Thus there is need for a machine which does not have this defect—the winch dyeing machine described later.

Jigger dyeing machines have also been redesigned so that they can now be had fully enclosed (to conserve heat) and automatic as regards their performance. By use of press-buttons the running of the fabric backwards and forwards through the dye liquor can be started and also so that it makes a pre-set number of passages through this liquor and then stops. The tension on the fabric can be controlled accurately and the dye liquor can be automatically maintained at any desired temperature. Additional devices ensure that the fabric runs straight and free from creases.

Winch dyeing machine The winch machine has been mentioned already (p. 231) in connection with scouring. The same machine is

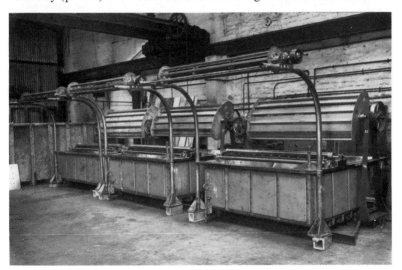

Figure 4.22 Three plain motor-driven stainless-steel winch dyeing machines. (Courtesy John Dalglish & Sons Ltd)

very suitable for dyeing all fabrics which must not be excessively pulled in the length. These include knitted tubular fabric, rayon and cotton woven crêpe fabrics, and fragile fabrics of all kinds.

Dyeing in the winch machine follows much the same course as scouring. The fabrics are threaded up side by side and the vat is filled with dye liquor. Then the winch is set rotating and the fabric is thus kept moving and so it becomes uniformly exposed to the dye liquor. In this machine the fabric absorbs dye whilst it is in the dye liquor for it lies in this fully submerged most of the time. It is almost free from tension and is thus more receptive to the dye. From time to time the dyer makes additions of dyes and dyeing assistants through the perforated partition in the front of the machine. After the usual period of about 1 to 2 h he succeeds in bringing the fabric to the desired shade. Then follows a rinsing with cold water and the fabrics are then taken out, hydro-extracted, and sent forward for drying. Hooded winch dyeing machines are preferred—they reduce wastage of heat and allow more rapid dyeing and better penetrated dyeings (Figure 4.21).

Padding processes Sometimes it is required to tint a large number of pieces of the woven fabric to a very pale shade. In this case a convenient method is that of simply running the fabric through a padding mangle provided with a trough containing dye liquor. The fabric passes through the trough of liquor, then is squeezed in the mangle, and is straightaway led over a set of drying cylinders. Experience in this method enables the dyer to keep up the supply of dye liquor to the trough so that the shade produced on the fabric is maintained constant. Tinting can be carried out very easily by this method at a high rate. Sometimes, in order to ensure uniformity of shade and superior dye liquor penetration, two such padding mangles are employed in series.

This padding method can be combined with the jigger method in the case of thick fabrics when it is necessary to squeeze the dye liquor into the fabric and where several such squeezes and time for dye absorption are necessary. The fabric can be run backwards and forwards fairly slowly through the dye liquor and a mangle, the fabric being drawn from one roller to another as in jigger dyeing.

Machines for dyeing fabric while wound on a perforated roller There could obviously be many advantages associated with dyeing fabric while wound on to a perforated roller or beam within a closed vessel, allowing dye liquor to be pumped through the fabric in either an inward or outward direction, provided that the dye liquor is free from suspended undissolved dye or impurities which would become filtered off on the outside layer of fabric, and provided also

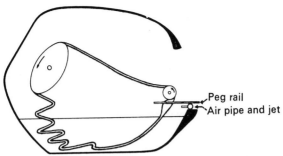

Peg rail
Air pipe and jet

*Figure 4.23 Device (I.C.I. Ltd) for attachment to a winch dyeing
machine to cause the tubular knitted fabric being dyed to balloon
while being drawn out of the dye liquor to pass over the winch thus
preventing it from having creases permanently fixed in it.*
    *The device consists of a horizontal pipe having holes at spaced
intervals and positioned near the peg rail along its length, which
keeps the fabrics being dyed separate and also positioned near the
guiding roller over which they then pass to the winch. The holes
are spaced so as roughly to be adjacent to each fabric. Compressed
air supplied to the pipe causes a jet of air to impinge on the fabric
so that some of it enters the fabric via its interstices and balloons
it either to remove any creases, or at least to shift their position
by the time they reach the winch and pass over it. Normally this
ballooning can take place without this device when the dye liquor
is maintained boiling; it is when one dye liquor is only hot that
ballooning does not occur sufficiently and this condition can occur
in the early stages of dyeing when the dyeing liquor is being heated
up to the boiling point with the fabric running over the winch.*
(Courtesy I.C.I. Ltd)

that the dye liquor passes so freely through the layers of fabric that
the outer layers do not become more deeply dyed than the inner
layers. In recent years dyeing machines of this type have become
available and for knitted fabrics especially are quite satisfactory
(these are more open in structure than woven fabrics and so allow a
freer through-flow of dye liquor).

It is usual to use a machine which allows the perforated roller to
have fabric evenly wound on it apart from the machine and also so
that it can then be positioned within a pressure-resistant vertical or
horizontal cylindrical container in connection with a dye liquor
pump and other auxiliary equipment. This will allow the dye liquor
to be maintained (usually automatically) at any desired temperature
up to about 130°C or somewhat higher (the cylindrical container
must be suitably strong to allow the necessary liquor pressures to be
used) and the liquor circulation to be satisfactorily controlled. It is
also necessary that the machine should allow samples of the fabric
to be taken from time to time (for shade matching purposes) without

interference with the dyeing operation—without opening-up the machine.

Such machines are now in wide use for the dyeing of fabrics made of synthetic fibres—fabrics whose dyeing can be much assisted by dyeing at temperatures above the boiling point of water.

Many of these machines allow the dyed fabric to be washed and even dried (by hot air) before removal from the machine.

An important advantage of this type of machine is that it avoids unwanted creasing of the fabric and hinders excess width shrinkage of the fabric such as can occur by dyeing in jigger and especially winch machines.

New continuous dyeing processes Since the end of World War II, much progress has been made in dyeing fabrics by continuous methods so that full shades, as distinct from mere tints, can be satisfactorily produced. When dyeing is carried out in jigger and winch machines, it is necessary to process fabrics in lots of, say, 5–20 pieces at a time. This intermittent form of dyeing can involve the expenditure of much unnecessary time and labour associated with preparing the dye liquor, arranging the fabrics within the dyeing machine and, finally, cleaning the dyeing machine for the next dyeing operation if successive lots of fabrics have to be similarly treated but dyed to another shade. Thus new methods and types of dyeing equipment have been devised, particularly for dyeing long runs of fabric to the same shade at the same time. Excellent progress has been made in this direction in many countries (see Thermofix dyeing, p. 280).

In a dyeing process it is generally necessary not only to apply an aqueous solution of the dye to the fabric but also to arrange for conditions which ensure that the dye external to the textile fibres is caused to penetrate these and become fixed within them. If, say, a piece of cotton fabric is dipped in a cold dye liquor and is then mangled free from excess liquor and dried at a low temperature, for instance by hanging out in the open air, it will appear to be properly dyed. Really it is not dyed at all, because if the fabric is lightly washed, even in water, it will very rapidly lose its colour. In such a simple method of treatment with dye liquor followed by drying, the fibres acquire colour by dye particles adhering loosely to the surface. For the dyeing to be satisfactory, it is necessary either for the dye liquor to be applied for a suitable period while it is hot or boiling or for the fabric impregnated with the cold dye liquor to be exposed to hot, moist air, for instance by hanging it within a closed chamber filled with steam. These latter more drastic conditions swell the fibres and probably break down the dye into smaller particles so that these can more easily penetrate the fibres where they become firmly fixed when the fabric is finally dried and the fibres contract to their original non-swollen state.

The new continuous dyeing processes conform to the above conditions necessary for adequate dye penetration and fixation. Thus, in the Standfast process, a special type of apparatus is employed which consists of an iron U-tube about 5 ft (1·5 m) wide and which is nearly filled with molten metal, an alloy of bismuth, cadmium, tin and lead which melts at about 71°C. A small volume of dye liquor floats within one arm of the U-tube on top of the molten metal and floating on the metal in the other arm is a correspondingly small volume of washing liquor consisting of an aqueous solution of Glauber's salts. Fabric in open-width, pre-heated by running it over a few cylinders internally heated by steam, is led downwards through the dye liquor and the molten metal in one arm, and then upwards through the molten metal in the other arm. It emerges through the washing liquor fully dyed and ready for final rinsing and drying. In this operation the molten metal is maintained at 90–100°C or any suitable high temperature and the dye liquor is continuously replenished so that it always contains the same concentration of dye. Under these conditions the fabric picks up dye liquor floating on top of this molten metal, and while passing through the hot metal this dye penetrates the fibres and becomes fixed. The water carried in the fabric partly turns into steam so that the fabric, while passing through the molten metal, is subjected to a rapid steaming. It has proved satisfactory with some fabrics to dye these at a rate of about 100 yd/min (90 m/min) in this Standfast apparatus which produces shades perfectly uniform and of excellent fastness. Obviously this method of continuous dyeing has much to recommend it (Figure 4.24).

The Standfast dyeing process is being used in America, but alternative processes have been devised there which will therefore be in competition with the British processes. One early devised American process known as the 'Williams' process, uses hot oil instead of molten metal. Otherwise the two dyeing processes are governed by the same principles.

In the 'Williams' process the dye-impregnated fabric passes through a column of hot oil which serves to ensure penetration and fixation of the dye. As the dyed fabric emerges from the hot bath it is led through a mangle and washing machines which scour out the residual oil to leave the dyed fabric ready for drying. There is thus some wastage of oil by this process and it may cause some difficulties in purifying the dyeworks effluent but, in spite of this, the process has attracted American dyers.

Figure 4.24 Standfast hot-metal type of machine for the continuous dyeing of cotton fabric, in which the fabric impregnated with dye, has the dye made washfast, first by a very short passage through a molten metal alloy at about 71°C, and then by passage through a drying chamber. (Courtesy Mather & Platt Ltd)

Both of these processes have the advantage of allowing the rapid and satisfactory dyeing of cotton fabrics with vat dyes and the production of colours of the highest possible fastness. These processes are also being adapted to the dyeing of rayon fabrics and even syn-

269

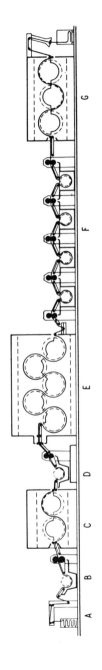

Figure 4.25 Continuous dyeing, washing and drying range for knitted fabric. Fabric in open width at **A** is impregnated with dye liquor at **B** and freed from creases as it passes over perforated drums at **C** while being dried with hot air. It is then impregnated with a solution of a softening agent at **D** and then steamed while passing over perforated drums at **E** for fixing washfast dyes. Then follows thorough washing at **F** by passing over perforated rotating drums each within a tank containing the washing liquor, and finally a drying at **G**. (Courtesy Fleissner Ltd)

thetic fibre materials using the appropriate types of dye. But the Thermofix (otherwise known as 'Thermosol') process (p. 280), in which the dye-padded fabric is heated for a few seconds at a high temperature to cause the dye to sublime into the fibres, is now very important and is widely used.

In all dyeing processes where the fabric is impregnated with a dye liquor and is then further treated by drying, heating or steaming to fix the dye just where it has been left in the fabric, it is obviously necessary that the apparatus for impregnation (often termed *padding*) shall ensure that the dye liquor forced into the fabric is uniformly distributed not only across the width but also throughout the length, so that when the dye is fixed the colouring is uniform throughout the fabric.

In the usual method of pad-colouring fabric this is run at a uniform speed through a padding mangle so that it first runs through a small trough of dye liquor and then between two or possibly three bowls under pressure to squeeze out excess of liquor. Thus for uniform colouring the pressure on the bowls must be readily and accurately adjustable and the pressure across their width must be uniform. Because the trough of the padding mangle is small its low content of dye liquor must be frequently or continuously replenished and this favours its having a constant dye concentration as is required to give uniform dyeings. Recently very much improved padding mangles for achieving these aims have become available. The pressure between the bowls is pneumatically controlled and the bowls are slightly off-set and not, as hitherto, one exactly vertically over the other, since this off-set position favours more uniform and thorough squeezing of the fabric saturated with dye liquor. The padding mangle bowls and the bearings in which they run are specially constructed so that they do not bend or bow under the high pressures employed. Such bending can allow more dye liquor to be left in the middle than at the edges of the fabric.

The dyeing of yarn

The dyeing of yarn may be carried out using a simple dye vat of the rectangular type as described for scouring purposes. In this case the vat is filled with dye liquor and in this are suspended the skeins of yarn threaded over rods which rest on the sides of the vat. The vat is only half filled along its length with yarn. Thus from time to time it is possible by hand labour to slide the rods to the vacant end and so draw the skeins through the dye liquor. At the same time the operative can shift the position of each skein on the rod so that the small upper part out of the dye liquor now becomes immersed. Thus even dyeing is assured.

In large yarn dyeing works use is made of more automatic machines. One type consists of a long upper framework from which project numerous porcelain rods each to carry about 2 lb (0·9 kg) of yarn in skein form. Underneath is a long comparatively shallow trough of dye liquor. At the beginning of the dyeing operation the skeins are hung on the rods and the trough is filled with hot dye liquor. Then the framework is lowered so that the lower half of each skein dips in the dye liquor. At the same time the rods are caused to rotate. Thus in this position the skeins are slowly revolved so that every part passes through the dye liquor. From time to time the direction of rotation of the rods carrying the yarn skeins is reversed. With such a machine the skeins are not so fully submerged in the dye liquor as in the hand dyeing method described above, but the constant rotation of the skeins ensures that the dyeing is uniform.

In the Hussong type of machine which has already been mentioned (Figure 4.26) in connection with scouring, the skeins are suspended from an overhead frame so that they can be lowered into a vat of dye liquor. Circulation of this liquor by means of a pump ensures even distribution of dye liquor among the skeins and so uniform dyeing is obtained.

Yarn can also be dyed in package form such as cheeses or whilst

Figure 4.26 Hussong machine for dyeing yarn in skein form. A block of skeins suspended from horizontal rods as shown is lowered into the adjacent vat of dye liquor; the fully immersed skeins are dyed there by a continuously circulating dye liquor which can be heated to any desired temperature. (Courtesy Samuel Pegg & Son Ltd)

wound on perforated bobbins or spindles, or rollers, and a considerable amount is now processed in this way. It is a method especially suitable for fine yarns which must not become entangled.

There is yet another special method for dyeing yarn to be used as the warp in woven fabrics. These single yarns are wound side by side on a roller from which they are drawn through a dye liquor, and at the far end of the machine are passed as a 'sheet' of closely spaced yarns over drying cylinders and rewound on another roller.

Sometimes skeins of yarn are linked together end to end and led chainwise through a dyeing liquor, but this method has only limited application.

Dyeing of loose fibres

It may be recalled that in converting loose fibres into yarn there is a stage at which the fibres have been brought into a parallel condition as roving or sliver. Yarn is made from this thick material by drawing it out and twisting. Sometimes it is desired to dye the roving (p. 151), and this can be done in a so-called 'top dyeing' machine which is often used for wool. The roving, which is about as thick as a finger, is wound into the form of balls, each of up to 10 lb (4·5 kg) in weight. These then resemble big pumpkins. Two or three of these are then placed one above the other in tall narrow

Figure 4.27 Standard yarn package dyeing machine in which dye liquor is pumped through the yarn. (Courtesy Samuel Pegg & Son Ltd)

273

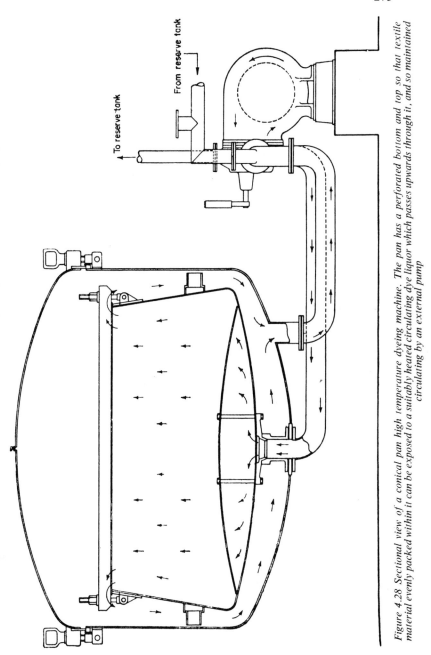

To reserve tank

From reserve tank

Figure 4.28 Sectional view of a conical pan high temperature dyeing machine. The pan has a perforated bottom and top so that textile material evenly packed within it can be exposed to a suitably heated circulating dye liquor which passes upwards through it, and so maintained circulating by an external pump

perforated cylindrical containers. A number of these containers are fixed within a dye vat. Then a pump is started and dye liquor is pumped up through the central perforated tube in each container and so outwards through the wool tops and into the dye vat. The dye liquor is thus continuously withdrawn from the vat to pass through the wool which becomes uniformly dyed. If desired the direction of flow of the dye liquor can be reversed by drawing the liquor inwardly through the wool and down the central tube.

After dyeing, the tops of roving are placed in front of a machine through which the rovings can be drawn, washed, combined and then further split up again. In this way any unevenness of dyeing is levelled out by a kind of mixing treatment.

An alternative method of dyeing roving is simply to run it through a vat containing hot dye liquor.

Quite considerable amounts of cotton, wool and other fibres are dyed in the loose state, since with this coloured material it is possible to spin special types of coloured yarn. A very convenient machine for the dyeing of this loose material comprises a large central container for the loose fibre which is evenly packed into it. The sides of this container are perforated and also there is a perforated central tube. This container is situated within a larger vessel to hold the dye liquor. Dyeing then takes place as might be expected. Dye liquor is pumped in either direction through the fibrous material until the required shade is obtained (Figure 4.28).

Dyeing of knitted materials

Men's socks　The dyeing of knitted garments, socks (half hose) and stockings (hose) demands special machinery. A very suitable machine for dyeing men's socks is the paddle machine. There are two types, of different shapes.

In the oval-type machine the dye liquor is contained in the space between the outer wall and an inner concentric wall. The socks are thrown into the liquor at the commencement of dyeing and are swept along by the large continuously rotating paddle together with the dye liquor. Thus there is a constant and very vigorous movement of both socks and liquor so that even dyeing is promoted. Steam can be blown into the dye liquor as it goes round and round the space and additions of dye can be made from time to time. Dyeing may occupy about 1 h, and some 100 dozen or more pairs of socks can be treated at the same time.

The horizontal paddle machine consists of a vat having a curved bottom and provided with a paddle which extends lengthwise in

Figure 4.29 Oval-type paddle dyer for hose and knitted garments. Can be used for dyeing, scouring, bleaching, moth-proofing and shrink resisting. (Courtesy Samuel Pegg & Son Ltd)

the middle of the vat. The dye liquor and socks contained in the vat are constantly moved as the paddle rotates on its axis and so even dyeing is ensured. The dye liquor can be heated as desired and the dyer has ample opportunity to add dyes when this is necessary.

There is, however, another type of machine which has no moving parts but which is quite satisfactory for dyeing various types of hose. All that is required to operate this machine is a good supply of steam at high pressure. The machine is specially shaped. In the bottom of the machine and slightly to one side is a long perforated pipe which is connected to the steam supply via an air injector. When the steam valve is opened the steam rushes through the injector and thus takes with it a fair proportion of air. Steam and air thus blow out through the perforated pipe in the bottom of the dye liquor and give this a circulatory motion which is quite vigorous. At the same time, the socks in the liquor are also swept round and round with the moving liquor. The steam brings the whole to the boil, and soon dyeing continues with a swirling mass of socks, dye liquor and escaping air bubbles. Even dyeing is obtained very easily. For some purposes it is not satisfactory to have the dye liquor absolutely at boiling point. To meet any objection on this point the machine can be modified so that the air is injected into the machine by means of a pump. It is then permissible to use the steam only as required for heating as the compressed air maintains the circulation of socks and dye liquor but, of course, without simultaneously heating the liquor.

Dyeing ladies' hose Men's socks are comparatively strong and can withstand a fair amount of rubbing and rough handling. On the other hand machines used for ladies' hose that are made of rayon or real silk must be exceptionally smooth and light in action, for it is very easy to damage the hose. In recent years, as ladies' hose have become finer and more gossamer-like than ever, so have dye machine makers been obliged to give this side of dyeing their special attention.

A dyeing machine formerly much used for this purpose is the rotary machine. This has been described for use in scouring (p. 233), but for ladies' hose it has to be specially designed. Generally it is made of stainless steel having a high degree of polish and tested so that there is no roughness at any point.

The hose are first turned inside out, to protect the outside at the expense of the inside which matters less, and placed in small mesh bags each capable of holding about a dozen pairs. Again this is a proved method for protecting the hose from rubbing and friction marks. The bags of hose are then packed in the compartments of the inner cylinder. Dye liquor is added to the outer cylinder so that it occupies the space between the inner and outer concentric cylinders and the covering flaps to both cylinders are then closed. Steam is blown into the dye liquor and the inner cylinder is rotated for a short period in one direction and then for a similar period in the opposite direction. This reversal is controlled automatically. This motion ensures movement of the bags of hose in each compartment and also a flow of dye liquor into and out of these compartments via the perforations of the inner cylinder. Dyeing is continued until the required shade is obtained. The hose are then rinsed in the machine with cold water and finally withdrawn to be hydroextracted and dried.

Ladies' hose, again in mesh bags, can be dyed satisfactorily in the horizontal paddle machine (p. 274) as described for men's hose. This machine is quite light in its action and seldom causes damage. On the other hand, the oval type of machine (Figure 4.29) is more vigorous and is only suitable for the tougher, coarser types of ladies' hose.

When ladies' very gossamer hose made of real silk were popular, it became necessary to devise even lighter forms of dyeing machines as the hose were so easily damaged. One form of machine was of a type on which separate stockings were hung downwards from rods into the dye liquor contained in a rectangular vat. The stockings thus received the least handling and movement, for the dye liquor circulated among them. Dyeing in this way somewhat resembled the dyeing of skeins of yarn.

In modern fully automatic machines for dyeing ladies' stockings, arrangements are provided so that about three dozen stockings can be tightly drawn over an equal number of closely spaced, vertical, highly polished metal shapes secured to a base plate and then enclosed with water-tight sealing within a tall bell-shaped container. Under substantially automatic control the enclosed stockings can be treated in succession at suitable temperatures with scouring, dyeing and rinsing liquors, and finally (after removal of the last liquor), be dried by circulating hot air. The dyed stockings are thus set to their 'finished' shape by drying on the shapes. Then the bell cover can be raised, the group of shapes moved out and the stockings taken off, while another group of shapes covered by stockings can be moved into position so that these can be processed in the same manner. Such a machine (Figure 4.30) allows stockings freshly taken from the knitting machines to be dyed and finished with the minimum of handling, and is very economical.

Other types of machines are available that operate in a similar way but that have individual differences.

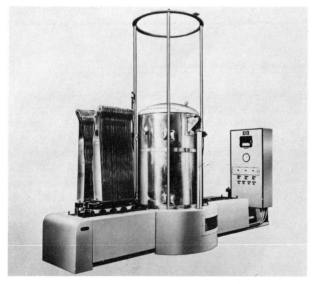

Figure 4.30 Machine for dyeing and boarding ladies' stockings. It can process four sizes simultaneously

The more delicate types of rayon and silk lace are usually dyed in quite small winch machines, the winches being turned by hand. Such laces are only about 10 to 20 yd (9 to 18 m) in length and can be dyed in a vat not much larger than $2 \times 3 \times 2$ ft (600 $\times$ 900 $\times$ 600 mm). Rayon and silk nets used for ladies' hats and as wedding veils are also dyed in these small winch machines.

It will be noted with the various scouring and dyeing machines mentioned above that the textile material (yarn, fabric or garment) and the treating liquor are maintained moving relatively to each other to secure a uniform treatment. When the textile material is being moved, say by drawing it forward over winches, rollers, etc. it can be subject to excessive tension that is able to cause dimensional distortion or possibly damage, especially when it is subject to rubbing. Thus to avoid these objectionable effects there is a trend in modern machine design to arrange wherever possible for the textile material (e.g. a fabric) to move while surrounded by the liquor moving with it. For this the liquor will be pumped and it will both propel the fabric forward while lubricating it wherever it comes in contact with the machine surface. Machines of this type now known as *jet* machines are described later (p. 349).

With some types of jet machines it has proved beneficial to inject air into the flowing liquor to prevent excessive compacting of the textile material being treated. However, sometimes this air can cause objectionable foaming which must either be omitted or be counteracted by the addition to the liquor of an anti-foaming agent.

High-temperature dyeing processes

It has previously been mentioned that, while it is usual to dye most textile materials in aqueous dye liquors at temperatures up to the boiling temperature of water (100°C), these conditions may not prove sufficiently drastic to allow the production of deep shades on synthetic fibre materials. It is further to be noted that while wool can be satisfactorily dyed in boiling aqueous dye liquors, it usually takes $1\frac{1}{2}$ to 2 h for the dye to be fully absorbed and produce a deep shade. Wool dyes more slowly than cotton and viscose rayon. For this reason it has not been practicable to dye wool fabrics by the continuous dyeing methods previously described for cotton and rayon. However, research has shown that at temperatures above 100°C wool and synthetic fibres absorb dyes very much more quickly and it is seen that the continuous dyeing of wool could be possible at these elevated temperatures. But such high temperature dyeing conditions could deteriorate the wool to an unacceptable degree. Thus, from the viewpoint of dyeing synthetic and other fibres there is much to be said in favour of using high-temperature dyeing methods. During recent years considerable attention has been given to this aspect of dyeing and rapid progress has now been made in devising new dyeing methods and apparatus.

Figure 4.31 Dyeing fabric in roll form at high temperatures within a closed vessel. The fabric is wound on a perforated roller which is positioned in the machine on a perforated vertical spindle, thus allowing the surrounding dye liquor to be pumped through the fabric roll in an outward or inward direction. This machine allows four rolls of fibre to be dyed simultaneously. A comparable but horizontal dyeing machine can be used

A much favoured machine for dyeing knitted and woven fabric allows the fabric, which is rolled on a perforated roller, to have dye liquor pumped through alternately inwards and outwards while positioned vertically or horizontally within a closed vessel that is capable of withstanding high pressures to secure high-temperature dyeing (Figure 4.31).

Difficulties have been encountered in introducing high-temperature dyeing processes. Not only had new forms of dyeing machines to be designed, but even new dyes made, for it has been found that many of those now perfectly satisfactory for application in boiling liquors decompose when applied at temperatures around 120 to 130°C. This decomposition not only results in a loss of dye but, owing to the formation of differently coloured decomposition products, it leads to the production of shades which are different from and often duller than those ordinarily produced. Further, some textile fibres, including wool and some synthetic fibres which have a low softening temperature or are liable to shrink excessively, are liable to be weakened at temperatures exceeding 100°C. Strongly built dyeing machines of various types are now available for dyeing all types of textile goods (yarn, fabric and garments) under pressure at up to 140°C by a continuously circulating dye liquor (Figure 4.32).

Such a type of dyeing machine has been developed for the dyeing of knitted garments made from most types of fibre, but especially polyester fibres, and it allows temperatures of up to 140°C to be used and thus makes the use of dye carriers unnecessary when dyeing synthetic fibre garments. It is a special type of kier having a false perforated bottom above which the garments are packed. The cover can be securely fastened to allow high internal pressure and the dye liquor is circulated continuously by means of a pump. A feature of the liquor circulation is that it throws the garments towards the centre and maintains them loosely packed, thereby avoiding the formation of permanent creases. For this the liquor is withdrawn downwards through the perforated false bottom to pass through the pump and exchange heater and return through a tube to just above the false bottom and thence upwards through the fabric (Figure 4.32).

It has also been found very satisfactory to use a modified form of the rotary washing or dyeing machine previously described for the handling of smallwares and hosiery, for with this machine both textile material and dye liquor can be simultaneously agitated to ensure even dyeing. Obviously, such a rotary dyeing machine has to be strengthened considerably to withstand the high internal pressure which is developed by heating the dye liquor up to 130°C.

Another method for rapidly dyeing synthetic fibre fabrics (especially those of polyester fibres) at high temperatures exceeding 100°C and in a continuous manner is known as a *Thermosol* (Thermofix) process (see p. 270).

In dyeing at these elevated temperatures it is often found that complete absorption of dye by the textile material is obtained within a few seconds, whereas at, say, 90 to 100°C this degree of absorption

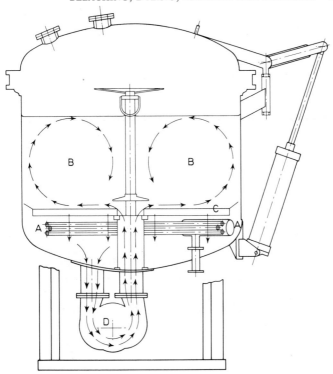

*Figure 4.32 Construction of the Pegg Toroid machine for scouring garments.
The scouring liquor is continuously circulated via a pump D through the garments
packed in the upper cylindrical vessel to return downwards through the perforated
bottom C and then through the steam heating coil A. The liquor circulation
shown at B is specially designed to tumble the garments to ensure their uniform
treatment. (Courtesy Samuel Pegg & Son Ltd)*

would occupy at least several minutes. In the case of synthetic fibre
materials these drastic dyeing conditions make possible the dyeing
of really deep shades using dyes which, under ordinary conditions
of dyeing, would be absorbed to only a moderate degree. However,
there is one snag. At the high temperatures employed, synthetic
fibre materials tend to shrink excessively and become slightly weaker.

In the application of dyes to synthetic fibre materials which are
reluctant to absorb them up to 100°C the dyer has to decide whether
or not he should employ a dyeing temperature above 100°C, say at
110 to 130 °C, or use in the dye bath a dye carrier (p. 115) and dye at
100°C by a normal dyeing process. Whatever the choice dyeing can
be more expensive and/or troublesome. Hence the continuous
search for new dyes or new fibre modifications to allow the use of

conventional dyeing methods. Recently attention has been given to dyeing in organic solvents instead of water but here again both advantages and disadvantages are encountered.

Dyeing with reactive dyes   The present availability of reactive dyes (p. 102) enables very fast-to-washing shades to be produced on mainly cotton, linen and rayon materials in a very simple manner so that these dyes (they become chemically combined with the cellulose fibres) have recently acquired great importance. It is a distinct advantage of reactive dyes that they can often be applied at relatively low temperatures, i.e. considerably below 100°C. Their application depends on bringing the fibres and dye together in the presence of an alkali such as sodium carbonate or bicarbonate, or sodium hydroxide, when combination of fibre and dye takes place rapidly and according to the temperature employed. The alkali acts as a catalyst to accelerate the chemical combination. Since at the same time the dye can react similarly with any water present a certain amount of the dye is lost or wasted by such reaction. These conditions of dyeing have, therefore, to be adjusted to favour combination of the dye with the textile material rather than with the water. But it is necessary at the end of the dyeing process to wash the dyed material thoroughly to remove the loosely held water-soluble colouring matter resulting from reaction between the dye and the water.

Various methods for applying reactive dyes are in use and a description of two will suffice to show the principles involved. In the first a cotton fabric is padded with an aqueous liquor containing the dye and the alkali, say, sodium bicarbonate or carbonate. This liquor is maintained cold to secure its stability. The fabric is then steamed for a few minutes or it is batched on a roller which is then stored at room temperature overnight or for a shorter period at a higher temperature within a closed chamber suitably heated by steam when this is desirable. The roll of fabric is periodically or continuously rotated to counteract any drainage of dye liquor to the lower part of the roll since this could lead to uneven dyeing. It is also wrapped with polythene film to prevent uneven evaporation. The storage period is necessary to allow all the dye to be utilised. Then follows thorough washing, whichever method of colour development is used.

In the second method the fabric is padded with a cold solution of the dye and is then, with or without intermediate drying, led through a hot solution of sodium carbonate or caustic soda to cause immediate combination between the dye and the fibres. This dyeing process can be made a continuous one.

Suitable methods are now available for producing fast shades on wool and nylon using reactive dyes. They are applied by methods similar to those used for acid dyes.

Dyeing two colours simultaneously

Such, then, are the main methods for dyeing solid shades on various kinds of textile material. It must be recognised, however, that by using mixture fabrics it is possible to produce two- and three-colour patterns by these same dyeing methods. We have seen that the different fibres have different affinities for dyes and it is generally possible to find a number of dyes which will colour one fibre and not another. Often, for example, the dyes applicable to wool and silk will be useless for cotton or viscose rayon. The disperse dyes used for acetate fibres fail to dye cotton or wool to any appreciable extent.

Thus, if a fabric is made with a pattern using, say, viscose and acetate threads, then it is easy to dye this in a jigger or other machine using a dye liquor containing a mixture of dyes which will colour only one of the fibres and leave the other unstained. The acetate or disperse dyes leave the viscose rayon unstained but dye the acetate fibres, whilst the viscose rayon dyes colour the viscose rayon but leave the acetate fibres unstained. In this way two-colour patterns can be produced.

This method of producing multi-coloured patterns is frequently used in the production of lace goods, since these can be made with three and even four different types of textile fibre. Four colours is about the limit to the variation which can be produced in this way.

Nylon polyester and acrylic fibres are now being made with added substances which differ as regards their dye-attractive properties (p. 118). Carpets made with these fibres can be dyed in one operation to give multi-coloured patterns.

PRINTING

From plain dyeing it is now convenient to turn to printing as a means of producing more colourful effects on fabrics and also on yarns, although to a lesser extent. The printer makes use of ordinary dyeing machines but, in addition, he uses special machines by which he can print the coloured patterns.

Block printing

The simplest method for printing a pattern is that of using so-called 'blocks'. Block printing was known to the Chinese some 2 000 years ago. It consists of carving a thick block of wood so that a pattern stands out in relief. This raised pattern can be smeared with a colour paste and then pressed upon fabric. In this way the coloured pattern is transferred to the fabric.

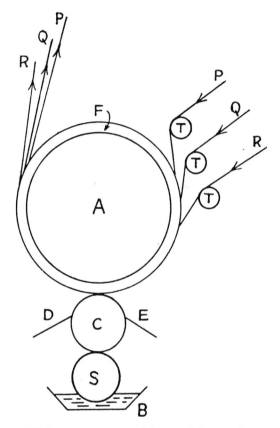

Figure 4.33 Diagrammatic view of the essential parts of a roller printing machine. The pressure cylinder A is covered with several thicknesses of fabric F to make it resilient. P is an endless thick blanket, to give further resilience which is protected from colour staining by the 'black grey' cotton fabric Q. P, Q and R (the fabric being printed) are guided by rollers T. The colour paste in trough B is picked up by the 'furnishing' roller S, which transfers it to the printing roller C. This impresses the colour on the fabric R which moves forward, together with the blanket P and the black grey Q, as the cylinder A rotates. A doctor blade E scrapes excess paste off the roller C, and another doctor D scrapes off lint and loose impurities picked up by C from the fabric

Cylinder (roller) printing

Block printing can be made to give very fine patterns and the method is used today where exclusive patterns are required, but it has largely given way to roller printing. A Scotsman named Bell is credited with having devised the first roller printing machine in 1783. Since then the principles of the machine have changed only a little, but refinements have been introduced so that its mechanical performance is now much improved. It is the roller printing machine which is used all over the world for producing printed fabrics in quantity, and its construction and operation can be understood by reference to the illustration in Figure 4.33.

The printing machine   The essential parts of the machine are seen to comprise a large central cylinder around which passes the fabric

Figure 4.34 Roller-type fabric printing machine. (Courtesy Mather & Platt Ltd)

to be printed. This cylinder rotates with the moving fabric. Against the fabric and this central cylinder a number of colour-printing rollers press. Each roller contributes one colour. Thus if there are five colours in the design then there must be five colour-printing rollers. Each of the rollers is made of copper and engraved on it is that portion of the pattern which it contributes. Thus, one roller may print all the leaves of a floral pattern in green, another all the roses in red, yet another all the stalks in brown, and so on. In an engraved roller the pattern is scooped out of the copper surface and not raised above the surface as with printing blocks.

Working with each colour-printing roller is a colour-furnishing roller which rotates in a small trough containing the printing colour or paste. A long fine 'doctor' or steel blade presses against one side of the colour-printing roller. There is also a lint doctor on the other side.

It will be appreciated that since the central cylinder is made of cast iron it has but little resilience and affords a poor support for the fabric being printed. To mitigate this it is usual to have a thick blanket between the printed fabric and the cylinder. This blanket is expensive and would soon become soiled by colour marking off from the fabric being printed, so, to overcome this further difficulty,

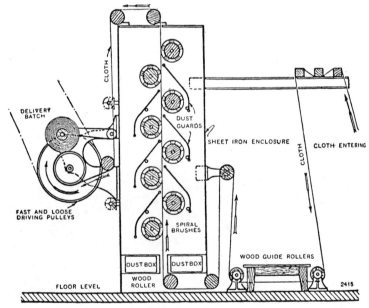

Figure 4.35 Machine for brushing loosely embedded impurities from cotton fabrics before printing. (Courtesy Mather & Platt Ltd)

it is customary also to have a cheap washable grey cotton fabric between the blanket and the fabric being printed.

It may be mentioned here that cotton fabric is usually sheared before printing so that its surface may be sheer and clean for reception of the printing colour pastes. Frequently the fabric is both brushed to remove loose fibres, motes, etc. and sheared. The shearing machine used for this purpose is similar in construction and action to an ordinary lawn mower (Figure 4.35).

Method of printing The driving power is switched on to draw the fabric forward and also to drive the printing rollers, and indirectly the colour-furnishing rollers.

The printing rollers (positively driven), by their pressure against the central large pressure cylinder, cause this to rotate at the same peripheral speed and carry the fabric forward with it. There should be no slip between the fabric thus being printed and the surface of each printing roller otherwise the coloured pattern would become blurred.

The colour-furnishing roller picks up colour paste from the trough in which it rotates and impresses this colour on the printing roller. This colour paste fills up the engraved parts and the doctor knife scrapes off excess colour. At this stage the colour-printing roller is free from colour on its surface except where the engraved pattern is filled with colour paste.

As the colour-printing roller comes in contact with the fabric and is pressed against it so colour transference takes place and that particular part of the pattern is impressed on the fabric. Then, as the colour-printing roller continues its rotation, the lint doctor scrapes off any pieces of lint, loose fibres or threads which may have rubbed off the fabric being printed. Thus the surface of this roller is maintained clean.

Each colour-printing roller continues to impress its part of the pattern on the fabric as long as the machine is in operation. Naturally, great care has to be taken to adjust the colour-printing rollers so that they 'register' or contribute their part of the pattern at the correct times and places. The machine has various devices to assist this and the differential-geared device is very ingenious and effective.

Meanwhile the fully printed fabric continues to move forward to the point at which it leaves the central cylinder. Thence it is led through a drying and steaming chamber to fix the printed colours whilst the 'back grey' is led away to be washed for further use. The blanket is endless and continues to go round and round the cylinder until it becomes sufficiently soiled to warrant its removal for washing.

Printing machines of this kind can be provided with up to fourteen colour-printing rollers and so they are able to produce patterns in the corresponding number of colours.

As described above the machine prints only one side of the fabric. If it is required to print both sides then the central cylinder, with its accompanying printing and furnishing rollers, is duplicated. Thus the fabric is first printed on one side and then passes on to be printed with the same pattern on the other side. Colour paste can penetrate thin spun fabric sufficiently to give a coloured pattern on both sides of the fabric without requiring such duplication.

Steaming to fix printed colours   The steaming process which normally follows printing must be considered part of the printing process, for without it most printed colours would be so loose that they would be largely removed in the first wash. It is necessary that the fibres of the fabric should be swollen by the hot moist steam to allow the printed dyes to penetrate and be absorbed by them. This can be understood when it is recalled that most dyeing operations are carried out in a hot or boiling dye liquor.

The steaming chamber is large enough to hold several pieces of fabric, say about 500 yd (450 m) in total length. Within the chamber are two rows (upper and lower) of closely spaced rollers (Figure 4.36). The fabric enters at one end, passes over an upper roller and

Figure 4.36 Machine for fixing printed colours by steaming. (Courtesy Mather & Platt Ltd)

then downwards to pass under the corresponding lower roller. Then it travels upwards to pass over the second upper roller and again downwards to pass under the second lower roller. Thus the fabric travels in

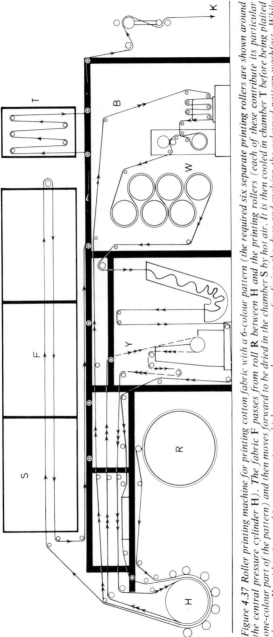

Figure 4.37 Roller printing machine for printing cotton fabric with a 6-colour pattern (the required six separate printing rollers are shown around the central pressure cylinder **H**). The fabric **F** passes from roll **R** between **H** and the printing rollers (each of these contribute its particular one-colour part of the pattern) and then moves forward to be dried in the chamber **S** by hot air. It is then cooled in chamber **T** before being plaited down at **K**, either for washing or steaming, whichever may be necessary for fixing the dyes and making the coloured pattern washfast. While passing through the printing machine, the fabric **F** is underlaid by a cotton 'black-grey' fabric **B** to protect the thick blanket from colour staining that underlays the fabric **F** at a lower level. The blanket is resilient and thus assists the level printing of fabric **F**. Both the black-grey and the blanket move in a closed circuit, which in the case of the black-grey includes washing and drying at **W**. (Courtesy Woods' Engineers (Ramsbottom) Ltd)

this up and down manner until it reaches the far end of the chamber and goes out. The object of these rollers is to allow the fabric to take a long time to travel through the chamber, many times longer

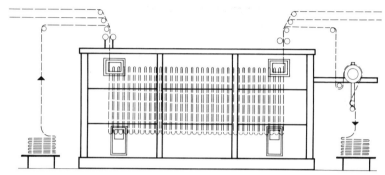

Figure 4.38 Modern Festoon steamer with a cloth capacity of up to 350 yd (318 m) per end. The cloth enters through steam-heated rollers, is looped over rods by a festooning mechanism and is conveyed to the exit arrangement by means of an endless chain carrying the rods. Plaiters draw the cloth from the chamber and pile it on to stillages. During passage through the steamers the cloth is freely suspended over rods which are slowly rotating, thus eliminating marking-off and ensuring that there is no tension applied to the fabric. (Courtesy Woods' Engineers (Ramsbottom) Ltd)

than if it went straight through. The longer the path of travel the longer the exposure of the fabric to the action of the steam and the better the fixation of the printed dyes. Not only does the steaming assist fixation of the printed dyes but it also assists the action of the discharge or resist pastes printed upon the fabric.

In many cases it is better to dry the fabric before it enters the steaming chamber. This is done by passing it over a number of steam-heated plates. The fabric moves close to these plates but just does not touch them. After drying and steaming so that the coloured pattern is fully developed the fabric must be well washed to remove all loose colour and residual chemicals. Sometimes this washing involves a hot soaping, particularly when the dyes printed are sufficiently fast to withstand this.

So-called 'flash-ageing' has recently come into use for developing the colour of printed fabric. It replaces the steaming chamber method. In flash-ageing the wet printed fabric is run over the hot surface of a large drum while being pressed thereon by a continuously running blanket. Steam is developed during the drying and held in by the blanket so that the fabric is thus subjected to a short rapid steaming treatment in a simple manner.

Flash-ageing can also be carried out by running the fabric rapidly through a Vaporloc reaction chamber (pp. 238 and 392).

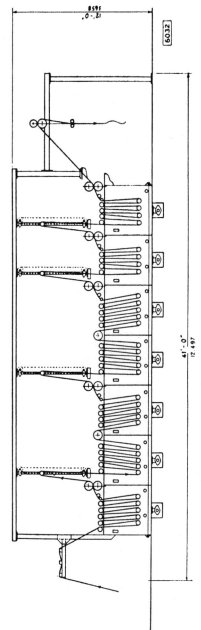

Figure 4.39 Side view of a controlled tension (open-soaper type) machine for washing fabric after printing. The fabric, in open width, passes through the compartments containing suitable detergent liquors, and then finally through a plain water rinse. In passing through the vertical tension-control devices between the compartments, it passes around a pair of spaced rollers, each held apart by spring tension. If the fabric tension is excessive this automatically pulls downwards the upper floating roller temporarily, thus relieving this excess. Such a machine is generally known as an open soaper. (Courtesy Mather & Platt Ltd)

Often it is quite difficult to wash the printed fabric rapidly and yet remove the pasty thickening agents and non-fixed dyes in it using the open-soaper as shown in Figure 4.39. Hence the reason for improving this machine by using rotating brushes to brush the fabric as it passes over the top guiding rollers as shown in Figure 4.40. Such brushing can be very effective.

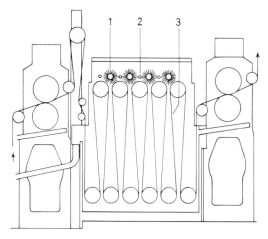

Figure 4.40 Method of more thoroughly washing fabric after printing, showing the new Benninger brushing compartment for washing machines. 1, brush; 2, spray pipe; and 3, cloth run

Discharge printing The method of printing so far described is simply adapted to produce a multi-coloured pattern on a white fabric. It could also be used for producing similar patterns on slightly tinted fabric, but certainly not on deeply coloured fabric, for then the ground colour would interfere with the coloured pattern printed upon it. As everyone knows, many of the most attractive fabrics are those which have coloured patterns on a 'ground' colour (as the original all-over colour of the fabric to be printed is termed). How can these patterned effects be obtained?

Before answering this question it will be useful to consider another kind of pattern—a white pattern on a ground colour. To produce this the following method is used. Firstly, the fabric is dyed all over with a ground shade, say brown. Then it is printed, not with a colour paste but with a paste containing chemicals capable of bleaching the brown colour to a white. Such a paste is called a *discharge* paste, and the method of printing is known as discharge printing. When the brown fabric is printed with the discharge paste, this does not usually act in the cold but only acts when the fabric is dried and passed for a few minutes through a chamber heated by means of steam which also keeps the air moist. The discharge paste then destroys the brown ground shade to give a white pattern in the printed parts.

This discharge method can be elaborated still further. There will be some dyes which are not destroyed by the chemicals in the discharge paste, so one or more of these can be incorporated in the discharge paste. After the printing and subsequent steaming a coloured pattern, instead of a white pattern, will thus be produced on the brown ground.

Obviously these white and coloured discharge processes lend themselves to the production of a wide range of patterns on coloured grounds.

Resist printing   There is, however, an alternative method known as *resist* printing. Here the white fabric is first printed with a paste containing a substance which will prevent fixation of certain dyes subsequently applied to the fabric. It is then dried and padded all over with a colour paste containing one or more of these dyes. On drying and steaming the padded dye is fixed everywhere, except where the resist pattern has been printed. The result, after washing off non-fixed dyes, is a white pattern on a coloured ground. This result is thus similar to the discharge method just described, although obtained in a different manner.

The resist method may also be elaborated. There will be some dyes which are not prevented from being fixed by the resist chemicals so one or more of these can be incorporated in the resist paste. Then, in due course there will be produced a coloured pattern on a coloured ground.

Obviously, all resist and discharge printing processes must be followed by a thorough washing of the fabric to remove loose dyes and chemicals. By using these discharge and resist methods of printing it is obviously possible to produce the multi-coloured patterns on coloured grounds to which reference has been made.

Screen printing

With roller printing it is important that there should be a large amount of fabric to be printed with the same pattern, otherwise the return is not sufficient to pay for the preparation of the printing rollers and the setting up of the machine. It is quite expensive to engrave the rollers, and the rollers themselves, being of copper (frequently they are chrome plated after engraving to lengthen their useful life) are expensive, even without the added cost of engraving. It is also evident that it is more expensive to print a design in ten different colours than it is in, say, two or three.

In the past, when there was a great demand from abroad for our printed fabrics, it was easy to secure large repeat orders for more or less standard patterns and so the cost of roller printing per yard of fabric was quite reasonable. However, in recent years, this export trade has fallen off considerably and our printers have had to give more attention to meeting the demands of the home market.

Figure 4.41 A Buser rotary screen printing machine in which the fabric to be printed is temporarily stuck to an endless belt which carries it forward so as to pass under a number of flat suitably spaced printing screens each of which is for imparting to the fabric one of the colours in the pattern. With the commencement of printing the moving fabric is brought to a stop for a suitable dwell period under the rotary screens which automatically transfer dye paste through these on to the fabric below. Then automatically the screens are all simultaneously raised and the fabric is carried forward by the belt. Thus in passing from one end of the machine to the other the fabric receives the 3, 4, 5 or more colour applications according to the number of colours in the pattern. The fabric is then drawn off the belt to pass through a hot air drying chamber while the belt returns to the entry end of the machine via a washing unit. The success of such a printing process depends on the absence of any overlapping of the applied colours which would give the pattern clear-cut delineation. (Courtesy Barke Machinery Ltd)

Here a difficulty has been encountered. Fashion changes very quickly in these days and it has become difficult to secure orders for large runs of the same pattern. Printing costs have therefore risen. It is on this account that increased use has been made of another printing method which does not require a large initial outlay for each pattern. This method is known as *screen* printing.

Screen printing is really a development of stencil printing, where the letters and numbers are cut out of sheet, and colour dabbed through by means of a brush. It is a somewhat crude method but, by careful attention to detail, it is possible to produce some quite attractive coloured patterns on fabrics. It is a method which has long been employed to decorate knitted fabrics since these cannot be manipulated satisfactorily in the roller printing machine owing to their ease of distortion by stretching.

Figure 4.42 Machine printing carpets or other pile materials with coloured patterns in a continuous manner which is especially characterised in that the dye solution is not rolled or pressed through each flat screen on to the carpet underneath as is usual in screen printing, but is copiously applied and sucked through under vacuum. Excellent penetration of the pile with least disturbance of it is thus obtained with this special machine.
(Courtesy Bradford Dyers' Association Ltd)

An early development of this stencil process was that of using a spray gun such as is now widely used for colouring motor car bodies to distribute the colour, instead of a brush. It was the second development, however, that advanced much more the usefulness of stencil printing. This was the introduction of the silk or wire mesh screen. In this modern form of screen printing each ordinary stencil is replaced by silk or stainless wire mesh fabric fastened across a square or rectangular wood frame. The screen interstices are coated or filled with insoluble varnish of hardened gelatine film except where they are left open according to the pattern it is desired to produce. Such a screen thus corresponds to the colour-printing roller of the roller printing machine. A separate screen is required for each colour in the pattern.

The fabric is laid flat in open width on long tables extending the whole length of the printing room. An operative then starts at one end and lays the screen on the fabric. With a brush or spray gun he then forces the appropriate colour paste through the mesh pattern on to the frabric underneath. This done, he lifts the screen and lays it down in the next place where that part of the pattern occurs. Again he forces colour paste through on to the fabric. In this way he continues to the other end of the fabric.

Meanwhile, another operative has commenced with the second coloured portion of the pattern and follows the first operative. According to the number of colours so will there be a succession of operatives following each other until the fabric is completely covered with the pattern. The fabric may then be steamed as for roller printing. With some types of colour pastes this steaming can be omitted.

The screens are cheaply prepared and, although they do not last as long as an engraved copper roller, they are sufficiently durable to print the comparatively short runs of fabric which are required. The results obtained by screen printing can be just as pleasing and clear as those obtained in roller printing. Up to fourteen or even sixteen colours in the one pattern can be produced. Screen printing by hand has now been almost completely mechanised so that the application of dye pastes through the wire or silk screens can be achieved more rapidly by mechanical means. For these new screen printing machines the fabric is drawn forward while supported on an endless belt and is automatically stopped appropriately to allow colour paste to be passed on to it through a flat or rotary screen before again moving forward to the next printing station. Thus screen printing has in these days become a very useful method. Its special value lies in the fact that it enables new patterns to be produced quickly and without a large initial expense.

Mechanised screen printing

In a mechanised screen printing unit the fabric is temporarily and levelly secured to a blanket supported by a rigid base divided into a number of printing stations each of width equal to the repeat coloured pattern being printed. Above each station is positioned a printing unit which comprises a flat or rotary screen with means for extruding colour paste through it on to the fabric below.

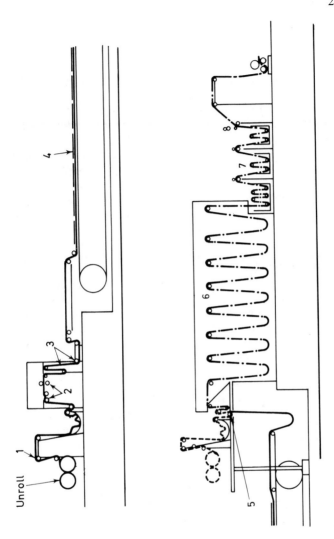

Figure 4.43 Essential features of a BDA screen printing machine for carpets, etc. The carpet (1) advances through a cleaning (brushing) unit (2), tensioning unit (3) and printing bed (4) on which the dye solutions, applied through screens, are sucked through the carpet. A dye-liquor padding device (5) allows the fabric to be given a solid colouring, and in this case, the fabric by-passes the screen printing section. A steaming chamber (6) makes the colours washfast before it enters a washing unit (7) and a suction slot device (8) to remove as much water from the fabric as is possible before drying. (Courtesy Bradford Dyers' Association Ltd)

At any particular time the printing unit may be lowered (simultaneously with all the others) to make contact with the fabric below so that colour paste can be transferred to it through the screen. The number of printing stations in use will be equal to the number of colours in the pattern and always the same coloured part of the pattern will be imparted to the fabric at any one station.

All the printing screens may be brought automatically and simultaneously into contact with the fabric and colour paste transferred to the fabric. Contact with the fabric will then be broken and automatically the fabric and blanket advanced one printing station forward, and another printing will take place as just described. In this way, by advancing one station at a time, all the different colours will have been applied and the colour pattern completed. The fabric will now be led forward apart from the blanket to pass through a steaming chamber to fix the colours and then be washed and dried.

An important part of such a mechanised screen printing machine is the manner in which the colour paste is transferred to the fabric. In one type of machine (so-called *rotary*) the screen is positioned around a hollow roller to the inside of which colour paste is supplied. Devices automatically move the roller over the fabric while the colour paste is extruded through the screen to produce a depth of colour as required by the pattern and so that the colouring is in perfect register with the coloured parts produced at the other printing stations. In the other type of machine (so-called *flat screen*) the colour paste is spread on a flat screen and rolled through it to the fabric.

Gradually these screen printing machines have been improved to be fully automatic so that, with great precision, completely satisfactory multi-coloured patterns can be produced at up to about 80 yd/min (70m/min) according to the kind of pattern being printed.

In the BDA Ltd screen printing machine, which in particular has an excellent reputation for colour printing tufted carpets having a thick pile, a copious supply of colour paste supplied through the screen to the fabric is sucked downwards through it thus ensuring excellent penetration even in producing deep colourings. Afterwards, the fabric is led through a steaming chamber to fix the colours and then further washed and excess water content removed by passage over a suction slot before drying (Figure 4.43).

Transfer printing

Recently a very old idea for producing coloured patterns on paper has been adapted for the colouring of fabrics and more especially those composed entirely or partly of synthetic fibres. It involves

Figure 4.44 Machine for 'transfer' colour printing fabric by running it sandwiched between a resilient blanket and a transfer paper carrying the coloured pattern around a hot cylinder, to cause transference of the pattern from the paper to the fabric. (Courtesy Hunt & Moscrop Ltd)

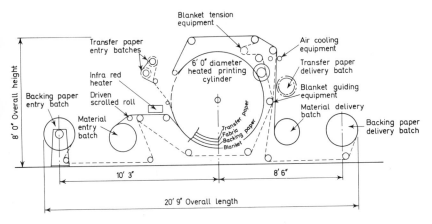

Figure 4.45 Passage of blanket, backing paper, fabric and transfer paper around the central internally-heated large-diameter cylinder on 'transfer' printing machine

heating at somewhat below its softening temperature (say up to 200°C), a nylon, polyester, acrylic or similar fabric in contact with paper carrying a colour-printed pattern (sublimable pigments are employed for this) so that very rapidly the pigments pass into the fabric thus transferring the pattern to the fabric in a very simple manner. This so-called 'transfaprint' method eliminates most of the many stages involved in roller and screen printing and may be carried out by the machines which have now become available at speeds of 25 yd/min (23 m/min). For example, no final steaming or washing of the printed fabric is required; it is a fully dry process.

Essentially such transfer printing involves passing into the machine in contact with each other the fabric being printed, the coloured-patterned transfer paper and a backing paper, so that this composite is compressed by a felt blanket while they all move together around and against a large diameter cylinder [say 6 ft d (1·8 m d)] which is internally heated either electrically or with oil fired burners. It is required that contact with the hot surface of the rotating cylinder shall be for about 30 s since within this period the transference of the pattern to the fabric will be complete. It is desirable to blow cold air against the composite leaving the hot cylinder where the components of the composite can be separated and wound on separate rollers. The felt blanket will of course be re-used many times and the backing paper is largely used to prevent colour being transferred to the blanket through the fabric being printed.

In the Fastran process (J. Dawson and Co Ltd) moist wool fabric impregnated with sulphamic acid is held (without drying) against the colour patterned transfer paper and a hot surface. The dyes, under these steaming (highly concentrated dyeing) conditions migrate (not sublime) into the wool at about 110°C.

Printing with pigments

Before leaving this section on printing, reference must be made to a development of very great interest. In the printing processes so far described, use is made of an affinity which exists between the dye and the material being printed. It may be that the fabric has to be pre-treated to have this affinity. Thus the fabric may be mordanted or in the case of wool fabric this may have to be first chlorinated. But, nevertheless, fixation of the colour is in the last resource dependent on the fabric having an attraction for the dye. This limits the dyes which can be applied and also restricts the manner by which they can be printed. The pigment printing procedure is calculated to remove these restrictions and requirements. It has originated in America and is known as the Aridye process.

This method uses pigment dyes, that is, water-insoluble pigments which under ordinary conditions are most suitable for the manufacture of paint rather than colour printing pastes. Such pigments often have a very high degree of fastness to light and other adverse influences, and in fact it has long been known that such pigments can be much faster than water-soluble dyes.

Such pigments are made into a colour printing paste by special methods, using various ingredients, but particularly a synthetic resin to act as a binder of the pigment to the printed fabric. The pigments are ground into exceedingly fine particles since this can much increase their colouring power and assist their ultimate fixation in the fabric.

In printing, the pigment colour paste containing the resin binder is applied to the fabric by any of the usual methods such as roller printing or screen printing and then the fabric is dried. At this stage the resin present to bind the pigment to the fabric in the printed parts is not in its most stable and durable form, so the fabric is now heated (not steamed as in ordinary printing) at about 150°C for a few minutes. This hardens the resin and makes it insoluble and a firm bond is formed between pigment and fabric. There is nothing more to do than to give the printed fabric a wash, and even this can, on occasions, be dispensed with.

When necessary the fabric can be padded all over with a pigment colour paste to give it a pale to medium ground shade and so it is possible to produce by this method practically any pattern which can be produced by the alternative methods. It is all very simple to operate.

It is true that pigment dyes have been used before in printing, but they were never much favoured. Natural gums were then mainly used for binding them to the fabric. The success now attained by the Aridye method is the result of using the synthetic resin as binder and also in the way of preparing the colour printing paste. It is claimed that the coloured materials obtained by this Aridye process and other similar processes have excellent fastness to washing and, according to the choice of pigment, this can be supplemented by excellent fastness to light, perspiration and other harmful influences.

Dye penetration

In all dyeing and printing processes, apart from those in which insoluble pigments instead of dyes are applied to the textile material, it is generally desirable that the applied dyes should penetrate the fibres to the maximum degree—dye left on the surface of the fibres is easily removed by washing and even by simple rubbing so that it can stain adjacent white materials. Furthermore, a dye which is fixed well inside a textile fibre is generally faster to light than a dye simply attached to the outside.

With hydrophilic fibres such as cotton, wool, etc. the wet swelling that occurs when these fibres come into contact with the aqueous dye liquor much assists dye penetration by ensuring that the fibre substance has increased porosity. But it has already been pointed out that hydrophobic fibres such as nylon, Terylene, Orlon, and even acetate fibres absorb very little water when wetted and so do not swell sufficiently to facilitate easy penetration by the dyes commonly applied to these fabrics. It is thus found that deep colourings are not obtained on such fibres unless some substance other than water is present to make the fibres sufficiently dye receptive. Thus has arisen the practice of applying with the dye a small proportion of a substance, generally termed a *dye-carrier*, which has the power of swelling the fibres—usually this is an organic substance and often it could, in a concentrated form, actually dissolve the fibres. Acetic and lactic acids, phenol, ethyl alcohol, and similar substances have proved very useful. This expedient has proved exceptionally useful in aiding dye absorption and fixation in the dyeing and printing of all the hydrophobic fibres such as acetate, nylon, Terylene, Orlon, Acrilan, Courtelle, etc.

After dyeing with the aid of a dye carrier it is important to remove this completely from the dyed fibres by thorough washing. Residual dye carrier can in some instances lower the light-fastness of some dyes, and it can also weaken the fibres or discolour them.

Mention has already been made of the use of high dyeing temperatures of 100 to 130°C to accelerate and increase dye absorption by hydrophobic synthetic fibres (p. 278).

Polychromatic dyeing

Many years ago a convenient method for random colouring textile materials including net and gauze fabrics so as to have a harmonious patchy blending of colours all-over was obtained by use of a spray gun. Recently this method has been improved and modernised so that the colouring is obtained by causing streams of dye solution to flow over a fabric under controlled conditions such that they intermingle and produce quite an attractive appearance. The new process has been devised by ICI Ltd and designated *polychromatic* dyeing. It has the advantage of allowing the production of multi-coloured patterns without requiring conventional printing machinery and therefore being able to satisfy cheaply a demand for such random uniquely coloured textile materials.

Figure 4.46 Showing a polychromatic dyeing machine which allows fabrics to be randomly coloured while passing underneath various dye solutions issuing from numerous dye jets before passing through a mangle. The dye solutions can fall directly on the fabric or have the opportunity of a mixing on an intermediate inclined plate extending across the machine and underneath the row of jets before dropping on the fabric surface, and before this is mangled. The inclined plate may be optionally horizontally or vertically traversed to cause various mixing of the dye solutions falling on the fabric surface. (Courtesy Sir James Farmer Norton & Sons Ltd)

The apparatus employed (Figure 4.47) comprises a padding mangle, through which cotton or other fabric prepared to be absorbent steadily passes, and a number of dye solution containers (each for a particular colour) say four or more. A stainless steel plane surface is mounted above the fabric and is inclined to direct dye solution falling on it to drop over its lower edge on to the fabric as it advances to the nip of the mangle. The different dye solutions are led from their containers to separately mounted jets on a horizontal bar positioned parallel and in front of the mangle and above the fabric. The dye solutions, falling from the jets on to the inclined plate and then on to the fabric below, become randomly mixed, and this mixing continues as the fabric passes through the mangle to aid their absorption by the fabric. Thereafter the fabric is steamed or otherwise treated in a conventional manner to effect fixation of the dyes in the fabric according to the type of dye and the type of fibre of which the fabric is composed. Both woven and knitted fabrics can be satisfactorily coloured in this way. Carpets, and yarns for the manufacture of carpets, can be similarly coloured.

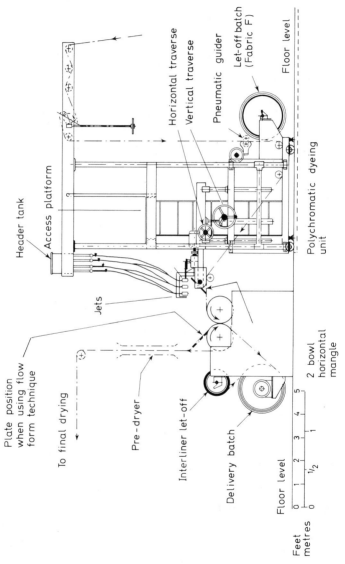

Figure 4.47 Constructional details of the polychromatic dyeing machine shown in Figure 4.46

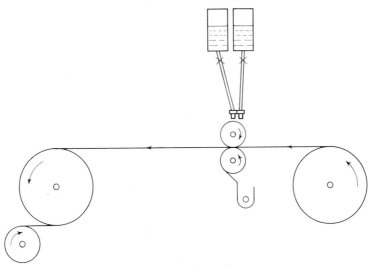

Figure 4.48 The falling of dye liquors on the moving fabric can be varied considerably, and hence the dyes are shown falling on a rotating roller on which it spreads before being mangled into the fabric. Alternatively, the liquors can fall on an inclined plate and then on to the fabric before mangling. These variations permit attractive colour mixing

Carpets and their colouring

Carpets originating in the far east, where they were variously used more for decorative purposes and for sitting, sleeping and even praying on, than as floor coverings, came to England during the 17th century where their manufacture was, and still is, concentrated around Wilton, Axminster and Kidderminster. Of the various types of carpet, Wilton and Axminster have been predominant up to about fifteen years ago when interest was taken in a completely new type now generally known as *tufted* carpet, coming mainly from America. This type is now very important and competitive with all others.

Wilton and Axminster carpets are loom woven and are thus very different from the tufted type which involve a special kind of needling process. Axminster carpets are made differently from Wiltons and the machines on which they are made impose scarcely any limit to the number of differently coloured yarns which can be used for giving multi-coloured patterns. By contrast Wilton carpets are more limited in such colouring and usually use not more than five different colours in a pattern. Tufted carpets are limited also in their colouring but these limits are being extended by using new methods for colouring them. These include ordinary winch dyeing, screen printing and the simple use of coloured yarns whose preparation can now be assisted by the use of highly sophisticated machinery.

Wool has long been preferred for carpets since it imparts resilience, durability and warmth. Other fibres such as cotton are used in other countries (but little in the U.K.) mainly because it is a cheap fibre. However, it lacks resilience and very readily becomes soiled. Synthetic fibres are now being increasingly used especially those of the nylon and acrylic type. Viscose rayon and, more important, those of the polynosic type are widely used because of their relative cheapness, i.e. Elvan of Courtaulds, and because special methods in their manufacture can be used to improve their suitability for carpets, for example, they can be made soil-resistant and anti-static. The exceptionally hard wearing properties of nylon fibres as compared with those of wool has led to the current practice of introducing about 20 per cent of nylon in wool carpets. However, all the synthetic fibres suffer from easy melting when a hot cinder drops on them. Rayon-tufted carpets are commonly referred to 'in the trade' as being made of 'fibre'.

Coloured carpets   Tufted carpets, as they are now widely made, are usually required to be coloured, that is with one solid colour all over, or to be multi-coloured to a pattern. It is possible to produce both types by the use of pre-dyed yarns but a carpet manufacturer usually prefers to delay the colouring as late as is possible so as to be able to meet quickly changing colour fashions without carrying large stocks of coloured yarn, some of which may ultimately not be wanted and which in any case could lock-up otherwise useful capital. It is thus often preferred to produce a multi-coloured pattern in the carpet by a screen printing process and lately this method has been much improved and is capable of producing a wide variety of coloured patterns, say with up to six different colours. However, there are improved methods in use for dyeing yarns immediately before tufting to allow the coloured pattern to be precisely produced when they pass forward to the tufting machine. The Alan Crawford (Edgar Pickering) method is one of these.

Another method for producing coloured tufted carpets has already been mentioned (p. 66). These carpets are made with selected mixtures of three different types of nylon yarn. All three types can be dyed about equally and normally with disperse dyes, but one type can additionally be dyed with acid but not basic dyes, while another can be additionally dyed with basic but not acid dyes. By dyeing such carpet finally in, say, a winch machine containing appropriate mixtures of these dyes under suitable controlled conditions coloured patterns can be obtained in the one dyeing operation which may either have contrasting colours or be of the tone-in-tone type.

The dyeing of tufted carpet in a winch dyeing machine is today more extensively used now that dyeing machine makers have given special attention to making this type of machine wider, and of more robust construction, than is usual when the machine is intended for use in dyeing ordinary lightweight fabrics, more particularly knitted fabrics. In dyeing carpet in this type of machine it is necessary that it should throughout the operation be kept in open width free from creases which could become fixed and spoil the carpet's ultimate appearance. In the dyeing process the carpet is continuously drawn up from the dye liquor in the front of the vessel containing the dye liquor to pass over an overhead oval or circular winch and then fall back into the dye liquor at the back of the vessel; this movement is continued until the carpet has acquired the desired shade when the exhausted dye liquor is run to waste and the carpet is then washed in the same machine. It is thereafter freed from excess water by passing it over a suction device and then dried.

In an alternative one-run method the carpet is run through a padding mangle to saturate it with dye liquor and is then run through a closed chamber where it is steamed for a few minutes and finally washed by passage through a series of washing units before drying.

When coloured yarn is used for producing carpets having a random type of colouring, use is often made of so-called 'space-dyed' yarn. As may be anticipated this method requires yarn to be coloured differently at spaced intervals along its length so that when the yarn is used in the carpet a multi-coloured pattern is produced corresponding to the spacing of the colours in the yarn.

Space dyeing is carried out in various ways and a number of different types of dyeing machine are available some being relatively simple while others are very complex and indeed may be controlled by electronic devices. The days have long since passed when the simple method used was to tie a tight knot in the middle of a skein of yarn and dye one half at a time by dipping it in different dye liquors to obtain differently coloured halves, and so rely on the tightly drawn knot to resist dye liquor penetration, in order to obtain yarn dyed in two colours with white spaces between them. Obviously the coloured yarn thus produced could be further modified by over-dyeing it.

In one space-dyeing process—a relatively simple one—carpet yarn is knitted to give a fabric which is then printed with an appropriate multi-coloured pattern. The fabric is then unravelled to give a corresponding yarn irregularly and differently coloured along its length.

In another simple method a sheet of parallel yarns closely spaced is led from one pair of rollers to another pair suitably spaced so that a thickened dye-solution extruding device can be drawn repeatedly from one side to the other while the yarn is passing forward at a steady rate from one roller to the other. More than one dye solution can thus be applied.

One further space-dyeing machine consists of a printing roller on which are parallel raised bars of varying width but equal thickness spaced at various distances. These bars may be padded with foamed rubber of other dye-solution-holding material. This printing roller rotates in contact with a lower plain roller so that when a sheet of parallel and closely packed yarns similar to a warper's beam of yarn (p. 177) is led between these rollers, colour bands are transferred to the yarn at distances apart corresponding to the spacing of the bars to which dye solution is continuously supplied.

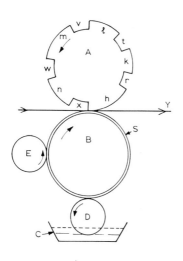

Figure 4.49 The essential features of this yarn space-dyeing apparatus are a roller (A) having a non-absorbent surface such as rubber and slotted as shown which runs under pressure contact with roller (B) having an absorbent surface such as a felt covering. Roller (B) runs so as to be indirectly supplied with dye liquor from trough (C) via roller (D) which rotates partially submerged in this liquor. Roller (E) presses against roller (B) to remove excess of dye liquor.

All the rollers are wide so that a sheet of parallel yarns can be led between rollers (A) and (B) and thus have dye liquor pressed into them by the upstanding parts h, k, l, m, n, etc. but not at the cut-out parts r, t, v, w, x, etc. thus space-dyeing the yarns according to the width and spacing of the upstanding parts of roller (A).

It is likely that the sheet of coloured yarns will require to be subsequently steamed to fix the applied dye so as to be fast to washing

Somewhat similar in construction is a space-dyeing machine comprising two rollers rotating in contact with each other. One roller has grooves cut helically in it while the other has similar grooves similarly positioned but being the mirror image of those in the upper roller so that on rotation these grooves perfectly register to coincide. The grooves are supplied with a thickened dye solution so that as a sheet of yarn passes between the rollers diagonal bands of colour are transferred to it.

In yet another machine for space-dyeing a sheet of parallel yarns to be used, say, in the manufacture of colour patterned tufted carpets, the yarns are drawn from wound cones to pass through a special type of printing machine which can be controlled to apply a dye solution in selected parts of the yarns, say as colour bars across them, perhaps at spaced distances. Afterwards the sheet of yarns is carried forward in loose small folds, supported on a moving endless stainless steel or aluminium belt, through a steaming chamber to fix the dyes within the fibres. It is then washed by passage through three units, in each of which the yarn, while passing around a perforated drum, has the washing liquor removed by being sucked through it. This is further aided by mangling between each pair of units. The yarns are finally dried while passing through a perforated drum-type of dryer before being wound on to a beam, the yarns still being maintained parallel to each other as originally.

By printing with liquid ammonia solutions of dyes, synthetic fibre yarns can be instantaneously spaced dyed.

The novel polychromatic dyeing machine recently developed by chemists of ICI Ltd now being made under licence by Sir James Farmer Norton & Sons Ltd and other textile machine makers and shown in Figure 4.46 can also be used for the space-dyeing of carpet yarns (although its main use is for dyeing fabrics) with a number of different colours. In this machine different thickened dye solutions are allowed to flow on to a moving sheet of yarns from several dye jets suitably spaced across the width of the sheet; the bar holding the jets can be given a traversing motion so as to spread the dye solutions falling on the yarns while alternatively other means can be adopted to make the impact of the dye solutions on the sheet of yarns irregular.

Much more sophisticated methods for producing multi-colour patterned carpets may now be mentioned. The first uses the Stalwart printing machine which is really an elaborate roller printing machine (p. 285). It comprises a system of separate printing rollers each engraved to hold the dye paste as continuously supplied to it, and to transfer ultimately this to the carpet in areas where it has to produce a particular part of the coloured pattern. When all the rollers have been arranged in register so as not to produce colour overlapping in the carpet this is run through the machine, say at 5 yd/min (4·5 m/min). The carpet is then led through a steaming chamber to fix the applied dyes and then passes through washing units before finally being dried.

Recently one of the largest British manufacturers of tufted carpets, Edgar Pickering (Blackburn) Ltd, has selected the Crawford Yarn Patterning System as devised by Alan Crawford and developed by Mohasco Industries in America as being of exceptional merit for producing coloured yarns in a substantially one-run process to allow tufted carpets to be made at relatively high speed with any design (at present 5-colour). It operates in two stages where the yarns drawn from a creel containing the separate yarn packages are first colour printed, steamed for dye fixation, washed, dried and wound on beams ready for passing to the tufting machine. On its way from the creels to the printing machine the sheet of yarn is divided into a suitable number of groups and the dye solutions are separately supplied to spring-loaded plungers (each covered with a dye solution holding-pad) over which the yarn groups pass. The plungers are mounted on a metal roller and they are brought into or out of action (automatically by an electronic device) as required by the colour pattern to be produced in the tufted carpet. The emerging groups of differently coloured yarns are then led through a steaming chamber and finally washed, dried and wound on a number of beams, convenient for tufting through a special control system, to produce an accurately coloured pattern in the resulting carpet. It has been reported that such a yarn colouring system is exceptionally satisfactory but is very costly to install.

A multi-tone colouring of polyester pile fabrics devised by ICI Ltd is based on the observation that when the exposed tips of polyester fibres forming the pile of a fabric are exposed to a mechanical polishing treatment so that they become heated to, preferably, $40-70°C$ above any previous heat treatment to which they have been subjected, this treatment modifies their dye affinity so that in the subsequent dyeing of the fabric they may acquire a depth of colouring which may be greater or less than that of the non-polished parts according to the temperature conditions to which the fibres have previously been exposed. If the fibres are crimped then the polishing can remove this either wholly or partly to increase their lustre.

By heat-setting a polyethylene terephthalate sliver-knitted pile fabric at $150-200°C$ and then surface polishing at $200-240°C$, the polished parts of the fibres lose some crimp and dye to a deeper shade than those parts of the fibres nearer to the pile base and which have escaped polishing. By contrast, if the heat-setting is effected at $100-120°C$ and then the pile polished at $160-190°C$ the polished fibre tips dye to a lighter shade. In either case the dyed pile has an attractive multi-tone appearance.

Polishing is effected by exposing the pile surface to one or more heated elements mounted on a rotating heated cylinder or a heated flat-plate, and it is possible by suitably arranging their positions to produce colour patterns.

It is preferred in the subsequent dyeing process to apply at the boil, with the aid of a dye carrier such as diphenyl, disperse dyes which have poor migrating power such as CI Disperse Orange 30, CI Disperse Blue 122, and mixtures of these.

For otherwise producing colour patterns in carpets screen printing methods can be used but they have to be adapted so as to ensure that the dye solution applied to the carpet can fully penetrate the dense fibre tufts—it is an advantage in using pre-coloured yarns that it is easier to secure such good dye penetration.

Figure 4.50 shows a machine for the continuous dyeing of tufted carpets.

FINISHING

When yarns, fabrics and garments have been scoured, bleached, dyed and printed they generally have to be 'finished'. This term indicates those final operations which are necessary to bring the textile into a presentable attractive condition. Many a manufacturer has said that it is the finish which sells the goods.

The finishing of yarns is not of particular interest to the ordinary reader, for most of the dyed yarn is used by weavers and knitters and only reaches the public at a later stage when it is in the form of fabric or garments. Of course, there are hand knitting yarns, but there is not much finishing about these except for the twisting of the skeins or winding them into balls of knitting yarn to be sold directly over the counter to home knitters. With fabrics and garments much finishing is needed, so we will consider these more closely.

Fabric, after wet processing such as scouring, dyeing or printing, will generally be in a distorted condition. It is almost impossible to run fabric backwards and forwards through dyeing machines, to mangle it, or even dry it without pulling it in length and making it correspondingly narrower. This applies particularly to knitted fabrics, which are particularly susceptible to distortion. One of the most important functions of finishing, therefore, is to straighten the fabric and bring it to the required dimensions. Generally in the trade it is necessary to finish the fabric to a standard width, say, 36 or 48 in (0·9 m or 1·2 m). With knitted garments, men's socks and ladies' hose these have to be brought to standard sizes and shapes.

It is also to be noted that the wet-processed textile material brought from the bleacher or dyer to the finisher is wet, and so drying forms part of the finishing process. Sometimes this drying can be carried out at the same time as the fabric is straightened

312

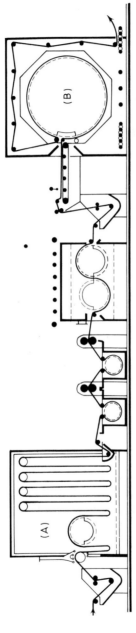

Figure 4.50 Machine for the continuous dyeing and curing of needled felt and other non-woven fibre-bonded fabrics. In a one-run process the needled felt is padded with a dye liquor and led through a steaming chamber **A** where steam is blown through it while it passes over a perforated drum while moving forward in festoon form. It is then washed free of excess and loose dye and dyeing auxiliaries, which are used in the dye liquor to promote maximum dye absorption and penetration of the felt, while passing around two perforated drums (with a mangle between and also after) which allow the washing water to be sucked through the drums and then back into the surrounding water. It then passes through a perforated-drum intermediate drying chamber in which hot air is sucked through the felt while around the two drums. In this way it is dried so as to retain 25–30% moisture before passing through a padding mangle which impregnates it with an aqueous dispersion or solution of the fibre-bonding agent (say a resin precondensate). Then follows the final stage in which the felt is led through a larger perforated drum dryer to be both dried and subjected to a high temperature which promotes curing (polymerisation and insolubilisation of the agent) and fibre-bonding. Features to note are (1) the free overhang of fabric around the perforated drum in steamer **A** to accommodate any temporary over- or under-running of the felt, (2) the intermediate mangle in the washing section to increase considerably the effectiveness of the washing and assist compacting the needled felt to protect it from damage; and (3) that the felt leaving the final drum dryer does not immediately pass out of the drying chamber but makes a circuit inside it over guiding rollers, thus allowing more time for completing the cure. The same machine can be used for dyeing tufted carpets with the following modifications: (1) the intermediate drying and application of a fibre-bonding agent are omitted (by-passed) since fibre-bonding is not required and (2) the direction of rotation of the final drying drum **B** is reversed so that the tufted surface of the carpet is uppermost and does not come into pressure contact with the drum surface. Hence flattening of the tufts is avoided. (Courtesy Fleissner Ltd)

and brought to its desired finished width and length, whilst at other times it must be dried first and then lightly damped for the final finishing treatment. The same considerations hold good for garments, socks and ladies' hose. In general it is important to remove the maximum possible amount of water from the textile material before it is dried. It takes about 1 lb (0·45 kg) of coal to generate heat sufficient to dry out about 10 lb (4·5 kg) of water, so drying can be an expensive business. Further, it is undesirable to bone dry a textile material, whether it be of cotton, wool, silk or other fibre. A textile fibre is in its most pliable and soft condition when it has a moisture content such as is natural to it in its air-dry state. If fibres are bone dried at high temperatures they are changed slightly so that they do not so readily pick up moisture from the air, and thus they never really soften down entirely. It is usually better to dry fabrics and garments in air at room temperature, but it is only seldom that this can be done in a textile works as such drying would be too slow and therefore too costly.

Removal of excess water from wet fabrics

There are a number of different machines available for removing the excess of water from a wet material, the choice depending on the type of textile material. If it is strong plain cotton fabric or any other material of similar robust character, then it is best to mangle it. The mangles used are either two-bowl or three-bowl machines. The former allows only one nip of the fabric whereas the latter type of machine allows two nips, one as it passes between the lowest and the middle bowls, and the other between the middle and the highest bowls. 'Bowl' is the term used for the rollers. Mangle bowls may be of wood, hard rubber, brass or even copper. The choice is governed by the hardness of the nip which is desired and the robustness of the textile material. In mangling it is possible to run at very high speeds, say up to 80 yd/min (73 m/min), or more. Generally the fabric is in open width but some mangles allow the fabric to run through in rope form. It is often dangerous to mangle rayon fabrics, especially those containing both rayon and threads of other fibres. Rayon is comparatively easy to cut in its wet state as may be easily verified by cutting different wet fabrics with scissors. So, if a hard cotton thread in the fabric crosses over a rayon thread then, under high pressure mangling, there is a risk of the cotton thread cutting right through the rayon thread. Such points as these have to be carefully watched. Also, it is impossible to mangle a fabric without stretching it in length. Thus crêpe fabrics which depend for their appearance on a small length shrinkage are not suitable for mangling. For such materials there are at least three alternative machines.

The first is the hydroextractor, which has been referred to previously. In this the fabric is placed in a perforated container rotated at very high speed and the water is thrown out by centrifugal action. This is a very useful machine and is used for all kinds of fabrics, garments, hose and yarns. One disadvantage about a hydroextractor is that the textile material within it is in a crumpled and crushed state. The high centrifugal force to which it is subject is equivalent to high pressure and permanent creases can thus be produced in susceptible fabrics. To avoid this, use may be made of one of the two other types of machine.

The second machine is also a kind of hydroextractor but the fabric is kept in open width. Essentially it consists of a horizontal cylinder on which 80–100 yd (73–91 m) of fabric are wound evenly. This cylinder is then rotated rapidly and the water contained by the fabric is thrown out by centrifugal force. The ejected water is confined by a hinged metal hood drawn over the fabric. This method of removing excess water from fabrics is not much favoured since it is very slow and it requires much labour to wind on and wind off the fabrics being treated. However, it is obvious that such a method keeps the fabric free from creases and disturbs its structure but little.

The last machine is most favoured. The essential part of this is a horizontal tube having a long narrow slot in its upper half extending the length of the tube. The width of the slot can be adjusted to coincide with that of the fabric which is led over it. The slotted pipe is connected to a vacuum pump and water ejector. Thus, as the fabric passes over the slot the air is drawn through it into the pipe, taking with it excess water from the fabric. This method is very suitable for fragile fabrics and cannot cause creases in them. A defect is that it is less efficient than a mangle and is somewhat slow. If the fabric is run too quickly over the slot then there is not sufficient opportunity for all the water to be drawn out of it.

Drying of textile materials

By one or other of these methods the stage has been reached at which fabrics, garments or yarns have been rid of excess liquor and are ready for drying. Quite a number of different types of machines are in use for drying, a feature to which textile engineers have had to give a great deal of attention.

Woven fabrics   For woven fabrics there can be no doubt that a set of drying cylinders is the most useful and convenient means of drying. It is true that by drying in this way the fabric is pulled somewhat in length and that the surface of the fabric may be polished a little by slipping on the cylinder surfaces, but nevertheless it is a method which is simple, rapid and efficient.

Figure 4.51 Fabric being drawn from a roller to pass through a padding mangle to be impregnated with a finishing (softening, etc.) liquor and then dried by passing around a number of drying cylinders internally heated by steam under pressure or optionally by hot oil. The bowed rollers shown in front of the machine (right) are freely rotatable. They are bowed to effect a weftwise spreading of the fabric passing over them and so remove creases. (Courtesy Mather & Platt Ltd)

The drying cylinders may be arranged either horizontally or vertically and a set may comprise any number up to thirty or even more. Each is of sheet copper or stainless steel and about 2 yd (1·8 m) in circumference. Through the inside of these cylinders steam is passed under a moderate pressure so that the surface of each cylinder is maintained slightly hotter than the boiling point of water. The cylinders are staggered so that as fabric travels over them from the first to the last it is in contact with the maximum area of the surface of each. Considerable heat is radiated from the cylinder surface and therefore the object is to have as much surface as

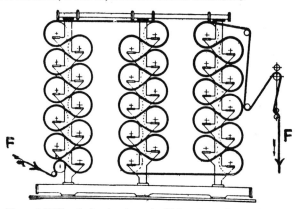

Figure 4.52 Sketch showing the path of fabric F round a set of vertical drying cylinders. The arrangement shown ensures the maximum area of contact. (Courtesy Mather & Platt Ltd)

possible covered with fabric. The special staggering of their positions ensures this. As the fabric passes from one cylinder to the next the side of the fabric in contact with the cylinder surface is reversed. Thus drying is very even. The rate at which fabric is led over such a set of drying cylinders is such that the fabric is just about completely dry when it is on the last two cylinders. If the fabric is run more slowly it is liable to be 'baked' and so have its handle harshened.

It is usual to have special devices in front of the first cylinder so that the fabric is freed from creases and is spread out to its full width. It is also necessary to keep a watch for double selvedges since these sometimes tend to curl in.

In a Jetcyl (Weston-Evans & Co Ltd) modification of a set of drying cylinders, it is arranged that hot air is continuously blown on to the fabric while passing over the cylinders which are enclosed so that the air can be re-circulated including reheating. Thus, with desirable increased efficiency of drying one cylinder can become as effective in drying the fabric as the three which are ordinarily employed (Figure 4.53).

In quite a different type of fabric-drying machine the fabric in open width is led over and in intimate contact with one, two or more large perforated rotating drums within a heat-insulated enclosure. Hot air is continuously circulated within the enclosure with arrangements which cause it to be sucked through the fabric into the drums and thence out of these for reheating and recirculation. The hot air passing through the fabric holds this close to the perforated drum surface and at the same time dries it. This so-called 'suction-drum' type of drying machine has been actively developed

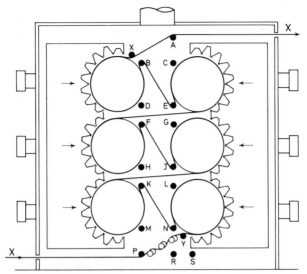

Figure 4.53 The essential features of the Jetcyl type of fabric drying machine. While the fabric is passed in open width over a number of drying cylinders that are internally heated, e.g. by steam, a mixture of air and steam is directed radially on the surface of the fabric. It is therefore heated from the cylinder surface and by the air and steam so that moisture evaporating in the fabric is rapidly removed from it. It is claimed that a drying cylinder operating under these conditions can be three times as effective as a cylinder over which the fabric passes in simple contact. The cylinders are enclosed as a group in a suitable housing so that the heating can be made as efficient as possible. An important feature of the rapid drying which is thus made possible is that irregular migration of any substances (dyes and finishing agents) within the fabric being dried is largely prevented (under conditions of slow drying such substances migrate to any parts of the fabric which are being more rapidly dried than others and this can lead to uneven colouring or finish). (Courtesy Weston-Evans & Co Ltd)

in recent years since it is efficient and highly economical. It can be used for all types of woven and knitted and even non-woven fabrics and allows the drying to be effected under the substantially tension-free conditions which are often desired. Suction-drum drying machines can also be positioned at the end of machines used for washing loose fibres such as raw wool and fibres in sliver form so that these textile materials can then pass directly through the drier and emerge fully dried.

Incidentally, it may be mentioned that just as fabric can be dried in suction-drum dryers by drawing hot air through it, so can fabric be impregnated with a bleaching, dyeing or finishing liquor by having the drum immersed in this liquor and drawing it through the fabric. Fleissner Ltd have been especially successful in devising new machines based on the above-mentioned principle and have made them available for use in many wet processing and finishing operations. Drum-steaming machines are now available in which the fabric has steam drawn through it while passing over perforated rotating drums. Thus, in a continuous process for dyeing knitted tricot fabric, perforated suction drums are used for impregnating the fabric with the dyeing liquor and drying it. Afterwards, this fabric, impregnated with a softening or other finishing agent, is steamed by using suction drums to fix the dyes, washed and finally dried.

Where it is not permissible to flatten the surface of the fabric, or to stretch it in length, it is better to use the festoon dryer. This consists of a large chamber which may be of brick, asbestos sheet or iron plate on an angle-iron framework. The air inside this chamber is kept very hot by means of internal steam-heated pipes. Fabric in open width is led into the chamber at one end near the top and is then formed into loops or festoons extending nearly to the floor. Such festoons of fabric are carried to the far side of the chamber on moving rods and there drawn out fully dry. During this drying the fabric is quite loose and is free to contract if it should have a tendency to do this. Such a drying machine is very suitable for crêpe fabrics for these will have been previously shrunk to give them their characteristic pebbly surface.

Brattice dryers for hosiery   Yet another form of drying machine is known as the *brattice* dryer. This is especially suitable for the drying of knitted fabrics and for ladies' hose or loose garments. It allows the material being dried to shrink to its natural size and shape and is a very useful machine in a hosiery dyeworks.

This machine also consists of a large chamber in which hot air is rapidly circulated. The problem for designers of this type of drying machine is to secure the longest path of travel of the textile material through it, so that it can be all the more thoroughly dried and at the highest possible rate. The general solution to this problem is to carry the hose or garments through the chamber on a long endless brattice made of perforated metal strips fastened to side chains. The garments are thus carried backwards and forwards two or three times within the chamber before emerging at the far end. During travel the hot air blows up and between the textile material and dries it with the minimum of disturbance.

Where large quantities of textile goods have to be dried each day such a machine as this will be kept running all the time, and as the goods arrive at the entry end an operative will throw them on the brattice so as to form a thin even layer; another operative will attend to them as they come out at the other end quite dry.

The brattice machine is also very useful for drying the knitted

tubular rayon fabric frequently used for making into ladies' underwear. This fabric cannot be dried on a set of drying cylinders as it
would become too distorted and stretch in length, but it can be
loosely but regularly plaited on to the brattice so that it will be
carried through the drying chamber free from length tension.

This same tubular knitted fabric, whether it be made of rayon or
cotton, can also be dried on another simple form of apparatus in
which advantage is taken of the fact that the fabric is tubular. Here
the fabric is threaded over a vertical tube, through which is blown a
strong current of hot air. The fabric is led upwards and drawn off
the tube to be gripped by a pair of rollers some 10 or 15 ft (3 or 4·5 m)
above the tube top. Thus the tubular fabric is closed as it is drawn
through the rollers to form a kind of long sausage. The hot air can
only escape by blowing outwards through the fabric and in doing
this it dries the fabric. It is found that rayon fabric can be dried quite
quickly in this way but, of course, there must be a good supply of
hot air. Thick wool fabric is dried in this way too, but it may have
to be sent through the machine twice since wool holds more water
than rayon. There is usually an internal stretcher placed within
the fabric between the top of the vertical tube and the pair of drawing
rollers. In this way the fabric can be dried to any desired width and
so be almost finished ready to be made up into garments.

It is to be noted that with the drying machine just described
the hot air which is blown only outwardly through the fabric escapes
into the surrounding air and a considerable amount of heat is thus
wasted. More recently such drying machines have now been modified
so that as the fabric moves forward through two adjacent separate
zones, while at the same time passing over a special form of stretcher,
the hot air is, in the first zone, blown through the fabric from the
outside to the inside and then in the second zone this same air
passes outwardly through the fabric. Thus the hot air passes through
the fabric twice instead of once to give a double more efficient
drying. A further advantage of this type of drying machine is that
it is not necessary to pile the fabric around a vertical tube through
which hot air is blown upwards—the fabric can be drawn directly
out of the truck in which it is brought from the scouring, dyeing or
other wet processing to pass upwards through the drying and stretching part of the machine and then through a finishing calender thus
to effect in one operation both drying and final finishing. The fabric
so obtained is at once ready for making-up into garments. A calender
finish is comparable to a hot ironing of the damp fabric (Figure 4.54).

The hosiery finisher sometimes finds this machine convenient for
drying knitted fabric which has not been tubular knitted but which
has had its selvedges sewn together and has hence been made tubular

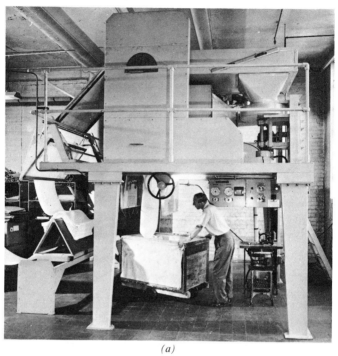

(a)

Figure 4.54 The machine shown in (a) allows wet-processed knitted fabric that has been freed from excess retained water to be dried with hot air in one run, and then to be

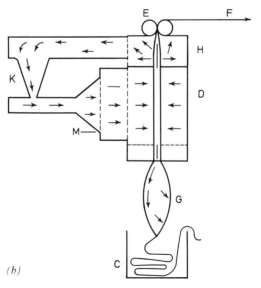

(b)

calender finished to set it to a desired width and to give it a smooth texture. It is then either brought into roll form or pleated into folds.

In (b) the fabric F is drawn upwards in rope form from truck C through the drying chamber, which has separate sections D and H, to the calender E. The air, which is continuously circulated by fan K, is heated while passing through the steam-heated battery M, and it then passes inwardly *through the fabric while this passes through section D. Some of this air passes downwards within the fabric and balloons it at G to assist it to be fully extended as it passes outwardly through it, and the drying is thus completed while it is held to the desired width, by passing it over the second internal stretcher. The dry fabric is then led through the calender E to obtain its desired finish (it is optional to steam the fabric to assist the finishing here).*

The air passing outwardly through the fabric H (mixed with a suitable proportion of fresh outside air) returns to the fan K for recirculation through the heating battery and dryer.

An important feature of this machine is that the hot drying air passes through the fabric both inwardly and outwardly, thus allowing greater efficiency and considerable heat economy. It is also an important advantage to effect both drying and finishing in the one-run continuous manner

temporarily for the purposes of dyeing. After drying, the sewing thread can be withdrawn and the fabric once more becomes flat.

Drying of lace goods   Finally, there is another method of drying which is much used for lace goods. This method involves stationary machinery which is very simple but, nevertheless, very satisfactory. The machine consists of two parallel iron frames along which are closely spaced pins. The damp lace is stretched between the frames with its edges impaled on the pins. A cross-bar, also provided with pins, is positioned at each end of the fabric so as to keep the ends straight. Then the frames are moved outwards until the fabric is just stretched to the desired width. At this stage wafters or fans overhead are started, and since the machine is positioned in a warm room the lace material soon dries by reason of the warm air being wafted around it. Such a method of drying involves a good deal of hand labour but is best for expensive lace materials.

The removal of excess water and drying are operations common to the finishing of most kinds of textile material. We now come to a variety of finishing operations all of which have to be adapted specially to the kind of textile material being processed. In some way or other the methods and machinery used are decided, not only by the nature of the material, such as woven or knitted fabric,

Figure 4.55 Knitted fabric entering a stenter to be brought to a controlled desirable width and free from length stretch

yarn, garments or hose, but also by the type of fibre present. For this reason there has been a tendency in the past for works to specialise. Thus some will finish cotton fabrics, others silk and rayon hosiery, others will deal with lace, and yet others with wool fabric. Some dyers and finishers are also specialists in yarn and have little to do with fabrics. So, in dealing with finishing proper, it will be most convenient to make particular reference to these divisions of the trade.

Finishing fabrics to desired dimensions on a stenter

Dealing first with cotton fabric, one of the most important operations in finishing is to bring it to the required dimensions of width and length. Of course, width is the more important for plain fabric.

The clip stenter Such cotton fabric may have been dried already or it may simply have been mangled. Let us consider the dry fabric first. To bring this to the right width it must be led through a stentering machine. This may be from 20 to 90 ft (6 to 27 m) long and is so arranged that the fabric is carried through it by two moving chains of clips, one on each side. The clips hold the selvedge firmly so that it cannot slip out of their grip. Each clip is 3 to 4 in (75 to 100 mm) long, and they almost touch, end to end (Figure 4.56).

The fabric, having been pulled in length during dyeing and drying, is generally somewhat too narrow. So it is arranged that the

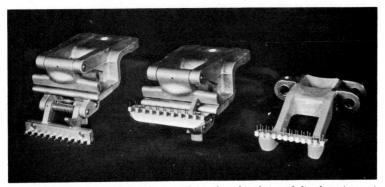

Figure 4.56 Types of simple pin plates (right) and combined pin and clip plates (centre) for holding the selvedges of fabric while passing along a stenter for drying and bringing to a desired width. At the left is shown the combined pin and clip plate opened so that the pin plate is lowered to be inoperative to allow the clip to be used. (Courtesy Mather & Platt Ltd)

clip chains diverge from the entry end for about a quarter or slightly less of the total length of the machine. Thus, as the fabric is carried forward, gripped on either side by a succession of clips, it is gradually widened. After attaining the required width the clip chains travel the remainder of the length parallel to each other and so the fabric is set to that width.

Figure 4.57 Cotton fabric being led through a brushing machine to remove surface adhering impurities and then over a stenter to bring it to a uniform width preparatory to subsequent treatment where it must be of uniform width and often in roll form. (Courtesy Mather & Platt Ltd)

In order that this stretching of dry fabric may take place easily it is either slightly damped in a damping-machine before being fed on to the stenter or it is steamed just before it is gripped by the clips. Then once it is stretched to width the moving fabric passes over gas flames or hot air heaters so that it is dried and so set at this width. At the far end of the stenter each clip releases the fabric and turns round to come back to the entry end to grip and carry forward fresh oncoming fabric. Meanwhile the fabric is drawn forward and plaited down into a folded pile of cloth or is wound on a roller.

The hot air stenter   The above is a description of the simplest form of stentering machine which is now only seldom used. The modern machine undertakes both the drying and the stentering so that the fabric is dealt with in one run at a rapid rate. The drying is done mainly or entirely by means of hot air, but it may be assisted in the beginning by a few drying cylinders. Hence the whole range of machinery is frequently referred to as a 'hot air' stenter. Some very efficient and up-to-date machines of this kind are now in use and they are widely used for finishing rayon crêpe fabrics as well as other fabrics.

At the entry end (Figure 4.58) there is a small two-bowl mangle of light construction and provided with a trough for holding any liquor with which it might be desired to impregnate the fabric. Then follow perhaps 6–10 drying cylinders. Next is the stenter proper. The greater part of this is enclosed in an asbestos sheet housing or tunnel. At the far end the stenter frame just extends out of the tunnel so that the fabric can be led off for plaiting or batching.

The stenter frame is mostly enclosed because in this way it is possible to effect drying of the fabric by means of hot air with the smallest waste of heat. As the fabric is carried forward within the tunnel or housing, jets of hot air are blown through it downwards and upwards, and also along it. In the short time that this hot air is in contact with the fabric it is scarcely possible for it to become fully saturated with moisture and so it can be used again. The object of the tunnel is to confine this air and enable it to be drawn off through the powerful fans to be led back through the heaters and blown again on the fabric. By circulating the air again and again in this way quite a large saving of heat is secured. So that the air never becomes fully saturated with moisture and therefore useless for drying, a small proportion of fresh air is continuously mixed with the recirculated air.

Thus the hot air stentering machine can be worked in this way. The dyed or otherwise prepared fabric mangled in the last operation

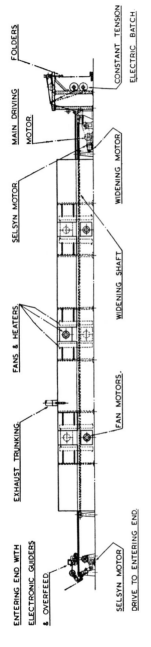

Figure 4.58 Sectional view of a single-layer hot air stenter showing the passage of the fabric

326

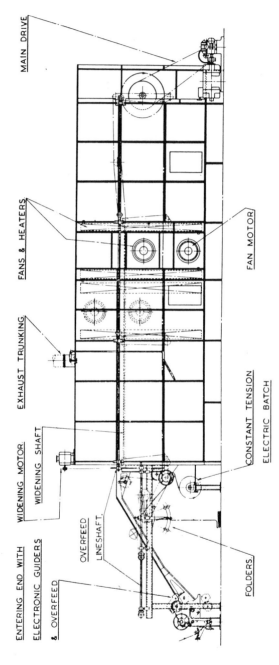

ENTERING END WITH ELECTRONIC GUIDERS & OVERFEED

WIDENING MOTOR

WIDENING SHAFT

OVERFEED LINESHAFT

EXHAUST TRUNKING

FANS & HEATERS

MAIN DRIVE

FAN MOTOR

CONSTANT TENSION

ELECTRIC BATCH

FOLDERS

Figure 4.59 Sectional view of double-layer hot air stenter in which the fabric makes a double passage through it to economise on floor space and reduce heat losses

Figure 4.60 Pair of fabric guiders operating to centralise a fabric passing forward to enter either a finishing machine or to be neatly rolled on a roller where it is essential that the fabric runs a straight course throughout. Each guider as shown consists of a pair of rollers aligned at an angle to the fabric which runs between them. A sensitive feeler or an electronic 'magic eye' detects when the fabric departs from its straight course and it then causes one or other of the pair of guiding rollers to rotate and (because of its inclination to the fabric length) it pulls the fabric sideways to correct its off-course running. As shown here, the right-hand guider is pulling the fabric to the right. (Courtesy Mather & Platt)

is led through the two-bowl impregnating or padding mangle for the purpose of incorporating in it some softening or other finishing ingredient. Thence it passes over the small number of drying cylinders to be half-dried. Then the fabric enters the stenter and is there stretched to width and set and dried at this width. All this can be done at rates exceeding 100 yd/min (90 m/min) with some types of fabric.

We have previously mentioned that rayon crêpe fabrics can be finished very satisfactorily in this machine. However, a modification of the clip chains is required for this purpose. From the nature of the clips it is seen that once the fabric selvedges have been gripped it is not possible to lengthen or shorten the fabric. Thus the stentering rate must be equal to that at which it leaves the stenter. This does not matter much with a plain cotton fabric but in the finishing of crêpe fabric it is usually necessary to bring each fabric to a specified length. It may therefore be necessary for the fabric to be led in at a rate which is different from that at which it leaves the stenter. Usually the fabric has to be led in faster than it comes out.

To enable this to be done plates carrying pins on which to impale the fabric selvedges are used instead of clips. Then the fabric can be fed in more quickly and impaled in a crinkled state on the pins by means of brushes, one on each side of the machine. As the fabric is carried forward it contracts in the drying and the selvedges thus straighten and lose their crinkle. The amount of contraction can be accurately controlled.

Mention should here be made of a peculiar modification of the stentering machine which makes it suitable for the finishing of cotton voile fabrics. These fabrics are required to be somewhat wiry and crisp in handle and the small mesh which characterises this type of fabric should be clear and well formed. This is attained by arranging that as the fabric is carried along the stenter frame the clip chains move slightly backward and forward with a jigging motion. This motion jigs the fabric and frees any of the interlacing threads that may be sticking to each other.

Stenters for flat knitted fabric In recent years stenters have been modified and improved so that these can now be used for straightening, drying and setting to width and length the various kinds of flat knitted fabrics. The trouble about wet flatknit fabric, or indeed any single thickness of knitted material, is that the selvedges curl up

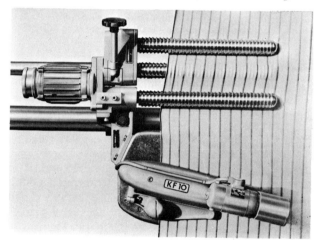

Figure 4.61 Guider/uncurler device for knitted fabric being stentered. Each selvedge becomes uncurled as it passes through a group of three spirally grooved rods. The fabric is also centred by each side of it passing through an inclined 2-roller device which, in response to fabric feelers, causes it to be drawn to one side or the other

tightly into rolls. In the ordinary way the clips of a stenter would fail to deal with these; stenter clips can only satisfactorily grip the flat clean-cut selvedges of a woven fabric. If the clips are replaced by pin plates then there is almost as great a difficulty in impaling the curled selvedges on the pins so that they will hold there. A further difficulty is that flat knitted fabric is so prone to distortion that it is hard to guide it uniformly into the stenter.

However, new devices now added to the entry end of the stenter enable these difficulties to be overcome. Firstly, pin plates are used instead of clips and a special device uncurls each selvedge just before these selvedges reach the pin plates. Then revolving brushes can at this instant impale the straightened selvedges on the pins. Automatic feelers are also placed, one on each side of the stenter machine, and these press against the sides of the fabric as it advances towards the uncurling device and the first pin plates. If the fabric is not moving exactly towards the middle of the stenter frame then one selvedge will touch a 'feeler' and this will instantly set in motion a device which temporarily moves the entering fabric to one side so that the fabric is then in the middle. If the fabric again moves to one side or other the 'feelers' react and the fabric is moved again in the

Figure 4.62 Another view of knitted fabric entering a stenter (Figure 4.61) through uncurling and centralising devices, and also a rotating circular brush for impaling the fabric selvedge on the pins. The rotation rate of this brush can be increased (as desired) to assist the lengthwise closing up of the fabric to ensure that it is shrink-resistant in subsequent washing

appropriate direction. Thus the stenter frame is automatically adjusted so that the fabric enters it centrally. Electronic guiders are now used instead of feelers to guide the fabric on to the pins. Now in wet processing it generally happens that knitted fabric is pulled in length considerably. This means that in finishing on the stenter a means must be available to close it up in length. This can be accomplished by means of a special device to feed the fabric on to the stenter frame more quickly or slowly as is desired.

Since the stenter easily allows the width of the fabric to be controlled we now see that by using one of these special pin plate stenters it is possible to dry and set the flat knitted fabric to any desired dimensions of length and width even though it differs widely from these just before it is brought to the stenter.

Combined pin-clips are now used to allow simple change from pin-plate stenter to a clip stenter as desired—it is no longer necessary to have separate chains (Figure 4.56).

Finishing of tubular knitted fabric Tubular knitted fabric cannot be finished on a pin plate or clip stenter for obvious reasons, but an alternative simple type of machine has hitherto been much used for such fabric. The machine consists of a central upright metal or composition tube standing on the floor. At the bottom is a flat board capable of rotation. Above the tube is a vertical stretching device which is adjustable to different widths, to suit the width required in the fabric being finished. This stretcher, which consists essentially of divergent metal rods, points towards and almost rests in, the nip of a small, light, two-bowl hard-rubber mangle or calender above it. (A mangle used in finishing is generally termed a *calender*.) Often this machine is placed so that the lower part is on one floor whilst the stretching part projects through to the floor above, where the calender is fixed.

In operating such a machine the damp tubular fabric is piled down around the bottom of the tube. The upper end of the fabric is drawn upwards so as to enclose the stretcher and then pass through the calender. Now hot air is blown up through the central tube, so that it balloons out the fabric as the air can only escape through the fabric, the top end being sealed by the nip of the calender. The calender draws the fabric through it and thus upwards from the vertical tube and over the stretcher. In this travel the fabric just becomes dry as it passes through the calender.

As the dry fabric comes out of the calender it can either be plaited down in folds on a table or it can be batched neatly on a roller. When so batched it is often useful to run a paper tape (marked in feet and yards) with the fabric so that its length can be

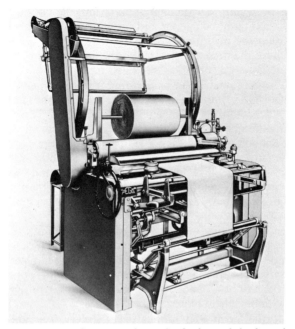

Figure 4.63 Medium-type calender for finishing tubular knitted fabric. The fabric is led first over an internal stretcher and then pressed between rollers heated by steam before being rolled. (Courtesy Samuel Pegg Ltd)

seen without further measurement. At this stage the fabric usually has to be calendered again to improve its consumer appeal. For this a light, medium or heavy type of calender will be used, according to the nature of the fabric. In this calendering the fabric is lightly steamed to increase the fibre plasticity. This assists its stretching to the desired width and also its setting with a smooth, soft handle in passing between the hot calender rollers. These characteristics are fast to ordinary wear and tear but not to wetting. The steel calender rollers are hollow so that they can be heated by steam or electricity. They are covered with two or three layers of fabric to make them more resilient and so that they do not glaze the fabric surface.

Mechanical softening of textile materials

With the exception of garments, men's socks and ladies' hose, which will be dealt with later, we have now considered the general processes of drying, straightening and setting the various kinds of fabrics at their required lengths and widths. At this stage it generally happens that the textile material is somewhat firm in handle and may not have the necessary lustre. So the next finishing operations must be designed to correct these faults.

Before leaving this aspect of finishing it may be mentioned that woven wool fabrics are dried and brought to the specified dimensions in much the same manner as cotton fabrics. There is, however, this difference. Wool fabrics are easily spoilt in handle by drying too much or at too high a temperature, so the use of drying cylinders is not so much favoured. Good use is made of stentering machinery, and curiously enough in the wool finishing works this machine is usually referred to as a *tenter* rather than a stenter. Another practice favoured by wool fabric finishers is that of using tenter machines in which the fabric goes backwards and forwards at different levels (Fig. 4.59) two or three times within the hot-air chamber instead of straight through once as previously described for cotton and rayon fabrics. Otherwise the clips and pin plates used to grip the selvedges are made on the same principles. Wool materials are left with the best handle when dried by hot air rather than by contact with hot metal plates.

Most textile yarns and fabrics become very firm and boardy when dried under tension. It seems that under these conditions the fibres and threads are given a special rigidity. This is easily broken down by jigging the fabric in the hands or by bending it one way and then another, but in a finishing works the most convenient way is to run the fabric through a calender. This can bring to the fabric a surprising degree of softness. At the same time it flattens the fabric surface and so makes it more lustrous, thus adding considerably to the appearance of the fabric.

Calenders for cotton fabrics   The calenders used for cotton and rayon fabrics are fairly heavy in construction. The bowls may be from 1 to 2 ft (0·3 to 0·6 m) in diameter and a great deal of pressure can be applied to these either by way of screw-down devices, by weighted levers, or by hydraulic pressure (Figure 4.64).

In its simplest form a calender will consist of two compressed paper or cotton bowls running against each other; it thus resembles a big mangle. Other calenders of a simple type will have a brass or copper bowl running against a paper or cotton bowl, but there are other much more elaborate calenders where several bowls of different composition are mounted above each other. Thus, in one run through the machine the fabric can receive several nips and so be flattened more and more.

Figure 4.64 Three-bowl, woven fabric calendering machine

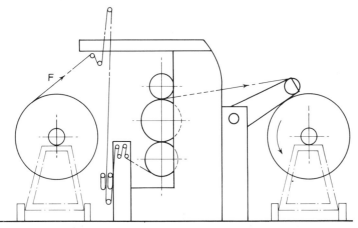

Figure 4.65 Path F *of fabric travel in the 3-bowl woven fabric calendering machine*

By simple calendering through these machines a kind of lustrous skin is formed on the fabric surface and this, coupled with the softening, gives cotton fabric an attractive character. Rayon fabrics are not calendered so heavily since they are less robust than cotton or linen, but silk fabrics often have a fairly heavy calendering.

There is one point about calendering which is most important. Fabrics must never be run between two metal bowls. If they are, there is grave danger of damaging the fabric by cutting or chafing it. If one bowl is softer than the other such damage seldom takes place. Hence the practice of running a metal bowl against one of paper or cotton.

Finishing to give lustre

Friction calendering Even by multiple calendering treatments as indicated above, it is sometimes not possible to secure the high degree of lustre required. In this case friction calendering must be used. In this method one of the bowls is driven at a speed greater than that of the other. The fabric between the two bowls is thus subject to a friction or polishing action. If wax or fatty substances are also present in the fabric it is easy to understand that the fabric surface becomes polished and lustrous to a very high degree. This is all the more accentuated if one of the bowls of the calender is of steel and is heated internally by steam or gas.

The Schreiner calender Whilst quite a high lustre can be secured by friction calendering there is another kind of calender which can achieve even more. It is the Schreiner calender. This machine has an interesting history.

It is primarily used for giving high lustre to cotton fabrics especially those of the sateen type. If we look at the surface of a piece of cotton sateen fabric under a fair degree of magnification we see that this surface is broken up in two ways. Firstly, there are the warp and weft threads interlacing each other. Secondly, there are the fibres twisted about each other in each warp or weft thread. This twist is shown by the fibres being aligned more or less parallel to each other but inclined at a small angle to the length direction of each thread.

Now, for a fabric or other surface to have a high lustre it must reflect most of the light incident upon it and this requires that the surface shall, in general, be very smooth. Any ridge or valley will cause a proportion of the light to be reflected into itself and be absorbed; this means a loss of brightness or lustre. It is evident then that for a fabric to be given lustre its surface must be flattened by high pressure calendering. In practice it is not found possible in this way to flatten it sufficiently to eliminate entirely the tiny ridges and valleys caused by the interlaced threads and fibres.

Some sixty years ago Schreiner hit on the idea of impressing all over the fabric surface a series of closely spaced parallel lines which would smother the irregularities caused by the interlacing of the threads and the twist of the fibres. Thus the fabric surface would be made smooth, except for the very evenly spaced parallel ridges and valleys produced by the impression of the lines. It was found that when the lines were spaced about 118 per cm then in spite of the ridges and valleys the fabric acquired a very high lustre when viewed from certain directions. The impressed fabric surface thus acquired a very attractive appearance. Plain cotton fabric finished in this manner had a handsome, lustrous, silk-like appearance and it soon became popular.

This method was patented but most unfortunately for the inventor he did not clearly specify that a large number of lines must be impressed. In this way he lost his patent rights and was the poorer by some hundreds of thousands of pounds, for he would undoubtedly have collected huge royalties during the life of the

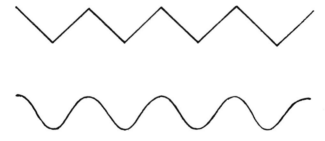

Figure 4.66 Section through two types of Schreiner lines impressed on fabric to impart high lustre

patent, so extensively has this method of finishing been carried out.

The method of schreinering fabric is comparatively simple, but a special machine has to be used. It is really a two-bowl calender, usually with a cotton or paper bowl as the softer one. The other bowl is of steel and is hollow so that it can be heated internally by means of gas, hot oil or electricity. The temperature of this bowl must be kept high and steam heating is not sufficiently powerful. The steel bowl has a highly polished surface engraved with closely spaced parallel lines. Fabric is run between the two bowls slowly and is thereby impressed with the lines, high pressure being obtained by hydraulic means. It goes in comparatively dull and comes out on the other side with a deep rich lustre.

Such an impression requires that the fabric shall be fairly plastic, and this is ensured by arranging that it is just slightly damp. Under the combined influence of heat, moisture and pressure most textile fibres become plastic or mouldable. The moisture is particularly necessary. Dry fabric fails to retain the impression upon it even if the pressure is very high. In ordinary calendering treatments it is better for the fabric to be slightly damper than usual.

Damping machines In order that fabrics may be given this necessary amount of moisture the finisher uses special machines. They are often termed *spray dampers*, for the earliest types created a kind of water spray through which the fabric was rapidly drawn. Nowadays the machines have been improved so that the water is practically in an atomised condition, like a fine penetrating mist.

One type of damping machine comprises a rectangular wood box at least as long as the widest fabric with which it might be called upon to deal. In the bottom of this box is a shallow layer of water and above this, but just dipping into it, is a rapidly rotating roller having projecting brushes, metal fins or spikes. As these briskly strike the surface of the water they make a spray or mist which fills the box. At the same time the fabric is led rapidly over a fairly wide slot in the top of the box so that it can pick up the finely divided particles of spray. The fabric is immediately wound on to rollers and allowed to lie for 2 or 3 h, or better still overnight, so that the moisture it has picked up can spread evenly throughout the fabric.

In some damping machines the water mist is created by forcing the water through a series of very fine jets. But, however the machine is devised, it is important that no large water drops fall on to the fabric, since these would not spread sufficiently and would eventually lead to irregular lustrous patches.

Other calendering and softening machines It is not possible to pass flat knitted fabric or even tubular knitted fabric through heavy calenders for it so easily becomes distorted. If the fabric goes through the calender nip with a crease then the pressure is usually so high that the fabric becomes cut at the crease.

Mention has already been made of a small steam-heated two-bowl calender for tubular knitted fabric, but there are two alternative machines of heavier construction. In both of them the fabric is not pressed between two bowls but is carried around a highly polished steel or sheet metal cylinder, being pressed in contact with this by a thick felted blanket. If the fabric is somewhat damp, drying

Figure 4.67 Medium type horizontal calender for finishing all types of circular fabrics, without the defect of bowing the fabric courses. (Courtesy Samuel Pegg Ltd)

in contact with a smooth metal surface, even under moderate pressure, is sufficient to give it a fair degree of lustre. With machines of this kind it is usual to have two cylinders so that in the one run both sides of the fabric can in turn be brought against a metal surface and be made equally lustrous.

Knitted fabric in tubular form can also be comparably calendered in a two-bed or hose press using a board of suitable width within the fabric as an internal stretcher (Figure 4.69).

Softening after lustring   In the calendering of all kinds of fabric the lustre can be made greater if the fabric is damped before passing through the machine. The best conditions for securing maximum lustre are to have the fabric wet and then press it hard until dry against a hot, polished metal surface. If it can at the same time be moved over the metal surface it will become all the more polished and lustrous.

There is one big snag in this finishing process. The damper the fabric and the hotter the metal bowl the more firm and boardy does the handle of the fabric become, so the finisher generally has to compromise. He really wants maximum softness and maximum lustre. Actually he secures high lustre with a moderate degree of firmness. In operating along these lines he sometimes finds it a good method to produce a high lustre with a fair degree of firmness of handle, and then pass the fabric through a machine to break down the firmness, whilst at the same time only destroying a small part of the increased lustre. There are two main types of machine for this mellowing of a firm handling fabric.

In one there are upper and lower rows of small-diameter rollers, each being covered by numerous small round-headed studs. The distance between these rollers can be varied as desired. The fabric is led through this machine so that it lightly touches the studded rollers and drags them round. In this way the fabric surface is lightly disturbed and it becomes much softer. The second machine is similar in principle, but the rollers are spirally grooved and are arranged around a cylindrical framework. As the fabric passes over these it suffers the same kind of breakdown and acquires increased softness.

All these considerations indicate how skilled is the art of the textile finisher. He has to use considerable judgment to secure the effects desired and it is surprising how high is the standard insisted upon.

Beetling of linen fabrics

The finishing of linen materials follows on much the same lines as for cotton goods, but linen materials are never brushed or raised to produce a hairy surface. It is to be remembered that the beauty of linen lies mostly in its high lustre and smooth thready appearance. So the finishing processes adopted for this type of textile material must in general have the effect of accentuating these properties.

There is one finishing treatment which is peculiar to the linen trade and this is known as *beetling*. It is carried out with a machine which is simple in construction but operates with considerable noise.

A beetling machine consists essentially of about forty heavy wooden fallers arranged closely side by side in a row and situated above a horizontal roller on which are wound many layers of the fabric to be beetled. Four or five lots of fabric are wound on this roller side by side so as to fill the roller from end to end, this being about 14 ft (4·3 m) long.

A mechanism lifts the fallers in rapid succession and then allows them to fall in the same order under the action of gravity and pound the roll of fabric beneath. At the same time the roller carrying the fabric rotates slowly and moves regularly a small amount from side to side so that the fallers do not continuously pound or strike the same part of the fabric throughout the whole of the beetling operation, which may take several hours.

After a suitable period the fabrics are unwound on to a second similar roller and this is placed beneath the fallers for a further period of beetling. This change ensures that the inside layers of fabric are now on the outside of the roll and thus the beetling treatment over the whole of the fabric is equalised. At the end of the operation the fabric has acquired a very lustrous appearance which most people associate with linen goods.

Instead of simple wood fallers it is possible to use fallers having a spring foot covered with leather; these fallers do not fall by gravity but pound the fabric as the result of spring action governed by cams. With a machine of this kind each faller will strike the fabric at the rate of up to 400 blows per minute.

Finishing of wool materials

In the finishing of wool, a number of treatments are employed which are used for cotton or other materials. Thus it has already been mentioned that cotton fabrics before being printed are usually sheared so as to give them a sheer surface free from projecting fibre ends which would interfere with the production of clear-cut colouring. On the same lines, many types of wool fabric, especially worsted materials which must have a sheer surface, are sheared. The machines used are very similar in construction to the ordinary grass lawn mower in which a rotating multi-blade cutter is employed. A number of these cutters are used in the same machine.

Wool fabrics are not calendered in the manner described for cotton and linen materials. When it is required to produce a smooth surface in wool fabric it is preferable to subject it to a hot pressing. One of the oldest and most favoured methods for achieving this utilises a hydraulic press. Suitably damp wool fabric is folded with layers of fibre-board between successive folds and the whole stacked vertically within a powerful press. When this is filled the press is closed and the lower end rises until a high pressure is exerted on the folded fabric, heat being simultaneously applied. Under these conditions a very pleasing surface is given to the fabric and it is easy to see the difference between the fabric before and after it has been pressed.

Naturally, this type of hot pressing takes considerable time and labour, so alternative machines have been devised in which the fabric can be handled more expeditiously and automatically. One favoured machine, known as a *rotary* pressing machine, has a central hot metal cylinder on which can be pressed upper and lower metal plates, curved so that they are concentric with the cylinder and can be pressed against it to any desired extent. The wool fabric is continuously led through this machine between the cylinder and the metal plates. The pressure on the fabric can be adjusted so that it can just be drawn through to receive a light hot ironing action which smooths its surface and gives it a pleasing finish.

Another pressing machine is on the same lines as a three-bed hose press often used for finishing ladies' hose. It consists essentially of lower, middle and upper hollow metal blocks which can be heated by means of steam from within. These blocks can be separated to allow several folds of the wool fabric, with fibre-boards between the folds, to be inserted between the lower and middle blocks and also between the middle and upper blocks. Then they can be brought together under suitable pressure and the folds of fabric are thus given a hot press finish.

A popular method for giving wool fabric a lustrous soft finish is that known as *decatising*. In this the fabric is wound around a perforated hollow metal cylinder the surface of which has been covered with a few layers of wrapping fabric to protect the wool fabric from stains. Steam is then blown from inside the cylinder through the perforations and thence through the fabric. Altern-

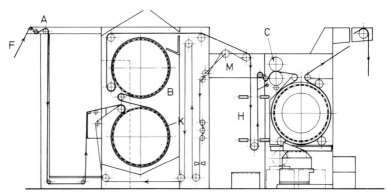

Figure 4.68 Continuous type of wool fabric decatising machine in which the fabric passing over perforated drums has steam blown through it from inside-to-outside and/or vice-versa. This gives it a pleasing, soft, smooth, lustrous surface and leaves it with low residual shrinkage of ½ and 1 per cent, in warp and weft respectively. In operation, fabric F is led over tensioning and straightening rollers A and is held against rollers at B by a thick endless blanket K which moves in a closed circuit. The fabric leaving the blanket at M is able to relax while passing over freely rising and falling rollers at H. It is then pressed between the hot rollers C. It finally passes over the third perforated drum so that it can be 'shock-cooled' by air sucked through the fabric and into the drum. (Courtesy V.E.B. Textilmaschinen)

atively the cylinder, with many layers of wool fabric wound upon it, can be enclosed within a chamber, the steam being blown in the opposite direction, that is, from the outside through the fabric and perforations and into the cylinder. This steaming treatment softens the wool fibres and, owing to the moderate pressure resulting from the tightness of winding, sets them to give smoothness and lustre.

A continuous decatising machine is shown in Figure 4.68.

Raising processes

A finishing treatment which is frequently applied to both woven and knitted cotton and wool fabrics is that known as *raising* or *brushing*. The object of this treatment is to change the smooth surface of the fabric to give a degree of hairiness which makes the fabric very soft to handle and at the same time warmer. Thus, wool blankets are raised, whilst cotton sheets are subjected to the same treatment with the object of making them warmer and more wool-like. Wool shawls and gloves are frequently brushed lightly while stockings and socks are brushed on the inside. Raised and brushed fabrics are much cosier to wear.

This finishing operation is carried out with a number of different machines, some quite simple whilst others are much more complex. In all cases the hairiness is obtained by scratching or plucking the fabric surface by application of fine points. Thus a very simple brushing machine may consist of a number of rollers over which are threaded a number of dry and hardened teazles. These rollers are caused to rotate whilst the fabric is drawn over them so that the sharp points of the teazles just enter the fabric surface and pluck out the fibre ends.

In the more complex machines the rollers are covered with wire carding cloth so that the points of the wire just enter into the fabric surface and have the same effect as the teazle points mentioned above. The usual arrangement of these rollers is on a cylindrical framework so that the whole can be rotated whilst the individual rollers are also kept revolving. The fabric is drawn over the rollers while being lightly pressed against them. Great care has to be taken in this adjustment for if the fabric is brought into too close contact then it will at once be torn or holed.

In these raising or brushing operations the degree of hairiness must be adjusted to suit the structure of the fabric or vice-versa. It is important that the loose projecting fibres be sufficiently anchored in the fabric that they do not work out in wear or loosen in washing. Neither must the effect be so drastic as to weaken the fabric.

Today, nylon bed sheets are very popular because they can so easily be washed and then rapidly dried for re-use. These are often raised to make them feel warmer and softer. These sheets are usually warp knitted. Non-woven cellulose fibre bed sheets are now being made.

Suede gloves for men are made with knitted cotton fabric which has been raised in a special manner so that the hairiness is short and very close. It is usual to produce this special raising by the use of rollers covered with emery powder rather than with teazles or wire carding cloth, both of which are too drastic in their action. Cotton flannelette is a well-known type of cotton fabric which is finished by a raising treatment.

Finishing of hose

We can now return to the finishing of knitted garments, ladies' hose and men's socks. A formerly widely employed and useful machine for this type of finishing, which is really just hot pressing, is the steam press. There are two-bed and three-bed presses and these differ in allowing either one or two layers of garments or hose to be pressed at the same time. Such presses are mentioned here from an historical interest viewpoint since they have now been re-placed by radically different machines which allow continuous processing.

As the name implies, a two-bed press consists of two beds, an upper and a lower one, which can be separated far enough to allow garments or hose on shapes to be inserted between them. The two beds can be brought together under pressure provided by a screw-down device or by a lever. The garments are thus set to shape. Each bed is rectangular and made of cast steel with the pressure surface highly polished. It is also hollow so that it can be heated internally by steam under a small pressure and the temperature of the metal surface brought to just above the temperature of boiling water (Figure 4.69).

In the three-bed press it is obvious that the same principles apply. In this case, however, a layer of garments or hose can be placed between the lower and middle beds whilst a second layer can be placed between the middle and top beds. This is an aid to more rapid production.

The shapes used for finishing garments, hose or socks may be made of wood or stainless sheet metal. For many years, and even today, wood shapes have been preferred, possibly because wood is more resilient and accommodating to slight inequalities of pressure.

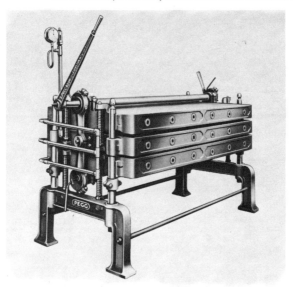

Figure 4.69 A three-bed steam press for finishing knitted fabrics and hose. (Courtesy Samuel Pegg & Son Ltd)

The pressing usually occupies 1 to 2 min, just sufficient to dry out the slight dampness of the garments or hose being finished. As soon as the garments are withdrawn from the press the shapes are taken out of them. The garments are then folded and stacked whilst the shapes are placed to cool ready for further use. A modern finishing method for dyeing and boarding (finishing to shape) uses the equipment shown in Figure 4.30.

Continuous hose finishing machines   It is now general practice to finish ladies' hose in continuous finishing machines. These machines have developed from the idea that the hose can be drawn on to shapes which can then travel through a chamber in which hot air circulates, and so be dried. All this is arranged in an elliptical route so that the shapes eventually return to the point where the hose were first drawn over them. Thus, on stripping off the finished hose, the shapes are ready for fresh hose to be drawn on to them. Two different types of machine are now available (Figure 4.70).

In one machine the metal shapes stand vertically attached and spaced out along a continuously moving endless chain. At the entry end operatives draw the slightly damp hose over the shapes which then travel forward through a damping chamber where they are thoroughly wetted. Almost simultaneously the shapes pass between

Figure 4.70 Automatic machine for finishing men's and children's socks. (Courtesy Samuel Pegg & Son Ltd)

pairs of revolving rubber rollers which smooth down the stocking on each shape.

After this smoothing treatment, which makes a good deal of difference to the final appearance, the shapes go forward to pass through the hot air chamber where they are dried. Thence they go through a cooling and conditioning chamber where they encounter cool, slightly damp air for the purpose of mellowing or conditioning the hose. The shapes now travel to the starting point where the hose are stripped off and fresh hose drawn over the shapes to take the same course as just described.

In the second machine there are a large number of regularly spaced rotating horizontal rollers which serve to carry forward the shapes placed flat upon them. The usual method of running this machine is as follows: firstly, the hose are drawn over the shapes, these being quite separate and not attached to any chain or belt. As soon as the shape is thus covered with a stocking it is laid flat on the first roller. This propels it forward to the next roller and so on right through the machine until it reaches the far end. In its course of travel the shape and the stocking are first wetted and then mangled by passage between two pairs of soft rubber rollers lightly pressed together. This mangling also serves to smooth out the stocking on each shape which travels toe first. The shapes then travel on the tops of the single horizontal rotating rollers and so pass through a chamber where hot air is circulated. This dries the hose. Then follows a passage through a cooling and conditioning chamber. Finally, the shapes reach the far end of the machine where the hose can be stripped off. The shapes are then allowed to slide down an inclined plane to the other end, ready for use with fresh hose.

It is to be noted particularly that neither of these machines applies pressure to the hose in the course of the finishing. The hose are free to contract in length and, in fact, this does occur, as can be seen by looking at the shapes during the drying stage. The tops of the hose gradually slide an inch or so down the shapes. In the early days of these machines this was brought forward as a defect of the method, for it will be understood that in the steam press method the hose cannot contract in length as they are held by pressure between the beds. However, the contraction seems to improve the appearance of the finished hose.

After finishing, fabrics of all kinds are taken to the making-up room for examination, packing and despatch according to requirements.

Some recent developments in dyeing and finishing

The open-soaper used for washing cotton fabric after bleaching and mercerising just referred to, has, over many years, been the most favoured machine for washing fabrics in open width following say dyeing and printing. It usually comprises up to five separate units, each of which is a rectangular vessel provided with an upper and lower series of rotating guiding rollers fairly closely spaced so that the fabric can pass forward over and under them to prolong its passage. Each unit will contain a solution of soap or some other detergent, an acid or an alkali, etc. but usually the last two units contain running hot or cold water to effect a final rinsing of the fabric. The mangles between each pair of units serve to pull the fabric forward and prevent it carrying over too much of the liquor in one unit into the next; devices can be present to control the passage of the fabric so that undue slackness or tension at any point is automatically corrected. The top of each unit is generally left uncovered but if a high temperature of the liquor is required then the unit must be fully enclosed to withstand the pressure involved.

In recent years much attention has been given to improving the efficiency of the washing. In older machines the fabric simply moved up and down thus mainly sliding through the liquor. Obviously such washing could be made more efficient by employing arrangements to cause the liquor to impinge perpendicularly to the fabric and thus be forced through it. All kinds of devices have been used to produce this effect. One method involves fixed vertical corrugated sheets of stainless steel or other corrosion-resistant material within each fold of the moving fabric to deflect the liquor to the fabric. In other devices the liquor is caused to have a swirling motion. Reference has already been made (Figure 4.40) to the use of rapidly rotating brushes to assist the removal of tenaciously adhering printing paste residues from printed fabric that is being washed in an open soaper.

Quite another type of aid to more efficient washing is to be found in a suction washing machine of Sir James Farmer Norton & Co Ltd. It is recalled that in the first type of mercerising machine described, the caustic soda being washed out of the fabric by water sprays above the stenter on which the fabric is passing, is drawn through the fabric into underneath troughs by suction. This manner of washing the fabric is very efficient, so the idea has been adapted for similarly sucking water through the fabric while passing through one or more washing units of an open soaper (Figure 4.71).

In such a modified open-soaper the washing unit (filled with water or other liquor) is divided into four similar and separate compartments as shown in Figure 4.71. Within each compartment an endless conveyor belt (4) consisting of nylon net fabric, moves in a closed circuit around guiding rollers and it serves to support the fabric being washed. This fabric F enters the machine passing over tension rails (1) and (2) to remove creases from it, before passing over a guiding roller to move downwards supported by the net fabric and then upwards through the water. In this path it presses against the suction boxes (3) and finally leaves the nylon conveyor belt to pass through the mangle (5) into the next compartment, and so on to the fourth mangle. It then leaves the washing unit to be plaited (folded) on a table. As the fabric passes over each suction box the surrounding washing liquor is sucked through it and returned via a small water-type vacuum pump thus effecting a very efficient washing of the fabric.

To prevent the fabric running too slack or tight the electric motors driving the mangles (which draw the fabric forward) are electronically synchronised. This same control can be used to run the fabric as free from stretch as is desirable and this makes it very suitable for washing all types of fabric including knitted fabrics.

In washing certain types of fabric, as for example, knitted fabric, it can be important that it should be subject to a minimum of

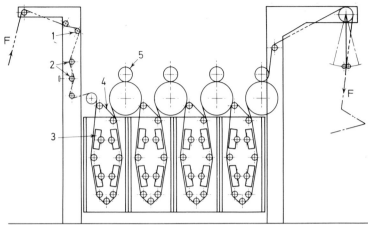

Figure 4.71 Knitted fabric suction washing machine that avoids excessive stretching. The fabric F passes over the tensioning rails (1) and straightening rollers (2) to enter the first stainless steel washing unit containing a washing liquor. Within this unit an endless Terylene mesh conveyor belt (4) moves around a series of guiding rollers to pass over four suction boxes (3) positioned at spaced intervals and it serves to carry and support the fabric over these boxes. At each box the washing liquor is continuously sucked through the fabric and the supporting Terylene mesh fabric into the box, and then out into the main bulk of liquor. Thereafter the fabric passes into the next washing unit via a mangle (5) positioned between the two units. At each box the fabric suffers flexing and this assists the loosening of any adhering solid impurities (e.g. printing paste residues) and their more complete removal by the washing liquor drawn through it. (Courtesy Sir James Farmer Norton & Co Ltd)

length stretch. If the washing machine comprises a number of rollers over which the fabric is drawn while passing through the washing liquor then it can be difficult to have the fabric tension-free, for then it tends to run with creases. A method for washing under these circumstances is to run the fabric around large per-forated drums instead of rollers using suction to draw the washing liquor through the fabric and drum perforations into the drum and then back from where it was drawn. In this manner the fabric is held against the drum by the liquor passing through it and can therefore be completely free from length tension as it is carried forward within an open-soaper type of washing machine.

This idea of treating fabric while held free from length tension to the surface of a perforated rotating drum has been widely exploited by the Fleissner Co, e.g. it is much used in the drying of fabric by hot air drawn through it (Figure 4.25).

A new type of machine is now available for securing the maximum efficiency in padding a fabric with a dye or other liquor (Figure 4.72), and is an alternative to the usual padding mangle in which the fabric is led one or more times through a lower trough containing the liquor, followed by a hard 'nip' in a mangle positioned above.

This machine allows padding to be carried out on the basis of the fact that the presence of air spaces between the threads and fibres of a fabric can be a serious hindrance to complete and rapid impregnation by the liquor, and that this hindrance can be completely nullified by first sucking out the occluded air under vacuum conditions and then immediately entering the fabric into the liquor so that it rushes in to occupy the spaces formerly occupied by air.

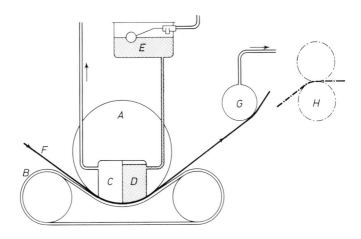

Figure 4.72 Principle of the Farmer Norton suction fabric impregnating unit. (Courtesy Sir James Farmer Norton & Sons Ltd)

The fabric F is led between a thick flexible endless rubber belt B moving in a closed circuit in which it passes for a short distance around a perforated cylinder A and presses upwards against it. Inside the cylinder is a section C connected to a vacuum pump and adjacent to this is a completely separate section D provided with a constant supply of, say, dye liquor E. Any perforations in the cylinder not covered by the passing fabric and rubber belt are suitably blanked off in order to prevent escape of vacuum or dye liquor.

In operation, the fabric F carried on the belt B is exhausted of occluded air while passing under section C so that it can all the more rapidly and completely suck in the dye liquor while passing under the adjacent section D. As the impregnated fabric then passes forward saturated with dye liquor it can pass over a suction slot G or through a mangle H to remove any excess of loosely held liquor.

Jet dyeing machines

Recent developments concerned with improving dyeing machines are to ensure not only level application of the dye but also to avoid wear and tear on the material being dyed. In the case of fragile or easily distortable knitted fabrics it is important also to avoid excessive length tension and rubbing. The high pressure of one layer of fabric upon another (especially with thermoplastic acrylic fibres) in a hot dye liquor can sometimes cause a peculiar lustrous *moiré* pattern which is objectionable and difficult to remove. Hence the demand for a type of dyeing machine which can avoid these faults.

However free and easy running the guiding rollers in an ordinary dyeing machine may be there is always a drag on the moving fabric which is undesirable, and it is to avoid this drag that a so-called *jet*-type dyeing machine has now become available in which all guiding rollers are dispensed with and the movement of the fabric is maintained by a specially controlled flow of the dye liquor. The fabric is fully immersed in rope form in the flowing liquor, which both propels it and protects it from rubbing. Sometimes air is injected into the liquor to keep the fabric in a loose buoyant state.

Reference is made here to a few types of these special dyeing machines as devised in various countries such as the U.K., America, Italy, Germany and Japan (see Figures 4.73 to 4.79).

The German Menzel machine (Figure 4.73) is about 20 ft (6 m) high and totally enclosed so that it will withstand the high pressure, say up to 40 lbf/in^2 (280 kN/m^2) required to obtain dyeing temperatures above 100°C. In operation the fabric F is continuously drawn upwards out of the dye liquor by the small rotating winch W positioned so that the fabric will drop into the tube T where it meets an inflow of dye liquor being maintained in continuous circulation by a small pump P. Thus the fabric passes down the tube to enter the main dyeing part of the machine and then rises steadily to pass again over the winch and repeat its cycle of movement. Air is injected into the dye liquor to form bubbles within the folds of the fabric to keep these open and not liable to become permanently fixed as creases, while facilitating easy circulation of the dye liquor within it and promoting uniform level dyeing.

Sometimes in a machine of this type excessive objectionable foaming of the dye liquor occurs and interferes with the dye liquor circulation. The movement of the fabric is so gentle in this machine that foaming is not experienced.

An improved jet dyeing machine of the Gaston County Dyeing Machine Co in America (Figure 4.74) has several novel features making it very suitable for the dyeing of many types of fabric but

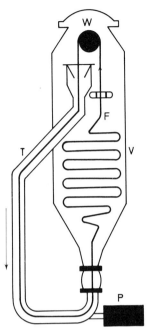

Figure 4.73 High-temperature jet dyeing machine with large diameter vertical accumulator vessel V that uses a winch W to raise the fabric F to the top. Maximum dyeing temperature is 145°C and the usual cloth capacity is approx. 100 kg, but a unit with a capacity of approx. 200 kg is available. (Courtesy K. Menzel Maschinen-fabrik)

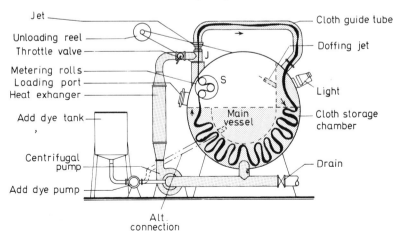

Figure 4.74 Improved jet dyeing machine suitable for the dyeing of many types of fabric, but particularly cotton. (Courtesy Gaston County Dyeing Machine Co)

especially cotton, cotton-polyester and acrylic fibre knitted fabric. It also allows the vat dyeing of cotton fabric. Throughout the dyeing period, which may occupy 3 h, the fabric is exposed in rope form to a highly turbulent dye liquor conducive to producing uniform, well penetrated dyeing, and is moved almost entirely by the flow of dye liquor which is at a high rate such that it can be entirely circulated within 30 s.

The machine is a long, horizontally mounted cylinder within which the fabric is continuously circulated within a closed cycle at a rate which is governed by the 3-roller driven device on one side of the cylinder at S through which the fabric passes. On leaving S and moving upwards it is caught up in a rapid dye liquor flow through the Venturi jet J (see Figure 4.74) coming from the pump and heat exchanger, so as to swirl it through the upper outside tube to the other side of the machine and fall down into the dye liquor in the form of loose folds while being sprayed with dye liquor issuing from the doffer jet at D. These folds move along the bottom of the machine to be drawn again through S and then follow the course just mentioned. It can be readily understood how, by always moving in a dye liquor flow, the fabric escapes any excessive tension and rubbing while acquiring a level colouring.

It is possible for excessive foaming to occur if impurities within the fabric lower the surface tension of the dye liquor; anti-foam agents may then be necessary.

Figure 4.75 shows a Pegg-Jet dyeing machine especially suitable for dyeing polyester warp- or weft-knitted fabrics in rope form at temperatures up to 140°C, while substantially free from length tension and protected against damage by rubbing.

The essential feature of this machine is that the fabric F initially piled in closed vessel D is continuously drawn upwards by the dye liquor coming from heat-exchanger A to pass through the jet chamber B. Thus this liquor and the fabric travel forward together (propelled by the moving liquor) through C to enter the lower part of vessel D. Here separation occurs to allow the fabric to pass upwards into jet chamber B while the liquor passes to pump P which again forces it upwards through heat-exchanger A to maintain a controlled temperature and flow to jet chamber B for recirculation. Throughout the dyeing process the fabric is propelled forward by the rapidly moving dye liquor, thus ensuring uniform dyeing and avoidance of fabric tension and friction.

Figure 4.76 shows a Platt-Longclose Ventura jet-dyeing machine for dyeing woven and knitted fabrics up to 140°C. The fabric is fully submerged in the dye liquor and moves through the tubular dyeing vessel V as an endless (ends sewn together) loop and in rope form. It passes through the venturi jet J, then through tube R lying close to the bottom of V and then returns in a loosely folded state under moderate compression within the upper part of V. It then

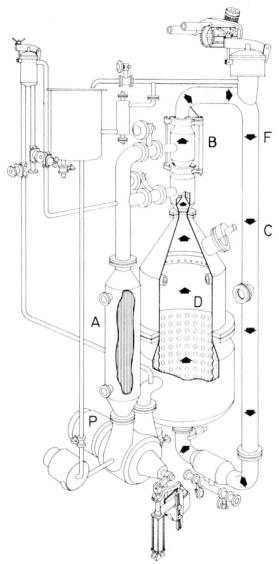

Figure 4.75 Pegg-Jet dyeing machine. (Courtesy Samuel Pegg & Son Ltd)

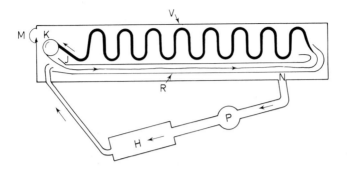

Figure 4.76 Ventura jet dyeing machine for dyeing woven and knitted fabrics in rope form at up to 140°C. (Courtesy Platt-Longclose)

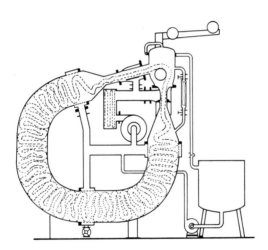

Figure 4.77 Jet and Twin-jet dyeing machine. A ring-shaped tube is arranged vertically and a tank containing a free running winch is located in the upper part of the ring. The machine operates at temperatures up to 140°C and is completely filled with liquor during dyeing; the liquor to goods ratio is quoted as 12:1. The liquor circulation is controlled through an adjustable valve in the suction return system. A second jet nozzle directed in the opposite direction from the liquor flow acts on the fabric as it leaves the lower reserve section of the tubing to provide a balancing and opening action. The rate of fabric movement is infinitely variable within the range 50–150 m/min. (Courtesy H. Krantz Maschinenfabrik)

354

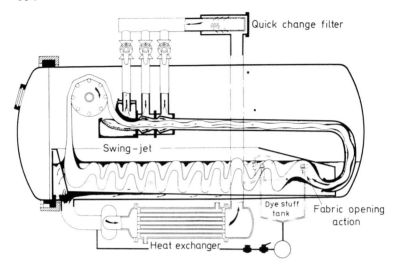

Figure 4.78 Japanese Uni-Ace Swing jet dyeing machine. The fabric is dyed within a horizontal, cylindrical, closed, dye-liquor containing vessel and is propelled by this moving liquor entering with it through the triple side-by-side jets. The fabric's forward movement is further assisted by a rotating winch positioned near the jets. A heat exchanger provided allows dyeing to be carried out at up to 140°C. (Courtesy Nippon Dyeing Machine Manufacturing Co Ltd)

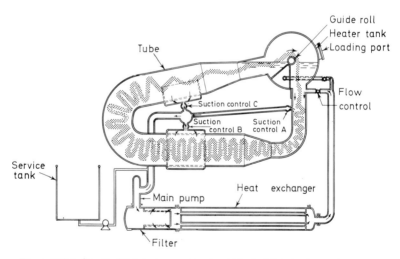

Figure 4.79 Italian Omli jet dyeing machine in which the fabric is always almost completely submerged within the flowing dye liquor which can be suitably heated while propelling the fabric forward. No winch is employed (only a guiding roller). (Courtesy OMLI)

drops over roller K into the venturi jet J and again passes along R while being pushed forward by the dye liquor to pass through the cycle repeatedly. The dye liquor circulation is sustained by withdrawing it from V at point N so that it passes through the pump P and then the heat exchanger H to boost it to the desired temperature before again reaching the venturi jet J.

Level dyeing is ensured since the fabric is continuously opened out in different directions by the moving liquor and while passing through · the venturi jet. Such opening out and avoidance of friction between the moving fabric and the surface of the machine with which it comes in contact is assisted by admitting a controlled amount of air near to the roller K so that it enters into the folds of the fabric.

Full submergence of the venturi jet within the dye liquor assists in preventing excessive foaming. The slope of dye vessel V beneficially influences the behaviour of the dyeing. The dyed fabric is withdrawn through the closable opening M.

Use of organic solvents in textile dyeing and finishing

Another recent development in dyeing and finishing is the use of organic solvents to replace water. It is anticipated that by such an innovation, processes could be accelerated and made more efficient and the necessary plant could occupy less space (a valuable feature in a textile works). There is also the incentive that by use of organic solvents it might be possible to solve pollution problems (now a matter of urgent consideration at all textile dyeing and finishing works), for in the recovery of the solvent, which would always be necessary to avoid costly losses, the residual impurities could be more easily collected and disposed of. Another consideration stems from the increasing scarcity and higher cost of water supplies.

Features which have to be considered in choosing and using organic solvents in textile processing include:
(1) fire risks; (2) volatility (the solvent must not be too volatile and yet it must be sufficiently volatile to allow its recovery by a distillation process); (3) cost of the solvent; (4) availability of enclosed plant to allow satisfactory processing—for example modern dry cleaning machines; (5) the marked smaller solubility of most dyes and finishing agents in organic solvents than in water; (6) toxicity of the solvents used in relation to health of the operatives.

Probably the fire risk must receive most consideration and because of this chlorinated hydrocarbon solvents and more especially 1,1,1-trichloroethane, trichloroethylene and perchloroethylene are the most used solvents. They have useful boiling points of 74°C,

87°C and 121°C respectively, are very stable in repeated distill-ation (for purification) and are inert, but in the presence of moisture they can decompose somewhat by hydrolysis (this hydrolysis is greater with 1,1,1-trichloroethane) to yield hydrochloric and some organic acids which can attack stainless steel.

The brand-named Markal process developed by ICI Ltd uses trichloroethylene and it has proved very useful for purifying raw cotton by removing natural waxy impurities as a first stage in preparing these fibres for dyeing, which demands that they are highly absorbent and reasonably white. It has been found possible to emulsify hydrogen peroxide in the solvent and thus additionally bleach polyester plus cotton fabric in a one-run treatment to be acceptable for dyeing, except for pastel shades. Since a starch size in cotton yarn (insoluble in organic solvents) can be its main im-purity, attempts are being made to discover alternative, equally satisfactory solvent-soluble sizes and thus make the solvent extrac-tion treatment satisfactory.

A foreseen advantage is that organic solvents very rapidly and completely penetrate textile materials even in the cold so that textile processing (scouring, dyeing, finishing, etc.) can be much accelerated. Many of the dyes in common use today are insufficiently soluble in organic solvents, but this is being remedied by preparing others having the desired solubility.

Dry cleaning machines are already available (sometimes specially modified) for processing with organic solvents and they allow such processing to be carried out at temperatures above 100°C, if desired.

Water-insoluble disperse dyes are more amenable than water-soluble dyes to dyeing from organic solvents and a number of processes suitable for colouring synthetic fibres have been devised using these dyes.

Organic solvents are likely to find much application in the finish-ing of textile materials. Thus in wash-and-wear finishing of wool (p. 419) a perchloroethylene solution of a partly polymerised isocyanate (available as the proprietary agent Synthapret LKF) can be applied to wool fabric and tumble dried in hot air to leave a thin film of the polymer on the fibres which by simple storage for 4 to 5 days at room temperature will further polymerise to a tough tenaciously adhering film to give the fabric a crease-resist finish which is fast to washing and to dry cleaning. This useful finish is obtained without requiring the wool to be fully pretreated with chlorine to make it non-felting.

Quite interesting is the fact that when polyester fabrics have aqueous solutions of proprietary agents such as Luratex and Permalose dried into them to confer anti-static and hydrophile properties on synthetic fibres it is necessary to subject them to a final high temperature to make these properties washfast. However, if they are applied from solution in perchloroethylene this final high temperature treatment is not necessary since the finish is already washfast. Somewhat similarly water- and oil-repellent finishes obtained by applying aqueous solutions or dispersions of silicones and fluorochemicals to fabrics give finishes less effective than those obtained by application of the same agents from an organic solvent solution.

If crease-resist finished cotton fabrics, produced by application of a cellulose crosslinking agent and an acid catalyst followed by drying, are finally cured (1) in the normal manner at a high temperature of, say, 150°C for 4 min or (2) cured in the vapour of trichloroethylene or perchloroethylene for 2 min at 87°C or for 3 min at 150°C, the fabric cured in the organic solvent vapour is left with a smaller loss of strength, although its finish is equally washfast to that of the fabric dry cured at the usual high temperature of 150°C.

It is evident that treatments with an organic solvent or with water can give significantly different results and at the present time there is much to learn about this.

High frequency radiation to assist dyeing

It can be advantageous in drying wet fabric to pass the fabric through an infra-red unit immediately in front of the stenter for partial drying—enabling the stenter to be run faster. The drying can also be effected by running the wet fabric over a few steam-heated drying cylinders but the manner in which heat is produced in the wet fabric is different. In the former method heat is generated within the fibres of the fabric whereas in the latter it is transferred from an external hot surface. The heating can thus be more uniform in the former case.

When a wet textile fibre is exposed to such high frequency radiation the molecules in both the water and the fibre oscillate synchronously with the frequency and proportional to its absorption which varies with different materials. The absorption is small for a completely dry fibre but high for water. The oscillating molecules generate heat and thus in the case of a wet fibre the water becomes hot and vaporises leaving the fibre dry. Such heating can thus be very efficient but the cost of the irradiation can be relatively expensive.

If textile material previously saturated with a dye solution is irradiated by microwaves that are essentially high frequency radiation and whose wavelengths are within 1 to 100 cm, then their absorption induces oscillation of the water, dye and fibre. It is possible that the dye may suffer some decomposition but if it is of a type which resists this then the irradiation creates a dyeing conditon, that is the fibres are in contact with a hot dye solution and thus the fibres should become dyed. It is along these lines that research is being carried out. Many advantages could stem from a dyeing process in which textile material such as loose fibre or woven or knitted fabric is impregnated with an aqueous dye solution and then carried along on a brattice (this could be made with a substance such as glass or polypropylene which is largely non-absorbent to microwaves) and irradiated to a degree which promotes dyeing at a useful rate.

It has been shown that irradiated wool fabric impregnated with a solution of a reactive dye can be induced to absorb up to 80 per cent of the dye present within 30 to 60 s—the amount is influenced by the type of reactive group in the dye and also by the character of the wool and of the fabric, i.e. flannel or serge fabric. Such a high rate of dyeing could help to make this method of dyeing practical and there is the possibility that additives could increase the rate, although equally some might lower it.

Drying of the fabric during absorption would have to be prevented for it is the water in the fabric which makes the heating possible. This research supplemented by the results of some bulk dyeings encourage the view that such a dyeing method could become practicable and economical on a large scale.

Continuous process for dyeing wool sliver

Wool is generally recognised to be a fibre which absorbs dyes slowly and so it is usually dyed in batches. By contrast cotton materials can be dyed continuously in which they are padded with a dye solution, heated (usually steamed) for fixation of the dye and washed and dried. All these stages can be affected by a one-run process. When a similar dyeing method is applied to wool in the form of sliver, various difficulties are encountered, i.e. a long steaming period is required and so-called 'skittery' (uneven) colourings are produced. In skittery coloured wool the fibres themselves are unevenly coloured along their length.

Progress has now been made to solve these difficulties and in particular to prevent skittery dyeing and make it possible to dye wool sliver satisfactorily in a one-run process. It would appear that such faulty dyeing in a continuous process where the sliver is impregnated in a padding mangle with a dye solution and then led through a steaming chamber and finally washed and dried, arises from the observed defect that on leaving the mangle the dye liquor does not form a continuous film covering each fibre but breaks up into drops with spaces between. Obviously this uneven distribution of the dye liquor will result in the individual fibres acquiring a varying depth of colour along their length. Following recognition of this broken film effect an expedient has been devised to overcome this skittery problem. It consists of adding to the dye solution a surface active substance and especially one based on sodium dioctyl sulphosuccinate which induces the dye solution to form a continuous film on each fibre. Further improvement is obtained by adding urea to the dye solution as a means for accelerating absorption of the dye by the wool so that the steaming period can be reduced.

Thus a satisfactory one-run process has become available in which wool sliver is padded with the specially modified dye liquor, run through a steamer and then washed and dried.

TEXTILE AUXILIARIES

Various kinds of machinery and chemicals used for the processes of scouring, bleaching, mercerising, milling, dyeing and finishing of textile materials have now been described. It has become possible in recent years to improve the efficiency of these processes by the use of special chemical assistants which are now marketed under the term *textile auxiliaries*. These assistants have indeed now become essential and no dyer or finisher would think of processing textile materials without first considering whether or not it was possible to make the process more effective or to carry it out more expeditiously by the use of a selected textile auxiliary.

Soap is one of the best-known auxiliaries since it has been used for many years. In washing operations the presence of soap in the detergent liquor promotes the spread of the liquor in and between the fibres while it also assists the removal of dirt from the fibres and then holds the removed dirt in stable suspension in the liquor so that it cannot be re-absorbed by the cleaned fibres. Further, if a small amount of the soap is left in the washed textile material this can remain softer than it otherwise would have been. The soap thus assists the washing process in four distinct ways.

Soap has some disadvantages. For instance, it is precipitated as an objectionable scum when used in hard water and it is liable to after-yellow during the storage of soap-containing fabrics and garments. Following a search for soap substitutes free from the defects, but having all the usefulness of soap, there are today many synthetic alternatives to soap and a large proportion of them are made from petroleum products rather than natural fats and oils; they can be used in hard water and often have superior wetting and detergent properties.

Arising from this search for synthetic detergents has come the discovery of many types of textile auxiliaries having specialised uses, so that today a special section of the chemical industry is given over to their production. It is likely that the manufacture of textile auxiliaries is now as important as the manufacture of dyes. Today there are some thousands of individual products which can be classified according to their uses thus:

Wetting agents   These are added to scouring, dyeing and other processing liquors to promote rapid penetration of the liquor among the fibres and so overcome the natural resistance to wetting which is shown by many textile materials. It is usual for the more *complete* wetting thus obtained to be accompanied by a more *even* wetting so that the processing is thus made much more satisfactory. Only selected wetting agents can be used in mercerising liquors for many are decomposed or precipitated by the high concentrations of alkali.

Detergents   These can often be used in hard water and sometimes even under acid conditions without losing their efficiency or forming a scum. They allow scouring operations to be carried out under conditions less harmful to the textile material. Most of them have good emulsifying power towards fats and waxes, combined with high wetting power, so that they are more efficient than soap in removing natural impurities from raw fibres or heavily soiled goods.

Softeners   Such products are today exceptionally useful to dyers and finishers, not only to give increased softness to ordinary articles but also to make soft those fabrics and garments which have been made harsher by the bleaching, dyeing or other treatments to which they have been subjected. Many of these softening agents (they can be regarded as fibre lubricants) are held by the textile fibres as tenaciously as dyes, so that the softening they produce is almost permanent. It is important not to use softeners which reduce the fastness of coloured goods to light or washing.

Dye-dispersing agents   These substances are added to dye liquors, to printing pastes and to the liquors used for rinsing dyed materials.

Their main purpose is to break down large dye particles or agglomerates of dye particles into smaller ones and thus facilitate their penetration into the textile material and absorption there by the individual fibres; these agents are also added to printing pastes for the same purpose. When present in a rinsing liquor the dye-dispersing agent assists the removal of any dye which remains loosely adhering on the outside of the coloured fibres.

Dye-carriers   These have been previously referred to (p. 302). They are organic substances such as benzoic acid, *para-* and *ortho-*phenylphenol and tri-propyl phosphate which when added to the dye-bath in the dyeing of hydrophobic fibres such as nylon, cellulose triacetate, Orlon and Terylene, assist dye penetration of the fibres either by swelling the fibres to make them more porous or by dispersing the dye into smaller particles. They are not much used in the dyeing of hydrophilic fibres such as cotton and wool since the water of the dye-bath swells these fibres sufficiently to aid dye penetration. By the use of dye carriers in a dye liquor the dyeing process can often be carried out under conventional dyeing conditions of temperature (about 100°C) and time thus making unnecessary the use of dyeing temperatures up to 130°C.

Fibre-swelling agents   These substances are also added (as dye-carriers) to dye liquors and also to printing pastes to assist dye penetration and absorption. Care must be taken not to employ them in too high a proportion or concentration otherwise the fibres may suffer serious weakening.

Metal-sequestering agents   The water used for dye liquors is liable to be contaminated with metal impurities such as those of iron, manganese, copper, etc. and when this is the case there is always the risk that the metal will combine with the dye absorbed by the textile material and change its shade (usually the shade is thereby dulled) and possibly lower its fastness to light and other adverse influences. To avoid this defect it is better to purify the water used, but where this is impossible for cost or other reasons it is often convenient to add to the dye liquor a small proportion of a metal-sequestering agent which has the power to combine with the metal and render it inactive towards the dyes used. Among the more important metal-sequestering substances are polyphosphates and ethylene diamine tetra-acetic acid.

Anti-foaming agents   Many wetting agents cause excessive foaming of the processing liquors to which they are added and this foaming can be a real nuisance. Thus, special auxiliaries have been introduced having the power to prevent or reduce this foaming without at the same time reducing the wetting effect. Selected silicones are useful anti-foaming agents.

Oil-emulsifying agents These have the power to emulsify fats, oils and waxes so as to give stable emulsions in water which can be used in the finishing of textile materials. Such agents can also be added to scouring liquors for the purpose of assisting the removal of oily or greasy impurities from fabrics and ensuring that these become evenly dispersed in the scouring liquor so as not to become deposited once more on the fabric during the scouring operation.

Moth-proofing agents These products applied to the textile material during finishing make the wool repellent to the moth larvae or act as a poison to them.

Bactericidal agents These are applied to counteract odour formation from perspiration in fabrics and garments. They are also used to prevent mildew and fermentation in some finishing compositions.

Anti-static agents These substances are often used on synthetic fibre yarns to give them increased electrical conductivity and so counterbalance their natural tendency to accumulate excessive amounts of static electricity during winding, weaving and knitting operations. It is usual to combine the anti-static agent, which generally has hydrophilic properties, with a fibre-lubricating substance to promote the freer movement of the fibres in the yarn. Some agents confer permanent and others only temporary anti-static properties. Synthetic fibre manufacturers can add an anti-static agent to the fibre-forming polymer before fibre spinning. Anti-static agents are available for spraying (in solution) carpets.

Dye-fixing agents Several of these products are now available and they are proving very useful. They are applied to coloured textile materials (often in the final rinsing liquor) to make the colours faster to washing. Many of these products are effective because they combine with the dye to form less soluble compounds. Recently introduced agents can chemically link dyes with fibres and thus be more permanently effective.

Anti-slip agents These are usually resinous substances which are applied to rayon and synthetic fibre fabrics of such loose construction that their very smooth threads easily slip over each other to cause fraying. The small proportion of the anti-slip agent which covers the surface of each fibre gives just that degree of roughness and fibre-adhesion which is sufficient to prevent thread slippage.

Rot-proofing agents   Fabrics which are exposed to prolonged damp conditions or are left for prolonged periods in contact with the earth deteriorate due to the action of bacteria and various types of micro-organisms. Rot-proofing agents based mainly on copper (sometimes mercury) compounds are applied to counteract this. Synthetic resins can be used as rot-proofing agents.

There are several other types of textile auxiliaries but it is not possible to deal with them here. The important point to note is that by the use of often quite small amounts of these special substances a textile process can be much assisted or the properties of the textile material improved.

CHAPTER FIVE

Colour and finish from the viewpoints of manufacturer and user

However textile materials are scoured, bleached, dyed, printed and finished it is important that they should be serviceable in the use for which they are intended. This proviso about their use is essential, for it is not always necessary for materials, yarns, fabrics and garments to have the utmost fastness of colour or the maximum strength and durability. Some materials are only handled lightly so that it is permissible for these to be fragile. Others are never washed and with these fastness of colour to washing is unimportant. Yet other materials are never exposed to sunlight and so dyes poor to light can be used if they satisfy other requirements.

Ladies' rayon hose provide a typical example. These are dyed with direct dyes which have only moderate fastness to washing. Yet the life of a lady's stocking is so short and the exact shade is relatively so unimportant that this defect of the dyes used does not matter. No manufacturer would dream of dyeing these hose in fast vat shades.

The properties of textile materials which have to be considered in relation to their use are:

(1) fastness of colour,
(2) lustre,
(3) handle,
(4) resistance to wear,
(5) shrinkability in washing,
(6) resistance to creasing,
(7) stiffness,
(8) waterproofing or water repellency,
(9) warmth of handle,
(10) rapid drying (after washing) with shedding of creases,
(11) resistance to soiling,
(12) flameproofing,
(13) resistance to bacteria and other micro-organisms,
(14) resistance to high temperatures (including hot ironing).
(15) retention of shape, size and appearance.

Fastness of colour

Dyers and dye makers have, over a period of at least half a century, given much attention to the fastness of coloured materials when exposed to various influences such as light, washing, perspiration, hot ironing, bleaching, acids, alkalis, rubbing and the like. In the early days of dye discovery and manufacture more attention was paid to increasing the number of dyes available. When a fair number of dyes were known and made, then came careful selection from the viewpoint of fastness. For instance, the first coal-tar dyes discovered by Perkin, which initiated the coal-tar dye industry, were vulnerable to light and often to washing. Basic dyes, the class to which they belonged, are now not as much used as formerly for the dyeing of goods made of natural fibres, except where the maximum possible brightness of shade is required, but unexpectedly they are proving very useful for dyeing some synthetic fibres such as acrylic fibres on which the resulting colours have quite good fastness to light.

Today a newly discovered dye is not put into production until it is shown either that it has a peculiarity of shade which is much desired, or that it has fastness properties superior to those of the dye which it is likely to supplant. The result of this rigid selection is that at least one hundred times as many new dyes are being discovered as are ultimately manufactured on a large scale.

Most works where dyes are made or applied have laboratories for the purpose of testing their products. A great deal of care is taken throughout this country as well as in others to ensure that the fastness of colours is sufficient to meet general and particular requirements.

Fastness to light   One of the difficulties of testing for fastness is to secure results within a reasonable period of time. Actually there is no better or more reliable test of a coloured material than to wear it, or otherwise expose it to the actual adverse influences which it will meet in use. But this may take many months, and a large number of samples may have to be tested in order to secure a reliable average result. Consequently a number of accelerated tests have been devised stemming from the first systematic method of testing developed by the French scientist Du Fay in about 1730. In many cases they can be relied upon, but in others there is considerable doubt as to their accuracy. Yet, so great is the need for rapidity in testing, that these doubtful methods have to be used until more reliable ones are devised.

It has been found that the violet end of the spectrum including the ultra-violet light is most harmful in the fading of coloured materials. Special fading lamps are now sold for testing light fast-ness in which the proportion of ultra-violet light is somewhat greater than that of ordinary sunlight. Other sources of illumin-ation for testing and comparison purposes have been made to resemble sunlight very closely. The great advantage of these fading lamps is that the exposure of a sample of coloured material can be carried out continuously for 24 h per day, and testing is thus made independent of the weather or season.

The Fadeometer, Xenon and Fugitometer, which are very useful fading lamps of this kind, are much used. The Xenon arc lamp is characterised by its light emission being especially similar to sun-light. In their construction attention has to be given to the temper-ature and the humidity of the surrounding air by the provision of cooling and air conditioning devices. It is found that too high a temperature of the exposed sample will affect the fastness of the colour irregularly, whilst an abnormally high humidity generally unduly accelerates the rate of fading. With a lamp of this kind it is possible to expose upwards of fifty samples at one time. Fading results obtained artificially in this way are usually checked in the long run by exposures to sunlight itself in different parts of the world.

Recently a new type of fading lamp proposed by Giles has been further developed by Parke-Davis and has proved to be as accurate as existing lamps but cheaper to buy and use (Microscal Ltd, London). It uses a mercury–tungsten fluorescent lamp instead of an ordinary mercury vapour lamp.

Dyes fade by the action of the sunlight decomposing them into simpler, less coloured or differently coloured substances. Some-times it is found that the decomposition of the dye also promotes decomposition and consequent weakening of textile material on which it is dyed. This has a special significance for coloured cur-tains made of cotton and cellulose rayon. Certain yellow and orange to red dyes (even vat dyes) have been proved to make cur-tains very weak in this way. Such dyes are now 'blacklisted' and must not be used by curtain dyers or dyers of other materials that are exposed to a considerable amount of sunshine.

Textile auxiliaries are available which can, by application to coloured textiles, increase their fastness to light. Such auxiliaries can act by being more reactive to light so that they reduce the harmful action of the incident light on the dye by themselves being decomposed first. By contrast, some auxiliaries, applied to coloured fabrics to ensure special effects such as resistance to creasing, can reduce the fastness to light.

Much research has been given to elucidating the manner in which various internal and external conditions attendant on the light fading of dyes in textile materials is influenced, and the reactions involved have been found to be very complex. Small amounts of oxides of nitrogen and sulphur which may be present in the surrounding air in industrial cities, and of ozone in more sunny localities, can have not only a marked but also an uncertain influence. Temperature and moisture conditions can materially govern the rate of fading and substances present in the textile material together with the dye can often produce abnormal fading in light. For example, the titanium dioxide much used as a delustrant in viscose rayon can markedly lower the light fastness of direct cotton dyes especially when the rayon is wet. If glycerine is also present (sometimes used in finishing treatments) then the light fastness can be further reduced. The presence of moisture can, with some dyes, accelerate fading ten times the normal under dry conditions. Another uncertainty encountered in exposure of dyed materials to sunlight is that the light incident on the textile material can raise its surface up to 50°C and thus accelerate fading or hinder it if a drying of the fibres occurs.

The crease-resist finishing of cellulose fibre materials (this is usually effected after dyeing) can reduce considerably the light fastness of several dyes and this effect can only be discovered by actual light exposure since it varies with the chemical composition of the dye.

As might be expected light fading varies with the chemical nature of the dye and also with that of the type of fibre present in the textile material.

Fastness to washing   In the case of tests for washing fastness it is usual not only to carry out standard washings of the coloured material and note the loss of colour, but also to wash this material with white fabric or yarn and ascertain if there is any staining of the white. Of course, washing tests of graded severity are used and they must be correlated with the type of material. For example, it might be useful to see if coloured cotton fabric kept its colour in a boiling soap and soda liquor, but this test would never be applied to wool goods, for in actual use wool materials would never be exposed to such drastic conditions. A so-called wash-wheel machine is frequently used for washing tests.

The wash-wheel machine has a metal frame which carries a number of small pots each containing a detergent liquor, a number of steel balls and the fabric under test. The frame is rotated, partly immersed in water heated to any desired temperature. The loss of colour and any staining of white fabric washed with the fabric sample can thus be determined.

Recently and with the increased use of domestic washing machines it has become more clearly recognised that the rubbing together of the fabrics and garments in the detergent liquor can have harmful effects.

One such harmful effect is that a coloured material can acquire a so-called frosted appearance. This arises when the dye is not uniformly distributed within each fibre but is concentrated in its surface thus leaving the interior white or nearly so. The rubbing in the detergent liquor can irregularly remove such surface located dye to expose the white interior of the fibre so that the all-over appearance gives the fabric the appearance of having a thin frost covering over its coloured texture. This same defect can result when the fabric consists of two or more types of fibre dyed to the same shade so as to present a solid colouring but with one type of fibre surface dyed or liable to a greater loss of dye than the other fibre.

This same rubbing can also cause a marking-off of the colour of one fabric on to another differently coloured fabric. Obviously this trouble can be avoided by not washing mixed coloured materials. But, it can also occur when the washed materials are taken out from the washing machine and (before rinsing thoroughly) allowed to lie especially when they are warm for then the dyes can bleed (if not sufficiently fast) in those parts where the fabrics are firmly pressed together. Residual detergent and especially ordinary soap in the fabrics can promote dye bleeding and hence the advisability to rinse early and thoroughly in cold water.

Changes of shape and shrinkage in garments due to felting of wool or relaxation of a stretched condition of cotton and rayon are discussed on pages 378 and 386.

Fastness to perspiration   The old recommended test for perspiration was to give a piece of the fabric to a stoker to use. However, this is not always possible and, moreover, coal dust is apt to interfere with the colour change produced by actual perspiration. Some years ago The Society of Dyers and Colourists investigated this problem of devising a reliable perspiration test for coloured materials. Men and women volunteered to dress in rubber suits and remain in a hot room so that they would perspire freely. The suits were arranged so that the perspiration could be collected and examined. Some very interesting results were obtained and people suffering from different diseases gave perspirations which differed considerably.

As a result of these tests it was established that with ordinary healthy people the perspiration was at first slightly acid, but that as it became stale it changed by bacterial action and became alkaline, so that it could even liberate ammonia. Most perspiration contains common salt. With this information, artificial perspiration tests have been devised in which the coloured material is treated successively with special acid and alkaline liquors whose effects closely resemble those of human perspiration.

The fastness of coloured goods to acids and alkalis and other chemicals is more of interest to those who wet-process materials containing coloured threads. Bleachers and dyers have to be familiar with this type of fastness.

Fastness to gas fumes Two curious special types of fastness concern especially dyers and users of coloured acetate fibre goods. It has been experienced that when dyed rolls of fabric have been stored in merchants' warehouses for several months occasionally the exposed edges of the fabric become very faded. This has been traced to the action of fumes from burning coal gas and in particular to the nitric oxide impurities which are not only found in burning coal-gas fumes but also in the atmosphere of industrial towns especially during the damp winter months. So today all dyes used for colouring acetate fibres, especially the blue and violet ones, have to pass a gas fading test which involves exposing the coloured acetate fibre material to the nitric oxide evolved from addition of hydrochloric acid to sodium nitrite within a closed vessel. The problem is not now so serious as it once was, for the dyemakers have been successful in producing blue and violet dyes not affected in this manner. It is a curious fact that dyes susceptible to gas fading when present in acetate fibres are usually unaffected when present in synthetic fibres such as nylon and Terylene.

Several products (mainly organic amines such as diphenyl ethylene diamine and methyl benzylamine) are available for aftertreating dyed acetate fibres to protect them against gas fading.

Some acetate (disperse) dyes have also been found susceptible to a curious temporary change of colour when exposed to strong sunshine. It was found that if an opaque object, such as a book or a penny, was placed on a piece of coloured acetate fabric exposed to sunlight for a few minutes and the object was then taken away quickly, the shaded piece of fabric appeared redder or otherwise different in shade from the remainder which had been exposed all the time. As this fabric continued to be left unshaded so it became equal in colour to the other exposed parts. Thus the patchiness of colour produced in this way was only temporary. However, this susceptibility to sunlight had to be considered as a defect. Dyes subject to it were termed *phototrophic* dyes and they had to be removed from those generally available to dyers. Yellow dyes used for acetate fibres were found to be more phototrophic than other colours.

Disperse dyes are characteristically water-insoluble so that they usually require to be applied in the form of an aqueous dispersion in dyeing processes. During dyeing the very fine suspended particles of dye accumulate as a thin layer on the surface of the acetate or synthetic fibres and then become absorbed to dissolve in the fibre substance. The molecules of disperse dyes must be free from water-solubilising sulphonic acid groups (these are present in many dyes used for cotton and wool) since these groups hinder their penetration and fixation in synthetic fibres.

As mentioned earlier (p. 270) disperse dyes can be applied to synthetic fibres by a so-called Thermosol process in which the textile material has a dispersion of the dye dried into it and which is then followed by a short high temperature treatment of about 150°C (or somewhat higher but below the softening temperature of the fibres) which allows the dye to sublime into the fibres and so become fixed.

Fastness to hot-ironing   It is generally necessary that colours should be non- or difficultly-volatile under conditions of hot ironing, because there is then a risk that volatile dyes would sublime into any adjacent white portions of the fabric. Also, if a dye is volatile at ironing temperature, there is a chance that it will sublime very slowly at ordinary temperatures and so stain white materials left near to it over a period of some months. Disperse dyes used for acetate fibres and the newer synthetic fibres such as nylon and Vinyon have to be especially selected from this viewpoint.

Fastness to rubbing   The last kind of fastness of colour which needs to be considered is that associated with rubbing. It is an interesting fact that some coloured materials can be very fast to light and even to washing and yet stain white material when rubbed on it. Dyeings which rub badly are usually those in which the dye particles are present more on the surface of the fibres than inside them. In applying some types of dyes it is a difficult problem to avoid this defect. The naphthol range of dyes is especially liable to this fault and dyers have to take special precautions in the dyeing.

Summary of dye fastness   If we review the fastness properties of the different classes of dyes it is seen that these may be summarised as follows:

Direct cotton dyes—mostly of good fastness to light and washing, but gradually become paler with hot soaping. Very widely used for colouring cellulose fibre materials where special fastness is not required. It is usually possible to select dyes having excellent fastness to washing or to light but not to both. Useful also for dyeing polyamide fibres.

Basic dyes—generally of poor fastness to washing and light on natural fibres, but show good fastness on acrylic fibres. Used where very bright striking shades are required.

Acid wool dyes—mostly of good to very good fastness to light and only good fastness to washing. Very widely used where special fastness is not required.

Mordant, metallisable and pre-metallised wool dyes—usually of very good to excellent fastness to light and washing.

Direct cotton dyes developed by diazotisation and coupling—this after-treatment much improves the fastness to washing but usually lowers the fastness to light. The shades are often dull.

Sulphur dyes—mostly of very good fastness to washing. Fastness to light varies from moderate to good. Newer types can have much improved fastness. Mainly used for cellulose fibres.

Vat dyes—these are used for producing colourings having the maximum fastness to light and washing as well as most other influences on cellulose fibres.

Indigosol and Soledon dyes—these are temporarily water-solubilised forms of vat dyes to aid their application. They yield the same colourings as ordinary vat dyes with the same fastness properties.

Naphthol or insoluble azoic dyes—the same high fastness properties as vat dyes on cellulose fibres.

Acetate (disperse) dyes—mostly of good to very good all-round fastness. Selected members have excellent fastness to light but many have only moderate fastness to washing. Much used for acetate and nylon fibres.

Reactive dyes—very good to excellent fastness to washing on cellulose fibres and often also very good fastness to light. They also give very fast shades to washing on wool and nylon.

Polymerisable (on the fibre) dyes—very good fastness to washing.

Lustre of textiles

The lustre of a textile material is one of its most important attributes. Generally public taste demands a fair degree of lustre in all materials but today there is an emphasis on a lustre which is subdued. Many rayon materials are sold having a matt appearance but on careful examination it will be seen that there is a degree of lustre which exercises a subtle influence and is unobtrusive.

We have seen previously how different textile materials may have a natural lustre and how this can be decreased or accentuated by the finisher. The important criterion of this artificial lustre is whether or not it is permanent. For example, mercerisation of cotton produces a lustre increase which is permanent, for it will withstand all washing and other treatments to which cotton goods are likely to be subject. Matt finishes and semi-matt finishes produced by the fibre manufacturers have a high degree of permanence.

The increased lustre produced in cotton, linen and wool fabrics by mechanical treatments including hot pressing, calendering (with or without friction) and schreinering is not fast to water. Wetting swells the fibres so as to disturb the fabric surface and the original lustreless appearance at once develops. However, this mechanically produced lustre can be made fast to wetting and light washing by having present in the fabric a resin pre-condensate such as dimethylol urea so that this is further polymerised to be insoluble by the hot pressing.

It is useful to remember that stretching and hot pressing can be relied upon to give temporarily increased lustre to a fabric or garment.

At the present time titanium dioxide prepared to be in its anatase (not rutile) form and extremely finely divided is the most used delustrant for textile fibres. Incidentally it is also used in household white paint.

Handle and draping properties

The handle and draping properties of textile materials are most important and a large amount of attention is given by the finisher to this aspect. Some materials are required to have a firm stiff nature but the majority, especially those made of the rayons and synthetic fibres, are wanted with a soft handle. In addition they should generally drape well.

A great deal can be done by the weaver or knitter to ensure these desirable characteristics. For instance, the yarns used should be of comparatively low twist so that they are naturally soft. Then in weaving and knitting a somewhat loose packing of the threads or stitches favours softness. But finally it is left to the finisher to secure these properties.

In scouring, bleaching and dyeing various chemicals are used and although these are comparatively harmless to textile fibres they do have some small degrading effect even when properly applied. If the wet processing is left in unskilled hands then the degree of degradation may be great enough to produce a definite harshening of handle. Thus, in dyeing wool with mordant dyes, the chromium or other metallic compounds introduced into the fibres as a mordant detract from the natural softness of the wool. The weighting of real silk makes it harsher. Strong bleaching of cotton materials makes them firm. Rayon materials are more easily harshened by these treatments for the reason that their fibres are more reactive. Wool made non-felting or unshrinkable by most processes loses in softness, but some processes are worse than others in this respect. The chlorination of wool fabrics so that they may be more receptive to dyes in printing is another means by which wool handle is depreciated.

Dyers and printers have come to recognise that they cannot carry out their various treatments without harshening textile materials to some extent, and so they have developed methods for correcting this. There are today a number of softening substances which are much used. In earlier days the main softeners were fatty and oily substances. Solutions of soap and emulsions of olive oil were once very much used. Later it was found that castor oil could be sulphonated to give soluble turkey red oil which can be used for softening. Since those days many new and improved softeners have become available, but the majority of them are still based on a fat or oil in a form which aids their satisfactory application.

Some of the early softeners did correct the harsh handle of the textile material, but the effect was only temporary as the oil was easily removed by washing. This type of softening merely misled the purchaser of the goods. Search has therefore been made for products which will soften and remain in the textile goods for a considerable period even when this includes repeated washing. Some substances which contain a fatty or oily residue have been produced which act as colourless dyes—they are absorbed into the fibres and there remain much as a dye does. Sometimes these special softeners can be added to the dye-bath used to colour the textile material. This is thus dyed and softened at the same time. It is important that these softeners be stable and that they do not discolour or decompose with storage. Otherwise softened white goods might go rancid and become yellower.

The use of the special high-bulk yarns previously mentioned (p. 135) has in recent years proved a very convenient method for ensuring that woven and knitted materials have a soft, full handle.

The softening of fabrics by mechanical means, as described in connection with calendering, is not permanent. The first wash is sufficient to dissipate such a softness once and for all. Several of the water repellent agents of the Velan type (p. 404) are also useful for conferring permanent softness.

While many textile softening agents depend for their softening action on the presence in their molecules of fatty or oily residues such as are found in natural fats and waxes, newer softeners consist of aqueous dispersions of polymerised hydrocarbons derived from petroleum fractions. These latter represented by polyethylene have the advantage of being very stable so that when present in a textile material they are not liable to promote fibre discoloration. They are also inert towards dyes and most of the finishing agents commonly applied to textiles. Thus polyethylene is much used for softening cotton fabrics subjected to crease-resist and durable press finishing—it can be added to the padding liquor containing the cellulose crosslinking agent.

Most of the softeners applied to textile materials have large molecules such that they cannot penetrate the fibres but are left adhering to the fibre surface and there act as lubricants to aid the sliding of the fibres over each other. Softness is generally associated with fibre lubrication.

However, it is recalled that the strength of a yarn is largely governed by the lateral adhesion of the fibres to each other and in general this adhesion is obtained by the twist normally imparted to a yarn during the spinning stage. Thus excessive softening of a yarn can decrease the fibre adhesion and thus weaken it, for yarn breakage largely results not from fibre breakage but from a sliding of the fibres over each other with a localised consequent attenuation of the yarn at the point where it breaks. In a similar manner excessive fibre lubrication can promote easy fraying of a fabric by allowing the yarns in the woven or knitted structure to slip over each other too easily.

Resistance to wear

The resistance to wear shown by a fabric or garment is governed by many factors such as the type of the textile fibre present, the nature of the weaving or knitting and the degradation which the material has suffered in its wet processing.

Any treatment of a textile material which has the effect of shortening the long chain molecules of which the fibres are made will decrease its durability. Actually it is impossible to convert raw fibres of any kind into manufactured fabrics or garments without taking away some of their natural durability. This is generally recognised in the textile industry and so attempts are made from time to time to apply special substances such as rubber and related highly polymerised bodies to textile materials to increase their resistance to wear. Increased attention has been given to this, and more durable fabrics can be produced but usually at the expense of other desirable properties. For example, much increased resistance to abrasion can be given to cotton fabrics by impregnating them with acrylate resins but they can make the fabric stiffer and non-draping.

During the actual wear of a garment while its surface is subject to rubbing the fabric itself is also subject to flexing as it accommodates body movement. This flexing must expose inside fibres to the rubbing in a manner not possible without the flexing. Some testing methods and machines fail to take this into account. Again, the roughness or smoothness of the back of the fabric can influence the behaviour of the face-side of the fabric to rubbing. Often it is quite difficult to recognise precisely the end-point of the test as for example when a break or hole is formed. If the fabric has been finished so that its surface is extremely smooth then skidding of the rubbing abrasive member may occur. It is obviously difficult to devise a testing apparatus which will give accurate wear values for textile materials of all kinds. It is concluded that the testing apparatus should be single-purpose designed as for instance a BTF wear testing machine devised by Courtaulds for measuring the wear of woven staple rayon fibre fabrics and the Ring Wear machine of Tootals for determining the loss of wear produced by a crease resist finishing treatment. A Wira wear testing machine is also available.

From a similar viewpoint it has been found difficult to finish a textile material with a wear-resist agent without modifying its other properties such as stiffness and draping characteristics. Only moderate progress has been made in finishing wearing apparel fabrics to have increased wear but it is otherwise with fabrics intended for industrial purposes when fabric stiffness is less objectionable.

One difficulty concerning wear is that of measuring this value. Obviously it is a tedious and prolonged precedure to ascertain wear values by having the materials observed during their normal use. It would be much better if the wear value of a fabric could be determined by subjecting it to rubbing under standardised conditions and noting how many rubs were required to produce the first visible signs of damage or wear.

Various types of wear machines have been devised to meet this need, but it is very difficult to imitate closely the actual conditions under which a fabric does wear out in everyday use. The result is that only indicative results can be obtained with these wear-testing machines. There is room for more reliable machines.

The principle on which those machines now available are based is much the same in all cases. The fabric to be tested is rubbed against a standard type of fabric. The rubbing may be backwards and forwards or circular. In either case the number of rubs is recorded to the time when the first hole is produced. In an alternative method the rubbing is continued for a standard period and then the loss of strength of the test fabric is measured. In yet another method the rubbing is continued for a time sufficient to allow a measurable weight loss which can indicate the wear value.

Wear and rubbing, by exposing the inside fibres of a fabric, can reveal a curious defect if the inside fibres have been incompletely dyed and are thus less deeply coloured than the outside fibres. Some dyes have a natural difficulty in penetrating thick and tightly woven fabrics and may even leave the inside fibres almost white. When rubbing brings these fibres to the surface they give the fabric an objectionable 'thin' or 'frosty' appearance.

The synthetic fibres vary among themselves in respect of their resistance to abrasion, but nylon is outstandingly good. It is now current practice to reinforce the heels and feet of men's wool socks with a proportion of nylon thread for the reason that they then resist wearing into holes. It is sometimes claimed that by reinforcing with about 20 per cent of nylon, hole-proof socks can be produced. However, an extraordinarily long life characterises socks made entirely of nylon. Wool carpets are usually reinforced by the presence of 20 per cent nylon and now 100 per cent nylon carpets are available.

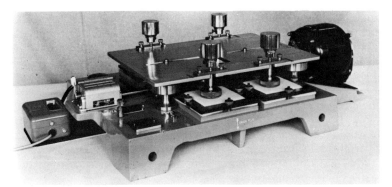

Figure 5.1 The Wira abrasion machine for testing the rubbing resistance of fabrics. Up to four specimens, 1½ in (approx. 38 mm) in diameter, can be tested simultaneously by rubbing against a standard abrasive surface with a varying motion including the warp and weft directions. (Courtesy Wira)

Shiny patches in worn wool garments

A disadvantage of a woven fine worsted garment is that due to wear it acquires a shiny appearance which is unsightly and difficult to remove. Wool technologists at Wira (Wool Industry Research Association, Leeds) have examined this behaviour of the wool and find that in the shiny patches those parts of the threads most exposed have worn thin so as to become flat and that in these parts wool has actually been rubbed away. It is this flatness which reflects light in a special mirror-like way when suitably viewed and it is responsible for the high shine (Figure 5.2).

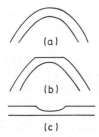

Figure 5.2 How shiny wool fabric is produced by wear, with loss of wool fibre by rubbing away at (b) and (c)

As will be mentioned later the application of resin to textile materials, especially those of cotton and viscose rayon, for the purpose of making them resistant to creasing and shrinkage during washing as well as for giving wash-fastness to mechanically embossed finishes, can have the effect of rendering the fibres more rigid and even brittle so that they are less wear resistant.

Shrinkability

Shrinkability in washing is a feature of textile materials to which a great deal of attention has been given in past years. This is because merchants of fabrics and garments have the idea that they can serve the public better by ensuring that the materials sold keep their shape and size throughout their useful life. This seems to be a sound proposition. The result is that today it is possible to purchase a wide variety of goods carrying a guarantee that they will not shrink in washing. It is interesting to consider this matter of shrinkage in somewhat more detail.

A textile fabric is made up of threads which are in a state of deformation for the reason that they are interlaced with each other. The threads are never quite straight but are bent 'this way and then that'. In the weaving or knitting of this fabric it is impossible to avoid straining or stretching the threads. A freshly-made fabric is therefore not stable; it is larger than it would be if the strains were relieved. When this fabric is wet processed by bleaching and dyeing, the wetting assists the fabric to contract to its natural size. As we have noted before, simple wetting of a strained textile fibre is the one reliable method for inciting it to return to its normal un-stretched state. So, if the dyed fabric were dried quite slack after finishing it would thus be brought to a condition when on further wetting it should not change at all as regards size and shape. How-ever, the finisher is obliged to bring the fabric into a smooth con-dition, and in order to meet this and other requirements it may have to pass through calendering or other machines, all of which tend to stretch and press out the fabric, particularly in length. So, under normal conditions of finishing, the fabric is left in a stretched state. In some cases this is perhaps something of an understatement for it has been known in the past for finishers to stretch fabrics inten-tionally with a view to producing fabrics longer or wider than is warranted by their structure.

However, with the modern revolt against fabrics and garments which in the shop appear to have a certain size but which after the first wash are found actually to be much smaller, the finisher is now called upon to arrange that he secures his effects of handle and appearance without stretching the fabric. In fact he is now called upon to conform to a general specification that the finished fabric will not shrink in washing beyond certain reasonable small limits. Consideration is given mainly to length rather than width shrinkage. These requirements involve the use of machines which allow a stretched fabric to be closed up in length to a degree such that it has no tendency to shrink in washing. Also the closing up must not be excessive so that the fabric in washing expands in length or width thus impairing the finish. Two types of machine were early developed and much used for this purpose. The fabrics produced are disting-uished by the trade names 'Rigmel' and 'Sanforized' shrunk.

British achievements with compressive shrinkage machines

About 1930 there arose a very insistent demand by the public for garments which could be repeatedly washed in the ordinary way and yet retain their original shape and size. At that time it was customary to buy a man's shirt or a ladies' skirt too large in order that in the first one or two washes it would complete its shrinkage and become stabilised at the desired size. The garment maker pressed the manufacturer and the dyer and finisher for fabrics which had already been brought into their shrink-resistant stabilised state. The problem became one for the textile finisher to solve and some method had to be found for shrinking completely, say, a shirting fabric (generally of cotton) following its earlier wet processing including scouring, bleaching, mercerising, possibly dyeing and a final drying. In all of these treatments where fabrics, each of 60–80 yd (55–75 m) in length (possibly several joined end-to-end), were run at a rapid rate through processing liquors and machines, it was obviously impossible to avoid length stretching, so that the final dried fabric could be 2 or 3 yd (2 or 3 m) longer than at the beginning of the wet processing. If this dry fabric could be wetted and allowed to lie completely free from length tension and similarly be dried it would largely contract to its original length thus losing its extension and becoming fully shrunk; the fabric would be in the state required by the garment maker. But no satisfactory method for carrying out this final shrinkage treatment was available on a large scale. Actually what often happened was that the extra lengths were cut off and distributed to the workpeople involved in the wet processing so that they could have them made up into shirts for their personal use. At that time this was an acceptable solution of the difficult shrinkage problem.

However, shrinkage was a serious problem which nobody appeared able to solve until it was put to Mr Alexander Melville, engineer at the Wrigley works of the Bradford Dyers' Association Ltd.

Melville decided that the bringing of a stretched cotton fabric into a fully shrunk state could not be achieved by any treatment which involved wetting the fabric, for before and in the final drying stage unavoidable stretching would occur. Further, wetting could destroy any special finish (softening, smoothing, etc.) previously conferred on the fabric. The shrinking treatment would have to be carried out on the dry fabric and sent to the garment maker without further treatment which could possibly involve length stretching.

It was at this stage that Melville had a simple idea which he put to the test. He bent a strip of rubber to a U-shape and then held pressed against its curved stretched surface a piece of plain cotton fabric while he allowed the rubber strip to straighten; he noticed that the fabric contracted as the curved rubber surface straightened and contracted. Hence the idea that fabric could be shortened by holding it against a contracting rubber belt and that this shortening could be fixed by pressing it with a hot smooth iron towards the end of the contracting stage. This shrinking of a dry fabric by contact with a contracting rubber surface and the application of a hot iron to fix the shrinkage under conditions such that the hot ironing pressure was not too high to prevent the fibres and threads in the fabric moving relative to each other to accommodate their closing-up, has become universally known as 'compressive shrinkage'.

Translation of the above idea into the now widely used Evaset machine and the later devising of a related Bestan machine which uses a more mechanical method for effecting fabric shortening, cost considerable money and effort. These machines, and also a modified type which allows patterned shrinkage of the fabric surface, are now solely made by Hunt and Moscrop Ltd. Melville, who has recently died, will always be remembered as the inventor of these machines.

Figure 5.3 shows the Evaset Mark II compressive-shrinking machine in which cotton fabric (slightly damped) passes through the shrinking unit and is then dried and its surface finally smoothed without dimensional change by passing through the Palmer dryer where it is held pressed against the hot surface of a large internally heated cylinder. The emerging fabric is then plaited down in a folded state having thus been made fully shrunk. The dampness of the fabric assists its shrinkage.

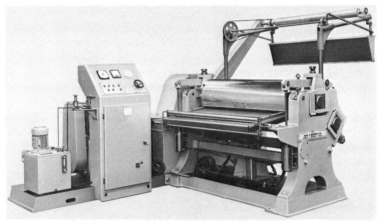

Figure 5.3 An Evaset machine. (Courtesy Hunt & Moscrop Ltd)

The shrinkage unit in the Evaset Compressive Shrinkage machine is more clearly shown in Figure 5.4. It comprises a thick [about 1 in (25 mm)] endless rubber belt R running around rotating rollers B and C so that in its travel path it presses lightly against the internally heated (250–350°C) rotating roller A. In operation

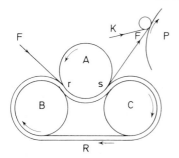

Figure 5.4 Compressive shrinkage unit of Evaset machine

the suitably damp fabric *F* makes contact with the outer surface of the belt at a point *r* where this is extended as a result of being thick and in convex contact with roller *B*. Immediately afterwards the fabric moves forward to be in light frictional contact with the surface of hot roller *A* at a point where the belt surface is contracting to become concave. Since the fabric adheres to the belt surface rather than to the surface of roller *A*, it now commences to close up lengthwise (without puckering, by reason of its light pressure by roller *A*) and so emerges at *s* to leave the nip between rollers *A* and *C* compressively lengthwise shrunk and heat-set by roller *A* to a degree controlled by the belt thickness and its convex-to-concave curvature change. At this stage the shrunk fabric is led through a Palmer calender *P* wherein it is pressed by a moving blanket *K* against the surface of a large, rotating, internally heated cylinder to give it a heat-stabilised pleasing surface and appearance.

Figure 5.5 shows the Bestan *confined passage* type of compressive shrinkage machine; so-called because the shrinkage is produced by leading the fabric through a confined passage under conditions such that it enters the passage at a faster rate than it leaves. This type of machine was devised to enable knitted fabrics to be shrunk to a much higher degree than woven fabrics. It was found that when a knitted fabric was passed through an Evaset machine to produce the highest possible shrinkage the emerging fabric had so great a resilience that it quickly self-extended to a degree that made it shrinkable in washing. The Bestan machine allows a much higher shrinkage to be obtained, say up to 50 per cent in length, and thus allows it to self-extend considerably afterwards and yet leave the fabric stabilised to resist shrinkage in washing (laundering).

The operational basis of this confined passage machine can be described by reference to Figure 5.6. The confined passage is between the lower surface of roller *A* and the highly polished upper surface of the fixed, internally heated (250–350°C) bed-plate *D* and extends from the nip between rollers *A* and *B* and that between

Figure 5.5 Bestan compressive shrinkage machine. (Courtesy Hunt & Moscrop Ltd)

rollers *A* and *C*. Because the peripheral speed of roller *C* is controlled to be somewhat greater than that of roller *B*, the forward movement of the suitably damped fabric *F* is retarded within the confined passage and thus the fabric is correspondingly closed up lengthwise. The moisture in the fabric and the raised temperature conditions facilitate this fabric shrinkage. The pressure of roller *A* on the fabric is minimal but is sufficient to prevent fabric puckering within the confined passage.

All the rollers are rubber covered but that on roller *B* is suitably somewhat softer than that on roller *A*, and this difference, coupled with the slower peripheral speed of roller *C*, gives this roller a stronger grip on the emerging fabric and peels this off from roller *A*.

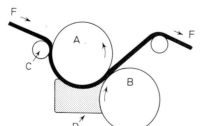

Figure 5.6 Shrinkage unit of Bestan confined-passage fabric-shrinking machine

At a rate of 40 yd/min (37 m/min) this Bestan 'Confined Passage' machine allows a high degree of fabric shortening and is therefore more suitable for finishing knitted fabric that has a tendency to extend somewhat afterwards. On leaving the shrinking unit the fabric can be led through a Palmer drying machine which also presses and sets the fabric so that it resists better any undue stretching during its making-up into garments.

With both the Evaset and the Bestan machines improved treatment of cotton fabric is obtained if this is first damped to contain some extra moisture, say a total of about 8–9 per cent. This moisture becomes effective while passing through the shrinking machine adjacent to a hot metal surface.

The modified Evaset machine just referred to allows the surface of a fabric to be pattern compressively shrunk, as for example, with seersucker stripes or with more complex configurations. For this the pressure roller or the rubber sleeve has a pattern designed surface and by adjustment of the pressure between these during the passage of the fabric the degree of fabric surface modification (embellishment) can be controlled. The pressure roller should be heated as already described with steam or oil and a convenient way of giving it a patterned surface is to key to its surface a number of suitably spaced rings (these will give warp striped effects on the fabric) or to cut a pattern in a bronze sleeve or shell and then secure this to the roller by a key. Alternatively a pattern may be cut in the rubber sleeve, and the pressure roller used unchanged. Intermittent and random designs are more conveniently obtained by cutting the rubber sleeve while continuous and geometric parallel patterns are more easily obtained by having the pattern on the pressure roller. Whether the patterning is on the pressure roller or the rubber sleeve the effects produced on the fabric result solely due to the surface contraction of the rubber sleeve and they differ from simple embossed effects.

As produced in a modified Evaset machine as just described, the patterned effects on the fabric will not be fast to washing unless they are set with the aid of a cellulose crosslinking agent or a resin precondensate as employed in crease resist finishing. For this purpose the cotton (or rayon) fabric is impregnated with an aqueous solution containing per 10 gal (45·5 litre) 8 lb (3·6 kg) of Fixapret CPNS (dihydroxydimethylol ethylene urea as a cellulose crosslinking agent and 0·6 lb (0·27 kg) of ammonium dihydrogen phosphate as a catalyst together with a softener and other additives). It is then dried at a temperature not exceeding 93°C so as to leave the fabric containing about 10 per cent of residual moisture (these drying conditions are used so as to avoid premature crosslinking occurring). The fabric is then passed through the modified Evaset machine and thereafter it is cured at about 150°C for 3–4 min or for 15–20 s at 175 to 180°C (a so-called 'flash-cure'). After this the fabric, with or without an intermediate wash, can be made-up into garments which have their seersucker, or other form of surface modified pattern, washfast.

Fabrics made with synthetic fibres or acetate rayon which are themselves thermoplastic are also amenable to this special processing but do not need the resin finish to make them washfast.

The processing by use of these compressive shrinkage machines has now been made easy and reliable but many problems were encountered from the time of the first idea of Melville. For example it was difficult to find a rubber manufacturer who could make the rubber sleeve and then it was found that the early sleeves wore out quickly so that a new one was required after finishing only about 80 000 yd (a week's work for one machine). Today with improvements in its manufacture it lasts for 1 000 000 yd. The surface of the sleeve must be free from flaws since these would make objectionable blemishes in the fabric surface. Then World War II interfered with development of the machine so that the patents covering it expired. Fortunately they were allowed to be extended but even then full patent protection was not secured.

By using such machines, fabric can be continuously shrunk in length at a steady rate so that it is left in a fully finished condition and such that it will not shrink more than 1 to 3 per cent during repeated washing.

The shrinkage in washing of cotton and other cellulose fibre fabrics can be much reduced by finishing them with a resin or cellulose-crosslinking agent by the methods described (p. 392) for making them crease resistant, but less than usual of the finishing agent is then required.

Sanforize shrink resist finishing process   While Melville developed his fundamental idea of shrinking fabric by holding it in contact with a shrinking stretched-rubber surface as just described, the competitive Sanforize process was concurrently and independently developed by Sanford Cluett in America. In this a thick wool blanket was used instead of the rubber belt used by Melville.

The fabric shrinkage unit in the Sanforize machine is shown in Figure 5.7. It shows the fabric *F* adhering closely to the initially convex surface of the thick blanket passing on the very small diameter feeding roller to the surface of the much larger diameter rotating Palmer drying cylinder. In this change the convex surface contracts lengthwise and the adhering fabric also with it. A hot highly polished stationary shoe smooths out any potential puckering arising from its shrinkage.

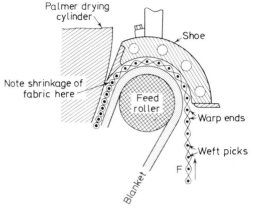

Figure 5.7 Shrinkage unit in the Palmer Sanforize machine

In its passage around the Palmer hot cylinder, pressed to it by the blanket moving with it, the fabric is dried and surface smoothed with simultaneous fixing of its length-shrunk state.

The operating conditions of blanket thickness and curvature change are arranged to shrink the fabric in length so that it just resists length shrinkage in subsequent washing. Resistance to width

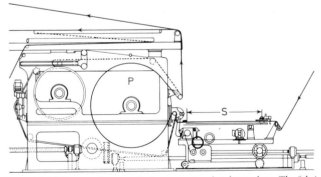

Figure 5.8 Main part of a controlled compressive shrinkage plant. The fabric, moving from right to left, is stretched to width by the short stenter S and then passes via the length-shrinkage unit (as shown in more detail in Figure 5.7) to the first Palmer machine P. It can then pass (a) directly to a plaiting device, (b) through a second Palmer machine where it is finished twice on the same side, or (c) through a fabric reversing device so that it can pass through the second Palmer machine and be finished on both sides. (Courtesy Sir James Farmer Norton & Co Ltd)

shrinkage is first secured by stentering the fabric to the desired width as it passes forward to enter the length shrinkage unit.

This Sanforize method is widely used but more recently the blanket has been replaced by a thick rubber belt somewhat similar to that used by Melville.

Unshrinkable (non-felting) wool   Wool materials could be closed up as 'Rigmel' or 'Sanforized' fabrics, but this would not ensure that they remain unshrinkable in washing. It has been noted before that wool fibres are unlike all others in that they have a natural tendency to close upon each other, a behaviour which is generally known as *felting*. So, if a wool fabric is relieved of all stretch it will still shrink in washing because the wool fibres then felt. To make a wool fabric really unshrinkable in washing it is therefore necessary to modify it in two ways: to change the fibres so that they do not felt and to close up the fabric so that there is no residual stretch left which would induce it to contract further. The well-known London shrink treatment is widely used to remove residual stretch in a wool fabric. It consists of damping the fabric and then drying it under slack conditions so that it is free to contract to its unstretched dimensions.

In quite early days of wool finishing, to reduce or eliminate its felting power the wool material was simply treated with a cold solution containing active chlorine, say an acidified solution of sodium hypochlorite. Although effective in making the wool unshrinkable in washing this treatment could seriously impoverish the wool by excessively thinning or even by removing the epithelial scales, thus exposing the less tough cortex of each fibre (Figure 5.9).

Searches for improved processes were guided by the requirements that the treated wool should be left soft (or amenable to softening after-treatment) and free from discoloration (say yellowing) but above all with full retention of its original durability (this includes resilience, strength and elasticity). Continued research has allowed these requirements to be largely satisfied, mainly by devising treatments which are substantially confined to modifying the epithelial scale layer of each fibre rather than by acting on the cortex or inside of the fibre.

It was early recognised that the felting properties of wool were largely associated with the epithelial scale covering (Figure 1.12a) of each fibre and especially to the manner in which their overlapping of the scales produced a so-called differential friction effect (DFE) with the fibres being rougher in the tip-to-root than in the reverse direction. In washing wool with repeated squeezing and relaxing in a detergent liquor the fibres interlock in the squeeze and do not completely disengage in the relaxing thus causing a progressive compacting (felting) of the fibres which is irreversible. Thus two main types of wool treatment have been devised to reduce felting— one by coating the fibre surface with a resinous film forming substance and the other by a chemical treatment to soften or thin the scales. By both treatments the DFE is much reduced either directly or indirectly.

An early American 'Lanaset' treatment involved applying a melamine-formaldehyde precondensate to the wool. Not only did it smooth the fibres, but in yarn and fabric it produced a kind of spot-welding of the fibres where they criss-crossed, and because of this latter effect produced an objectionable stiffening of the wool material. This method has been largely abandoned in favour of the use of other resins optionally applied under special circumstances as for example by a mild chemical pretreatment (with active chlorine, per-monosulphuric acid, etc.) to sensitise the fibres such that they more readily accept the applied resin and thereafter retain it more securely.

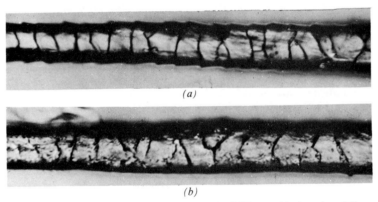

(a)

(b)

Figure 5.9 Enlarged photographs of (a) untreated and (b) over-chlorinated wool fibres. Although chlorination has rendered the wool shrink-resistant it has deteriorated the epithelial scales so that they no longer protect the interior of the fibre from destructive influences. As a result the fibre is less durable

Among the chemical treatments to reduce felting may be mentioned those aqueous solutions containing (1) a mixture of sodium hypochlorite and potassium permanganate, (2) sodium chlorocyanurate, (3) a mixture of peracetic acid and potassium permanganate, (4) permonosulphuric acid and (5) sodium hypochlorite under specially controlled pH conditions. Most of these treatments depend mainly on an oxidation of the epithelial scales by active chlorine leaving the cortex substantially unchanged.

The chlorination of wool occurs more slowly under neutral alkaline conditions than under acid conditions, and it leaves the wool much yellower. Sodium chlorocyanurate is also slow acting but leaves the wool white which is unchanged in subsequent treatments such as dyeing.

In the treatment of wool with liquors containing active chlorine there can be such rapid absorption by the wool of the chlorine that uneven treatment results. However, this can be avoided by adding to the liquor protein compounds which retard the chlorine absorption to give an even shrink-resistant finish.

Quite an interesting means for confining the action of the shrink-resist agents to the epithelial scales was at one time given early attention. It was based on applying the agent in an organic solvent which unlike an aqueous liquor produced no appreciable fibre swelling.

Such treatments (they modify substantially only the epithelial scale layer) were typified by those in which the wool material was treated with a solution of sulphuryl chloride in white spirit or of caustic soda in a 9:1 mixture of white spirit and butyl alcohol or in ethyl alcohol. In this way wool could be made shrink-resistant (non-felting) without appreciable impoverishment but there were the usual disadvantages attached to the use of the expensive organic solvents and so these have been abandoned. However, an alternative method, in which potassium permanganate dissolved in a concentrated sodium chloride solution has been much used—in this case the salt considerably hinders fibre swelling.

Another useful method is that in which wool conditioned to contain a limited amount (about 12 per cent) of moisture is exposed to the action of dry chlorine gas (the Woolindras process). It gives good results which derive from limited penetration of the fibres by the gas under the particular dry conditions employed.

At the present time quite a number of different processes are available for reducing the felting properties of wool and making them shrink-resistant in washing. If some harshening of the wool is incurred there are several softening agents readily available which can be applied to correct this.

One of the newer processes which is especially useful for producing non-felting yarn involves running very well scoured wool sliver (p. 160) through a relatively weak solution of sodium hypochlorite acidified to a pH of 2·5, then through a solution of sodium bisulphite to destroy excess active chlorine, rinsing it and then passing it through a solution or dispersion of a suitable polymer and drying it to retain about 15 per cent of water. At this stage the sliver is shrink-resistant due to the chlorination and to the deposition on the fibres of a film of polymer which tenaciously adheres to the fibres following high temperature drying. By continued insolubilisation during storage this polymer film will adhere even more tenaciously to the wool fibres. However, at this stage following drying, some of the fibres may be sticking to each other and to separate them so that the sliver can be converted into yarn it will generally be necessary to gill it (p. 159) which approximates to combing. The yarn thus ultimately obtained can then be used for weaving or knitting to give shrink-resistant fabrics.

In this treatment of the wool sliver the prechlorination is not sufficient to make the wool fully unshrinkable. Its function is to facilitate the deposition and fixation of the polymer (in this case, Hercosett, a polyamide–epichlorohydrin condensate, is used) to smooth the surface of each wool fibre and thus make the wool fully shrink-resist fast in repeated washing.

A small number of other suitable polymers are already available but improved types are being discovered.

It is important to recognise that the above treatments and others like them which make wool fabrics shrink resistant in washing only ensure that the fabric retains its original area during washing. They do not appreciably if at all ensure that any creases and pleats which have been intentionally introduced into a wool garment or fabric after a non-felting treatment will be retained after washing even if such pleats, etc. have been hot pressed to fix them. Yet today there is a definite call for wool fabrics and garments which have this particular useful washfastness.

A step forward in this direction came from the discovery that creases of moderate washfastness could be produced in wool fabric by first making it non-felting by one or other of the treatments already described, then drying into it a solution containing a reducing agent such as ammonium thioglycollate or sodium- or monoethanolamine bisulphite and then hot pressing any crease introduced into it. The resulting creases could persist through several moderate washings and this method has been used very largely to make more or less permanent the creases generally desired in men's trousers. However, yet superior washfastness has now become obtainable and is dealt with later (p. 420).

Resistance to creasing

An early discovery

When wool and silk are compared with cotton and rayon it is soon noticed that the animal fibres are more resilient and that the fabric made from them is very resistant to creasing. Hence the early incentive to discover a method for making cotton and rayon similarly non-creasable.

As is now well known such a method was discovered by Foulds, Marsh and Wood of the Tootal Broadhurst Lee Co Ltd of Manchester, and their process in various modified forms is being used world wide to improve cotton and rayon goods in respect of creasing. The method, by a special modification, can be used for linen materials.

The Tootal method seems to be comparatively simple. It is that of impregnating the fabric with a synthetic resin. Actually the resin is formed from its components urea and formaldehyde within the fabric and this urea-formaldehyde resin was chosen mainly because it was white and not easily removed from the crease-resist finished fabrics by washing and commercial laundering.

In carrying out such a process it is usual to impregnate the fabric with a liquor containing an acid catalyst and a water-soluble formaldehyde – urea precondensate and then to dry it and finally subject it to a high temperature curing treatment (say for 3 min at 150°C) for the purpose of converting the precondensate into a more polymerised, stable, hardened and insoluble resin uniformly distributed entirely within the fibres. Any resin formed as a coating on the fibres must be removed by washing since this can confer more, rather than less, creasability.

This crease-resist process proved exceptionally useful at the time of its discovery for application to fabrics made of staple viscose rayon by giving them a desired fuller handle and better appearance in addition to valuable resistance to creasing. The process was not then shown to be useful for application to acetate fibre fabrics.

It was soon found satisfactory to use melamine-formaldehyde resins and other types of resin instead of the urea-formaldehyde resin first used in the Tootal process. Experience with most of these resins has shown that the crease-resistant fabrics also have useful dimensional stability so that they resist shrinkage and loss of shape in washing and in the case of rayon fabrics these better retain their strength in their wet state. It was also found inadvisable to bleach crease-resist finished fabric with a bleaching liquor containing sodium hypochlorite or other form of active chlorine without regard to the type of resin present, for some resins greedily absorb and strongly retain the chlorine and release it in the form of hydrochloric acid in any subsequent hot ironing (with moisture present) thus leading to yellowing and marked impoverishment of the fabric. Such resins are those derived from components that contain basic amino groups.

Finishing of cellulose fibre fabrics with resins and cross-linking agents

The introduction of the Tootal crease-resist process about 1930 met with considerable criticism arising partly from its extreme novelty and partly from faulty processing excusably arising from lack of experience. But it has now been largely perfected and has completely changed the course of textile finishing by revealing so many different ways in which resin-forming and cellulose-cross-linking agents can be employed to increase the usefulness of cotton, rayon, linen and other textile materials even including those in which non-cellulose fibres may also be present. Behind the present day production of so-called drip-dry, self-smoothing, minimum-iron, wash-and-wear, etc. fabrics and garments is the crease-resist process discovered by Foulds, Marsh and Wood in the research laboratory of Tootal Broadhurst Lee Co Ltd in Manchester, and incidentally, it is interesting to note that while this research and development was proceeding Whinfield and Dickson of The Calico Printers' Association Ltd on the opposite side of the road in Manchester were quietly discovering how to make that type of polyester fibre now known as Terylene.

In the intervening years research has revealed the fundamental principles which now guide the production of crease-resist fabric by the use of resins and cellulose crosslinking agents and these may now be outlined.

It is now generally agreed that as a result of a crease-resist finishing treatment the long cellulose chain molecules which lie more or less parallel to each other within cellulose fibres become crosslinked at spaced intervals along their length, thus giving the fibres an increased degree of rigidity and resilience which causes them to resist distortion (creasing) and to recover from any distortion temporarily produced. To obtain this valuable property the discoverers found that no resin or crosslinking agent (possibly polymerised) should be left adhering to the cotton fibre surface since this could make the fabric more creasable. It is essential to modify only the inside of each fibre.

While most crosslinking agents, for example the well-known dimethylol ethylene urea types, are water-soluble and can thus be dried into cotton fibres from their aqueous solution it is not generally possible to apply fully formed resins since these are usually water-insoluble and even if applied in a finely divided form as an aqueous dispersion the resin particles would be too large to pass into the cotton fibre and so would be filtered off to adhere merely to the fibre surface. In applying a resin it is thus necessary to dry into the fibres an aqueous solution of the components of the resin and then expose the fibres to conditions which cause these components to combine within the fibre to form the insoluble resin. However, such components apart from the fibre can be caused to combine (this is usually termed *polymerise*) incompletely to form a so-called resin precondensate which is water-soluble and whose particles are sufficiently small in aqueous solution to pass cotton fibres, where they can be induced to polymerise further and form the required insoluble resin as is produced by applying the resin components from aqueous solution.

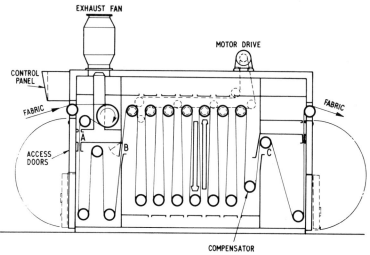

Figure 5.10 Construction of a Vaporloc apparatus for steaming (optionally heated) fabric passing through it. Various methods of heating may be used; for example, steam, electricity, gas, etc. Very efficient seals are provided at A, B and C to prevent escape of hot air or steam or the intake of air. (Courtesy Mather & Platt Ltd)

A crease-resist finishing process will thus normally involve impregnating say cotton fabric with an aqueous solution containing a resin precondensate or a crosslinking agent and a catalyst (most frequently an acid or acid-forming substance such as ammonium sulphate), drying under conditions such that no appreciable change (polymerisation) takes place except that the water content of the fabric is removed, and then subjecting the fabric to a curing stage in which it is heated for 3–5 min at 150–160°C to cause insoluble resin formation and crosslinking of the cellulose molecules. When a resin precondensate is used then some of this may take place in crosslinking and another part be left simply as insoluble resin in the spaces between the molecules.

In the above process the crosslinking occurs when the cotton

Figure 5.11 Cotton fabric impregnated with a crease-resist padding liquor passing through a Vaporloc high-temperature vessel to cause the crease-resist agent to cross link the cotton molecules and give a crease-resist finish. (Courtesy Mather & Platt Ltd)

fibres are quite dry and in a fully collapsed non-swollen state and such conditions will leave the cotton fabric resistant to dry creasing much more than to wet creasing.

However, if the process is carried out by allowing the wet impregnated fabric to lie for several hours at room temperature or moderately higher then the crosslinking occurs while the cotton fibres are wet swollen, and under these conditions high resistance to wet creasing is obtained accompanied by only a moderate degree of resistance to dry creasing.

Resistance to both dry and wet crease resistance can be obtained by applying both the above types of treatment in succession. The same result can be obtained also by drying the impregnated cotton fabric under mild conditions which leave about 10 per cent of moisture in the fabric (at this stage good *dry* crease resistance is obtained with the fibres only partly collapsed and while leaving some unreacted crosslinking agent) and then winding the fabric on a roller and allowing this roll to lie in its damp condition for say 18 h at room temperature (the residual crosslinking agent in this period reacts to give *wet* crease resistance).

A useful modification of this combined process is to add to the impregnating liquor a substance which strongly retains water during the drying stage and thus better preserves the wet swollen state of the fibres to allow production of a superior wet crease resistance during the last stage with the fabric lying in roll form for several hours. A further useful modification is to add also a substance which can promote an even greater fibre swelling than is produced by water itself.

So far mention has been made of only acid catalysts but alkaline catalysts can also be used provided that the crosslinking agent can be activated by them. Alkaline catalysts can be advantageously used for producing wash-and-wear finishes in the following manner.

Cotton fabric is padded with a liquor containing a monomethylol acrylamide crosslinking agent of formula

$$CH_2\!\!=\!\!CH.CONH(CH_2OH)$$

and zinc chloride or magnesium chloride as an acid catalyst. The fabric is then dried and cured for 4 min at 150°C to produce dry crease resistance. Then the fabric is impregnated with 6 per cent caustic soda (catalyst) solution, brought into roll form and allowed to lie at room temperature say for 10 h covered with polythene sheet to prevent uneven evaporation from the fabric and thus produce wet crease resistance additional to the dry crease resistance already produced. The caustic soda fabric also swells the fibres.

In another process for producing self-smoothing cotton fabric, based on a recently modified Tootal process, the fabric is padded with a liquor containing one of the commonly used crosslinking agents which require an acid catalyst (for example dimethylol ethylene urea) an acid catalyst and a special cellulose crosslinking agent which is the sodium salt of tris-(beta-sulphato-ethyl) sulphonium and is then dried and cured at about 150°C. In this curing the sulphonium agent remains inactive while the other agent effects dry crease resistance. The fabric is then run through a cold or warm solution of caustic soda, which as an alkaline catalyst activates the sulphonium agent and causes it to effect crosslinking of the cotton fibres while in a wet swollen state, so that wet crease resistance is produced additional to the dry crease resistance already produced. This last crosslinking requires only a few seconds for its completion. Finally the fabric is washed.

It is impossible to give here particulars of the many variations which have been tried with the aim of securing superior combined dry and wet crease resistances in a cotton and rayon fabric so that it has a drip-dry, self-smoothing or wash-and-wear finish which allows the fabric or garment to be washed and then dried by hanging on a line or with hot air in a tumbling machine without acquiring random creases which would finally require hot ironing to remove them. Obviously drying with hot air in a tumbling machine where the fabric is random creased and in folds is a more effective test for a wash-and-wear finish than is line-drying.

But while highly satisfactory wash-and-wear finishes can thus be obtained there is another difficulty encountered in crease resist finishing which remains to be overcome. It is that as a result of the crosslinking the finished fabric or garment can lose up to 40 per cent of its original strength and also even more of its resistance to wear when subject to abrasion. These losses arise largely from a small degree of embrittlement of the fibres resulting from the presence of insoluble resin or crosslinking. Softening agents such as polyethylene and polyacrylate resins are usually added to the impregnating liquor to reduce fibre embrittlement or give fibre lubrication but such addition can then impair the resulting crease resistance. Many such methods have been tried to avoid this difficulty but with only moderate success. Pre-mercerisation of the cotton fabric can assist in the production of stronger crease-resist finished cotton fabrics and this expedient is much used. The mercerisation itself makes the cotton fibres stronger and it is this effect which contributes to a final stronger crease-resist finished fabric. It is claimed that mercerisation by liquid ammonia instead of caustic soda (p. 250) is more effective in protecting cotton fabric from such deterioration.

Reference has already been made (p. 390) to the damaging effect of hypochlorite bleaching on crease-resist finished cellulose fibre materials for which some crosslinking or resin precondensates have been used, more particularly those which contain their active methylol groups linked to imino groups, e.g.— $NH.CH_2OH$. This damage does not result, or is much reduced, when the methylol group is attached to a nitrogen atom, e.g.—$N.CH_2OH$.

From another viewpoint the crosslinking agent must be selected such that it gives no yellowing to the finished fabric. But so far it has not been possible to select the crease-resist finishing agent and the conditions of applying it so that an ideal crease-resist or wash-and-wear finish is obtained. Hence the intensive research already carried out and being continued to discover an ideal process. This research is world wide and has already cost millions of pounds.

In this connection it is most interesting to recall the use of formaldehyde itself in crease-resist finishing. The application of this simple cheap agent to cellulose fibres was employed by Eschalier, about 1905, to give rayon fibres increased wet strength and improved dyeing properties. This action is now known to be due to cellulose crosslinking. By application with an acid catalyst either using a high temperature curing process or one in which the fibres are wet, quite good crease resistance can be obtained. But so far it is an agent whose action is most difficult to 'tame' sufficiently to avoid adventitious impoverishment of the cellulose fibre textile material. It crosslinks cellulose molecules within cotton and rayon fibres thus:

$$\underset{\text{Cellulose molecule}}{\text{Cell OH}} + \underset{\text{Formaldehyde}}{\text{H.CHO}} + \underset{\text{Cellulose molecule}}{\text{Cell OH}}$$

$$\downarrow$$

$$\underset{\text{Crosslinked cellulose molecules}}{\text{Cell O—CH}_2\text{—O Cell}}$$

Formaldehyde is not much used as such in crease-resist finishing. Its action can be much better controlled when in the form of a methylol group which is present in nearly all the crosslinking agents now in use. A methylol group is formed by reacting an amino $-NH_2$ or $= NH$ group with formaldehyde thus:

$$-NH_2(\text{or} = NH) + HCHO \rightarrow\; — NHCH_2OH$$
$$\text{or} = N(CH_2OH)_2$$

Recently it has been found that by using a special dual catalyst containing magnesium chloride and sodium fluoborate better control of the crosslinking action of formaldehyde can be obtained to reduce the usual strength loss of the cotton.

So far attention has been given to the production of cotton and rayon fabrics having dry and/or wet crease resistance which leaves the fabric with its original flat, smooth appearance, fast to washing. Such fabric is satisfactory if it is to be used as such but if it is to be made up into garments or other articles which are required to hold not only their dimensions but also any folds, creases or pleats, etc. which are usually introduced on purpose by the garment maker by hot pressing, then the crease resistance which they have can resist the introduction of such creases and pleats. So fast may be the first produced crease resistance by a crosslinking process that it may objectionably require a chemical treatment to break down this finish in order to allow the introduction of these desired creases, etc.

To overcome the above difficulty it has thus become necessary, as for example, in the case of men's trouser fabric, to apply to this a crease-resist finishing liquor and to dry it without appreciable cross-linking so that it can be sent in a stabilised state to the garment maker. There, possibly after several weeks storage, it can be shaped into trousers and the desired creases are temporarily pressed into these before subjecting them to a high temperature cure which will permanently set them washfast, with the remainder of the trouser fabric set smooth and resistant to shrinkage in washing.

In a similar manner it can be possible for partially finished fabric to be made into ladies' dresses and other garments and have introduced pleats permanently fixed in them by a final high temperature cure.

The principle underlying such processing is that however the fabric is held distorted while the crosslinking takes place, so will the fabric accept such distortion as its natural condition and thereafter always in washing or other treatment retain the distortion (a crease or pleat).

Thus the problem posed to the textile finisher a few years ago was to devise a crease-resist finishing process which he could interrupt after the drying but before the curing stage, and yet leave the fabric resistant to premature curing during storage under cool conditions, while awaiting its use by the garment maker. This problem was solved by using special crosslinking agents and also special catalysts, and one process which is very widely used is the Koratron process which uses 1:3-dimethylol-4:5-dihydroxy-ethylene urea (or the closely related 1:3-dimethylol-4:5-propylene urea) and a zinc nitrate acid catalyst.

At the time of the introduction of this process it was unusual for a garment maker to have the plant necessary for high temperature curing such as will normally be present in textile finishing works where crease-resist finishing is carried out. But after some hesitation many garment makers have installed such curing plant and the Koratron process is now widely carried on in many different countries.

Several other comparable processes (they are generally referred to as *durable press* or *delayed cure*) have now been devised and for some of these claim is made that the usual objectionable loss of fabric strength and resistance to wear that are produced simultaneously can be considerably reduced.

One such recently developed process involves drying into cotton fabric (to leave it moist) a strongly acid liquor containing a polymerisable, but not crosslinking, methylated mono-methylol melamine and a crosslinking dimethylol-dihydroxyethylene urea or a methylol methyl carbamate, steaming at 82°C for 15 min to produce some wet, but not dry, crease resistance, and also insoluble melamine polymer to hold the fibres swollen. The fabric is then rinsed free from acid and padded with a zinc nitrate catalyst. After drying, the fabric is in a fully stabilised state so that it can be sent to the garment maker. After its conversion into shaped and/or creased and pleated garments as desired, these are high temperature cured to washfast-fix the pleats, etc. and confer dry crease resistance additional to the wet crease resistance. It is not then necessary to wash these garments and this is an important point since it would not be acceptable to the garment maker if such washing was necessary. It is claimed for this wet fixation process that it is attended with a moderately lower loss of strength but with a much higher abrasion resistance because the crosslinking within the fibres is more evenly distributed. These advantages arise from the melamine resin present holding the fibres swollen during their final crosslinking.

Crease resist finishing normally also confers some resistance to shrinking in washing but to produce such unshrinkability in cotton and rayon materials it is only necessary to apply about one-half of the resin precondensate or crosslinking agent required for obtaining satisfactory crease resistance.

It has earlier been mentioned that application to wool materials of a resin or similar polymeric substance (p. 389) can smooth the fibre scale surface and thus also reduce the felting properties of the wool so that it is less liable to shrink by felting in washing. In such treatment it is difficult to avoid producing resin on the fibre surfaces and this produces a degree of fibre-to-fibre spot-welding. This effect reduces the freedom of movement of the fibres such as takes place in felting and this also contributes to reducing the liability of the resin-treated wool to shrink in washing. An associated defect of such resin treatment is that it can reduce the softness of the wool material.

Turning once more to the high loss of strength and resistance to abrasion which cotton fabrics can sustain in crease-resist and durable-press finishing, mention must here be made that because of the difficulty of reducing these changes to negligible proportions the expedient commonly adopted is that of having present in the cotton fabric a proportion of polyester fibres. For various reasons the proportion is usually 33 or 65 per cent. These polyester fibres are substantially unaffected by the finishing process and in particular they retain their strength and also their thermoplasticity so that they enable a finished fabric to be produced which has suffered a smaller loss of strength and abrasion resistance and thus has superior durability. The polyester fibres can also contribute to the washfastness of any creases or pleats introduced.

A useful feature of crease-resist processes is that various

finishing agents can be added to the impregnating or padding liquor containing the resin precondensate or crosslinking agent without causing precipitation or reducing the effectiveness of the finishing. Such compatible agents include softeners and fibre lubricants, water-repellents, fluorescent whitening agents, dyes, bactericides, fillers, etc. On the other hand the crease-resist finish can leave the fabric with an increased tendency to pick-up soil and this defect can be enhanced by the inclusion of polyester fibres in the fabric as already mentioned. This soiling tendency is especially noticeable for greasy soil for this can be tenaciously held by polyester fibres. To some extent it can be counteracted by application of one or other of the soil release agents now available (p. 418).

Crease-resist and durable press finishes are always applied as a final treatment of fabrics and garments notably after dyeing. It is thus necessary to use in the previous processing substances, for example dyes, which are unaffected by the resin precondensates and crosslinking agents. Actually many dyes change shade and lose some of their fastness to light and these are thus not to be used. In connection with white materials it has been noticed that some fluorescent whitening agents such as might otherwise be used to improve the whiteness are made ineffective by the catalysts used to assist crosslinking in the crease-resist processing.

It is advisable whenever possible to wash thoroughly the cotton or rayon fabric after crease-resist finishing since residual chemicals can, during storage of the fabric, develop objectionable formaldehyde and other odours. Obviously there are many features of the finishing and the finished textile materials which require careful attention.

Considerable interest is now being shown in a washfast permanent press process which avoids the use of resins and cellulose crosslinking agents but allows washfast creases and pleats and a flat finish, in fabrics containing 80 per cent of a special polyester type of fibre and 20 per cent of cotton or rayon, to be obtained by a simple hot pressing in a 'hot-head' Hoffmann press at 170–200°C for a short period, dependent on the temperature and the fabric texture. The hot cure common to durable press finishing which involves considerable fabric impoverishment, is unnecessary.

Permanently stiffened and laminated fabrics

Stiffened fabrics The above methods of stiffening are all similar to the older method in that something is dried into the fabric. In late years quite a different method has been developed and the results have proved very satisfactory. This method involves the production of laminated fabric.

Stiffened fabric is to be understood as being fabric which has been treated chemically to modify the substance of which the fibres are composed, or treated to add something to coat or enter the fabric and/or fibres and there temporarily or permanently fix such substances as starch and gums. Thus permanent stiffness, softness and draping properties can be beneficially modified.

Temporary stiffness can be obtained by drying into the fabric such substances as starch and gums, while permanent stiffness can be produced either by insolubilising resin-forming components (usually by heating) within the fabric or by treating the fabric with solutions of fibre-swelling substances and afterwards further treating to remove these substances and to allow full collapse of the fibres. This leaves them in a hardened state which can usually be accentuated by high temperature drying.

All these treatments will be devised to suit the type of fibre present. Thus cotton fabric can be stiffened by a treatment in which it is treated with concentrated sulphuric acid followed by thorough removal of the acid, or it can be impregnated with an aminoplast resin precondensate and starch or a similar stiffening substance, and then heated to insolubilise both so that the stiff finish is washfast. The former acid method is not used much today.

Synthetic fibre materials can be treated with suitable organic solvents which swell the fibres to a high degree, wash out or evaporate the solvent and then dry at a high temperature. Thus phenol can be used for nylon and acetone for acetate fibres.

The methods mostly used are based on the formation of an insoluble resin within the textile material and it is necessary that the resin is substantially colourless.

At one time permanent stiffening, particularly of cotton fabric, was effected by padding it with a viscose (an alkaline solution of cellulose xanthate as used for the manufacture of viscose rayon) and then regenerating cellulose by acidification or heating to become strongly fixed within the fibres and fabric. Unfortunately by using the crude viscose the cotton fabric became stained yellow and acquired a sulphur odour, and for this reason has been abandoned. But recently the process has been revived and the defects avoided by using a viscose especially freed from alkali and sulphur impurities so as to be an aqueous solution of *pure* cellulose xanthate.

Laminated fabrics Laminated fabrics are produced by superimposing two or more layers of fabric and with the aid of applied adhesives causing the layers to become strongly bonded to each other. Such lamination can be to effect permanent stiffening, as for use in shirt collars and cuffs, as stiffening fabric for use in dresses and other garments and as backing for carpets and upholstery fabrics. It is not necessary for the fabric to confer stiffness although this is often desired.

A very interesting process for giving permanent washfast stiffness to men's shirt collars is known as the 'Trubenised' finish.

The process is based on the following observations. If a fabric made up of alternate threads of cotton and acetate rayon in warp or weft (or both) is moistened with a mixture of alcohol and acetone and then suitably hot pressed, it becomes stiff. What has happened is that the organic solvents applied have the effect, under the influence of heat, of gelatinising the acetate fibre threads. Then the high pressure squeezes them uniformly in and between the cotton threads as a jelly. On cooling, a fabric is obtained in which the cotton threads are securely cemented together by a firm film of cellulose acetate. If the fabric was made with acetate entirely in the warp or weft then the treatment described would produce a fabric having only a little strength in one direction. By having cotton threads in both warp and weft a strong fabric is ensured as cotton remains entirely unaffected by the process.

Obviously, by such a process it is possible to produce permanently stiff fabrics very easily. The nature of the final stiffened fabric can be determined by the proportion of acetate threads present. It is not necessary that they be present alternately with the cotton threads but there can be two or three or more cotton threads to every one acetate thread.

The 'Trubenised' process can also be extended to produce two-layer or multi-layer fabrics which are nowadays generally termed *laminated* fabrics. It is merely sufficient to superimpose the desired number of fabrics containing acetate threads and hot press them as just described. The gelatinised cellulose acetate acts as a bonding layer between the superimposed fabrics.

Here it may be recalled that some of the synthetic fibres are similar to acetate fibres in that they can be gelatinised or melted by exposing them to a sufficiently high temperature. So developments are also taking place in this direction.

Laminated fabrics are also being made by using synthetic resins as the bonding material. Some very tough board-like materials are now being made in this manner. They are not used for clothing purposes but in engineering and constructional work. In addition to being tough they are very durable and are practically resistant to all influences except boiling acids.

An important type of laminated fabric is made by bonding a

layer of foamed synthetic polymer to say cotton or other fabric to give a pliable thicker fabric which combines a desirable lightness of weight with a warm handle, excellent resistance to creasing and good dimensional stability so that garments made from it have a marked tendency to retain their shape during wear. Such finished foam-fabric laminated material is widely used in outer wearing apparel and it usually consists of one layer of foam bonded to one layer of fabric. But, also used for industrial purposes and for motor car seat coverings and for house slipper soles, etc. are fabrics in which a foam layer is sandwiched between two fabrics.

A favoured method for producing these laminated fabrics consists of leading the foam layer over a horizontal row of gas flames carefully adjusted to a uniform height so that the surface of the foam layer softens and melts, and then as this foam layer passes forward it is brought into contact with a fabric and pressed against this by passage between a pair of rollers cooled by water running through them. The emerging laminated fabric is immediately cooled by blowing cold air against it or by passing the fabric over a perforated cylinder through which cold air is sucked.

A polyurethane type of foam is much used and according to its method of manufacture its melting temperature may be about 195°C or 290°C. The heating conditions are such that if the foam surface is melted to a depth of 0·03 in (0·75 mm) up to 0·02 in (0·50 mm), satisfactory bonding can be obtained at a rate of about 20 yd/min (18 m/min), but of course these conditions of lamination can be considerably varied.

The synthetic resins commonly applied are either a polyurethane dissolved in an organic solvent or an aqueous dispersion of a cross-linkable acrylic resin. Such liquids can be applied to fabric by a spraying method or by pick-up from a roller carrying a thin layer of the liquor evenly distributed by the use of a doctor blade. After being brought together with the other fabric(s) the thus laminated fabric is compacted by passage between a pair of rollers and then heated to volatilise organic solvents (recovery arrangements are provided so that this can be again used) or to further polymerise and insolubilise the acrylic resin.

Many different machines have become available for carrying out a variety of such lamination procedures. When the laminated fabric is required for textile purposes then it usually is required to have good draping and washfast properties while additional desirable characteristics are resistance to soiling and being amenable to dry cleaning.

In an alternative bonding process a molten resin is applied by a printing method on to a backing fabric which is then led into contact with another fabric so that, under pressure while passing between rollers, bonding immediately takes place. Alternatively, a powdered resin may be applied and bonding induced at a suitably high temperature.

WATER AND OIL REPELLENCY

Water repellency

The production of waterproof or water-repellent fabrics is an important section of the textile trade both for garment manufacture and for farm, domestic and military equipment. For the rougher, heavier types of waterproof fabrics such as tarpaulins and waggon covers it is usual to employ a coating composition which comprises tar products since these are cheap and can be applied thickly. It matters little how much the appearance or handle of the fabric is affected by such compositions. But with fabrics to be used as garments, raincoats, mackintoshes and the like it is necessary to leave the fabric with a reasonably soft handle.

For many waterproof garments, it is most satisfactory to coat one side with a layer of rubber, but nowadays rubber is being replaced by synthetic resins and high polymers such as vinyl chloride resins. A characteristic of such proofed fabrics is that whilst being resistant to water they are also entirely impervious to air. They may also be adversely affected by grease or oil with which they may come in contact.

For raincoats it is necessary that the fabric should be permeable to air and yet be able to withstand fairly heavy showers of rain. This air permeability is important since it allows moist air from the body to pass through the fabric. Otherwise perspiration deposits on the inside of the garment as is found on wearing a rubber-proofed mackintosh.

In earlier days the greater proportion of waterproofed raincoat fabric was made by treating it first with a solution of an aluminium salt, for example, alum or aluminium sulphate, and then with a soap solution. In this way an insoluble, somewhat greasy aluminium soap was deposited in the fabric in such a way that this was readily permeable to air but not to water. Now aqueous emulsions containing both the aluminium salt (usually aluminium acetate or formate) and the soap (or a protein such as casein) are used. It is simply necessary to impregnate the fabric with such an emulsion and then dry. Such a single-bath method of proofing is obviously much better in many ways than the older two-bath method.

A widely used and cheap waterproofing composition for application to cotton and rayon fabrics consists of an aqueous dispersion of hydrolysable aluminium (or preferably a zirconium) salt and paraffin wax. It is simply dried into the textile material. Mistolene MK7, Cerol Z and Persistol are examples of such compositions.

Fabrics proofed with aluminium salts and soap (or wax) are not completely fast to dry cleaning and washing such as must be applied to raincoats from time to time. Thus new methods have been devised to remedy this defect, but they are more costly to apply. In devising these new methods it has been recognised that nothing better than a soap or fatty acid or similar oily compound has so far been found for producing water repellency. So the new developments have taken the line of preparing complex substances which have the required water-solubility while containing a fatty water-repellent component. When these, within the fabric, are heated to a high temperature they break up and leave this fatty compound either chemically combined or at least very firmly fixed to the textile fibres. Velan (ICI Ltd) and Zelan (Du Pont) are both products of this type. Organic isocyanates having higher alkyl groups, such as stearyl isocyanate, in their molecule can also be used in this way.

A disadvantage of the Velan and Zelan type of water repellent is that after fixation of this agent in the fabric by a baking treatment it is necessary to carry out washing and drying processes to free the fabric from by-products, more particularly pyridine which has an objectionable odour—(Velan PF is stearamido pyridinium chloride and it is the strongly hydrophobic stearyl residue left combined with the cotton fabric which confers water repellency). Hence the incentive to devise alternative agents which allows the thorough washing process to be avoided. Success has thus come by the discovery that very good water repellency can be obtained by applying paraffin wax together with a thermosetting resin. Phobotex FT (CIBA) is an agent of this type and is fixed in the fabric by a high temperature curing stage say for 5 min at 145°C with no necessity to wash afterwards.

It is useful to note that various water-repellents can be added to the padding liquors employed to produce crease-resist finishes on cellulose fibre fabrics and thus simultaneously secure additional water repellency.

The new processes take this form. The fabric is impregnated with a padding liquor containing one of these special substances. Then the fabric is dried at a fairly low temperature. After this the fabric is passed through a chamber heated to 120°C, or even up to 150°C, the time of passage being shorter as the temperature is higher. In this treatment the substance decomposes and leaves the fatty residue firmly anchored to, or combined with, the fibres. Thereafter the fabric is rinsed to remove by-products and then dried. Cotton fabric thus gains a water repellency and crease resistance which is fast to dry cleaning and a reasonable amount of washing. At the same time the handle of the fabric may be improved, for it so happens that these new products are also softening agents.

Before such water-repellent fabrics are used in the manufacture of waterproof garments it is usual to test them. The test commonly used is to allow drops of water to fall from a standard height on to the fabric and to note when the water first begins to penetrate, and also the amount which passes through the fabric in a specified time. The Wool Industries' Research Association has recently devised a so-called *shower testing* apparatus for such testing. It uses a water shower more closely resembling natural rain (Figure 5.12). There is also another test in which the fabric is subjected to a head of water under pressure and then the pressure required to cause the first flow-through of water is noted.

The water-absorptive properties of viscose rayon are now being reduced by special modification of the manufacturing process (see polynosic rayon—p. 41).

Water-repellent fabrics are now being produced by quite another method and fabrics of this type are being sold under the trade name of Ventile. The water repellency of these fabrics is obtained by the close weaving of selected cotton yarns in which the individual fibres lie almost parallel to each other so that on wetting the fibres swell so that the threads become sufficiently thick to fill up the fabric interstices and resist water passing through them. Such water-repellent fabrics are known as Ventile fabrics.

Silicones, especially a mixture of polydimethyl siloxane and polymethyl-hydrogen siloxane, are now much used for making textile fabrics showerproof. Such a finish is obtained by drying into the fabric (of cotton, acetate, nylon, etc.) an aqueous emulsion or organic solvent containing 1–2 per cent of the polysiloxane mixture and then heating (curing) the fabric at about 150–160°C for 3–5 min. The polysiloxanes thus become very firmly fixed to the fibres (especially if a titanium or zirconium salt is also present as a catalyst), thus giving the fabric a showerproof finish which is fast to washing and dry cleaning.

Since a high temperature baking treatment is not suitable for processing wool it is preferred to work wool materials at about room temperature in a dilute acidified aqueous emulsion of the silicon water-repellent agent, also containing aluminium formate as a

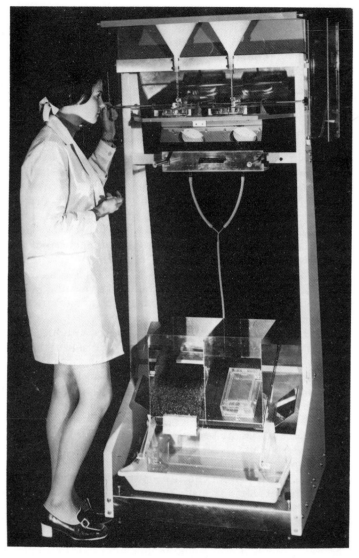

Figure 5.12 The Wira shower testing apparatus for measuring the water repellency of wearing apparel fabrics which has been found to give values comparable with the performance of the fabrics in actual use. It is based on allowing a uniform shower (total volume 500 ml) of well separated drops of water comparable in size to rain drops to fall on the fabric below, which is stretched flat over a ribbed glass plate and noting the time for the first 10 ml penetration, the amount of water absorption of the fabric and the total amount of water which passes through the fabric. (Courtesy Wira)

catalyst, and then dry at a moderately high temperature to complete fixation of the agent in the wool.

Oil and water repellency

Since about 1950 a number of organo-fluorochemicals have become available and some of these have been found useful for application to cotton fabrics to give them strong oil-repellent properties coupled with a fair degree of water repellency. They are relatively expensive and it has been found that it can be cheaper and yet satisfactory to apply them with an aminoplast such as a melamine methylol compound (resin-forming). This expedient also improves the simultaneously obtained water repellency and is employed in using the well-known oil- and water-repellent Scotchgard FC 208 made by the Minnesota Mining & Manufacturing Company in America which has pioneered very successfully in this field.

The organo-fluorochemicals are characterised in having complex molecules to which are attached along their length fluoro-carbon groups such as C_7F_{15}. The Scotchgard FC 208, first available about 1960, was believed to be a partially polymerised vinyl perfluoro acid ester obtained by reacting perfluoro-octanol with acrylic acid and whose long chain molecules had the formula

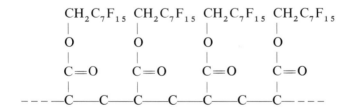

but in recent years research has indicated improved types which give oil and water repellencies better balanced.

A problem which has arisen from the use of these fluorochemicals is that the resulting fabric can have increased attraction for soil and so become more easily soiled in use and also more difficult to clean in laundering. This changed behaviour is considered to stem from the increased hydrophobicity of the finished fabric. Some success for the alleviation of this defect has been obtained by introducing into the hydrophobic fluorochemical a suitable number of hydrophile groups of the ethylene glycol type. The presence of these groups assists an aqueous detergent liquor as used in laundering to penetrate cotton or other fibres and so push out any oily soil which may be present; their presence can also similarly assist to prevent re-absorption of any soil that may be present in the dirty detergent liquor.

These hydrophile groups are inactive under dry or moist conditions when the fabric is, for example, exposed to a rain shower so that it then has its desired water-repellent properties.

Some fluorochemicals and other similarly effective agents are often added to the crosslinking padding liquors used in the crease-resist finishing of cellulose fibre fabrics.

Useful newer fluorochemicals which have become available and which confer both oil- and water-repellent properties include the copolymerisation product of ethyl perfluoro-octanoate with ethylene imine which have the formulae

$$C_7F_{15}COOC_2H_5$$
Ethyl perfluoro-octanoate

$$CH_2$$
$$|$$ NH
$$CH_2$$
Ethylene imine

and also the reaction product of THPC that is tetrakis-(hydroxy-methyl) phosphonium chloride, with 1-dihydro-perfluoro-octyl-amine. It is recalled that THPC is a flameproofing agent.

When the fluorochemical Scotchgard FC 208 together with the above-mentioned water repellent Velan, is dried into cotton fabric and followed by a curing (baking) at about 130°C for 5 min, the fabric acquires a combined water and oil repellency which with-stands repeated washing with soap. This type of finish has been found especially useful in the production of industrial protective clothing.

Warmth of handle

The next property of a textile material to be considered here is that of warmth of handle or heat-retaining power. This property is im-portant and wool and silk owe much of their popularity to the fact that they can be used for the manufacture of warm garments. Cer-tain fibres are recognised as yielding cold handling fabrics whilst others are expected to give warm fabrics. Silk and wool have al-ready been mentioned as being warm. Linen, cotton and the cellulose rayons are known to produce colder fabrics. So, most people have come to assume that it is the type of fibre which decides whether or not a fabric will be cold or warm. Whilst this conclusion is true to some extent it would seem from the most recent researches that it is far from being the whole story.

Quite a large amount of research has been carried out in the last twenty-five years to determine the features about a fabric which decide its heat-retaining value. All the results suggest that the different types of textile fibres have about the same heat conductivity. Expressed in other words, this implies that one fibre is just about as good or bad as another in transmitting heat. What is most important is the distribution of tiny air spaces in the fabric, for still air is one of the worst conductors of heat. Scientific opinion is that the warmth of a fabric is decided by the nature of the air spaces present and that the fibres must be considered effective in this only so far as they assist to hold the air spaces together. Now it is necessary to be quite clear about the function of the air spaces in a fabric. Perhaps it would be better to call them air pockets, for in order that they may be effective as non-conductors of heat they must not be open to the surrounding air. The air pockets must be like the spaces within a sponge to give a fabric warmth. If the spaces are such that wind or air can blow freely through them, then they detract from the fabric's ability to keep a body warm.

Thus in the ideal fabric the fibres are so spaced in the yarns which constitute the fabric that they occlude innumerable cells or pockets of stationary air which is not able to move out of the fabric. Also, the yarns of the fabric are required to be interlaced in such a manner that there are no large spaces through which air can blow from one side of the fabric to the other. Then when such a fabric is placed over a hot body, heat from this can travel outwards only, via millions of tiny stationary air pockets. Since air is a bad heat conductor the hot body loses its heat only very slowly.

If, on the other hand, the fabric is woven with hard twisted yarns made of fine fibres and these yarns are widely spaced so that air can blow freely through the fabric, then, when this is placed over the hot body, quite different conditions exist. In this case outside air can freely pass through the meshes in the fabric, strike the surface of the hot body and take away some of its heat, and then move outwards to be replaced by fresh cool incoming air to act in the same way. Also, since there are fewer stationary air pockets, heat can pass from the hot body by conduction through the closely packed fibres which conduct heat better than the air and so lead to rapid dissipation of the heat. A fabric of this type will be useless for keeping the body of a human being warm.

Now the reason why wool and silk materials are found in general to be warm is that the fibres are of such dimensions that they lend themselves particularly well to the production of fabrics full of the necessary tiny air pockets. Thus the extreme fineness of silk fibres and the crimped (waviness) of wool fibres can play an important part. To produce a warm fabric it is therefore necessary to study the distribution of the fibres in a fabric rather than the selection of a particular textile fibre.

Most types of fabric can be made warmer by raising their surface to give a pile which is non-conducting to heat because of its high air content.

Warmth of handle has now become an important subject for consideration by manufacturers of textile fabrics and garments, mainly on account of the high price of wool. Manufacturers have had to face the problem of producing wool-substitute materials which have the excellent warm-handling characteristics of all-wool goods. Even the substitution of but a small proportion of the wool by a cold fibre, such as cotton or viscose rayon in a wool fabric, can be easily detected.

Experience has shown that to produce warm fabrics composed partly or wholly of non-wool fibres, especially the rayons and synthetic fibres, it is necessary to use them in their cut-up or staple form. Yarn made from staple fibre is much bulkier and therefore warmer to the touch and a poorer conductor of heat than is continuous filament yarn. By proceeding along these lines several manufacturers now claim that they can produce fabrics and garments made entirely or almost entirely from rayon and synthetic fibres so that it is scarcely possible to distinguish them from all-wool materials. Wool has several qualities other than warmth of handle to commend it, so that it is unlikely that this fibre will ever be satisfactorily replaced, but it has to be admitted that there is an increasingly keen struggle between wool and the artificial fibres.

More recently, the use of so-called textured synthetic fibre yarns having 'Hi-bulk' characteristics has allowed warmer nylon, Terylene, etc. fabrics to be made (p. 135).

Pilling

Pilling is the name given to a defect of woven fabrics, but more usually knitted fabrics and garments, which is revealed during wear by the formation on the surface of the fabric of small balls or pills each consisting of a mass of entangled fibres which have apparently worked out of the threads which constitute the fabric. The presence of these pills gives the fabric a highly objectionable appearance and character such that during the past few years strenuous efforts have been made to make or finish fabrics so that pilling does not take place—unfortunately complete success has not attended these efforts and there is much textile material now being made which will be faulty because of its pilling tendency.

Pilling has become especially noticeable since the introduction of the synthetic fibres, and since these are highly resilient and have a very smooth surface it is generally concluded that it is just this character which favours their movement out of the threads. On the other hand many types of wool fabric are subject to pilling—in this case it is most noticeable with very loosely constructed fabrics and also with those which have been made shrink-resistant by a treatment which reduces the felting power (pp. 208 and 386) of the fibres (it also incidentally makes the wool fibres much smoother).

Careful investigation of materials subject to pilling has revealed that it results from fibre movement so that fibre ends protrude through the fabric surface and there become entangled to form pills (fibre tufts or balls) held to the fabric by relatively few fibres. As the material is further worn so do the anchoring fibres gradually wear and weaken until they break and the pill falls off. Meanwhile fresh pills are being formed. Thus at any given time the number of pills present is the balance of those forming and those falling off. This pill formation obviously leads to fibre loss from the fabric and a reduced resistance to wear.

Three expedients can be used to reduce pilling to the minimum. Firstly, the yarns present in the fabric should be tightly twisted and the fabric should be compactly woven—this reduces the opportunity for fibres to move relatively to each other and thus restricts fibre ends protruding through the fabric surface. Secondly, the number of protruding fibre ends in the fabric can be reduced in it as part of the finishing processing by singeing the fabric to burn off any present (p. 219) and also by shearing (p. 243) off the protruding fibre ends. Thirdly, the fabric can be given a treatment which causes the fibres to shrink, for in such treatment the protruding ends will shrink most and thus become well buried within the fabric surface.

In the case of fabrics made with a mixture of wool and synthetic fibres it is useful to subject them to a milling process (p. 252) under conditions which favour a movement of the wool fibres to the fabric surface sufficient to bury the synthetic fibre ends.

A novel method for preparing synthetic fibre fabrics resistant to pilling is to use fibres which have been specially treated so that each fibre is weak at spaced intervals along its length. This has the effect of promoting a more early breakage of the fibres anchoring the pills, so that at any time there are fewer pills present than when there are no especially weakened fibres.

The cross-sectional shape of synthetic fibres has been found to influence their tendency to form pills on fabric. A clover-leaf or other trefoil cross section hinders pilling, and such fibres are now being used to reduce pilling.

Some types of fibres which have very irregular instead of smooth round cross-sections yield fabrics less liable to pilling. Also the presence in the fabric of some bi-component (composite) fibres, which are self-crimping, can reduce pilling (p. 133).

Rot-proofing of industrial fabrics

Many types of canvas, ropes, cordage and netting materials used out-of-doors are exposed continuously to attack by bacteria, moulds, mildews and various micro-organisms. Copper compounds are well known to hinder the growth of such organisms and simple treatment of the textile material with, say, a solution of copper sulphate gives some protection. In recent years it has been found much more satisfactory from various viewpoints to treat with copper and zinc soaps, especially the naphthenates of these metals. Such products are marketed under various trade names. Copper-8-quinolinolate is an excellent anti-rot agent for protecting cotton during weathering or soil burial.

An interesting but not much commercially used method of giving cellulose fibre fabrics protection against attack by micro-organisms is that of partially acetylating or cyanoethylating them. For this the cotton or other fabric is treated with a mixture of acetic anhydride and acetic acid, using sulphuric acid as the catalyst, or the fabric is reacted with acrylonitrile in the presence of caustic soda. In either case only a relatively small proportion of acetic acid or acrylonitrile is induced to combine with the cotton fibres—a larger amount is unnecessary for the purpose in view and would certainly cause undesirable profound changes in the fibres. These treatments reduce the wet swelling of the fibres and thus hinder penetration by micro-organisms, and their activity.

A recent practical method originates with the discovery that a water-soluble precondensate of melamine and formaldehyde (similar to that used in crease-resist finishing processes) can be applied to cellulose fibre materials and there be fixed while further polymerising under wet (not dry) conditions. The fixed insoluble resin forms a barrier to micro-organisms and so the treated material becomes rot-proof. By addition of an anti-bacterial mercury compound the textile material is also made surface-resistant to mildew.

Nylon, Terylene and Ulstron (polypropylene) ropes and twines, etc. are now being widely used since they are naturally rot-proof. Ulstron ropes are light enough to float on water.

When desired it is possible to deposit bactericidal substances in fabrics and garments so that they counteract harmful bacteria and any micro-organisms which can, by their action on, say, perspiration, give rise to body odours, which can arise during wear or other uses. A difficulty which has been encountered but solved is to ensure that the bactericidal finish is fast to washing. It is very important that the bactericides used should be non-irritant to the skin.

Flameproofing

From time to time national interest is shown in requiring textile materials to be made flameproof, but still a large proportion are highly inflammable, probably because it is not easy and/or cheap to make them flameproof.

Many processes for making fabrics non-inflammable are available. They are not as yet widely used for fabrics employed in the home and for wearing apparel, for the simple reason that they make the fabric more expensive while taking away from it desirable characteristics such as softness, good draping and warmth of handle. With most processes it is necessary to deposit within the fabric a considerable amount, say up to 20 per cent or more, of a substance to hinder the fabric from catching fire when a light is applied to it. However recent legislation has compelled textile manufacturers to produce a proportion of flameproof goods such as children's nightdresses.

There is one other point about a fireproofing treatment. It should be fast to washing and dry cleaning if the textile material is customarily exposed to such treatments. Many fireproofing processes which use such substances as ammonium borate, ammonium phosphate, zinc borate, boric acid or metallic oxides such as antimony oxide, do not conform with this requirement.

Research, originally mainly in connection with the demand for fireproof army textile materials, has led to the discovery of fireproofing treatments which depend either on depositing within the fibres water-soluble substances or on combining selected compounds with the fibres; a fair degree of washfastness can thus be obtained. Thus the fabric may be treated with a mixture of antimony oxide, a chlorinated paraffin wax and a resin together with some zinc borate. The first ingredients hinder flaming while the zinc borate hinders afterglow. Alternatively, it may be reacted with an acid phosphate in the presence of urea whereby the phosphoric acid liberated combines chemically with the cotton or rayon fibres. But so far, these processes are used for special rather than for general purposes. Most of the synthetic fibres melt when burning so that drops of molten polymer fall from them. Synthetic fibres such as Verel, Movel, Rhovyl, Saran and Dynel whose molecules contain chlorine are substantially flameproof.

Considerable progress has recently been made in discovering new methods for making cotton and other cellulose fibre materials fireproof so that they not only resist ignition, but if ignited they quickly cease to burn when the igniting flame is taken away to leave a charred residue which does not glow. It is considered very important that a textile material should not only resist catching fire but that any charred remains should not continue to afterglow since this can be very dangerous too.

Following the ignition of a cotton or other cellulose fibre material, fibre decomposition normally occurs with liberation of combustible gases and vapours, at, say, above 300°C, and these then ignite with flaming above 350°C to liberate heat which produces yet more rapid fibre decomposition and flaming until ultimately there is only a solid carbon char left which can continue to glow. Easy accessibility to air promotes such combustion.

It is known that comparable decomposition of a cellulose fibre (but without flaming and even at room temperature) to yield a carbon residue can be promoted by dehydration of the cellulose by treating it with concentrated sulphuric acid thus:

$$(C_6H_{10}O_5)_n \xrightarrow{\;H_2SO_4\;} 6(C)_n + 5(H_2O)_n$$

Hence the stimulating idea that substances might be discovered that are more convenient to use than sulphuric acid, but capable (when incorporated in cotton fibres), of comparably dehydrating the fibres so that even if initiation of this change is required, a temperature up to say 250°C (i.e. below the temperature of 350°C mentioned above) could inhibit liberation of inflammable gases and thus inhibit flaming.

From this viewpoint a very efficient Ciba-Geigy proprietary Pyrovatex CP flameproofing agent has been made available. It is a cyanamide-phosphorus compound and when applied to cotton and other cellulose fibres under conditions which include a high temperature curing stage to fix it washfast it confers excellent fireproof properties. This product is especially useful for application to cellulose fibre nightwear, furnishing and industrial fabrics, so that they then satisfy the most stringent flameproof requirements.

Other recent flameproofing processes make use of new highly

complex phosphorus organic compounds such as tetrakis-hydroxy-phosphonium chloride (called THPC for short), which, when applied in the presence of a melamine-formaldehyde resin, and the fabric is heated for a few minutes at 140 to 160°C, polymerise within the fibres so as to become washfast and give practically complete resistance to fire and afterglow. Unfortunately the cost of such processing is fairly high and it can objectionally stiffen the fabric.

This process has recently been improved by a discovery which allows the melamine resin to be replaced by an after-treatment of the fabric (pretreated with THPC plus urea) with ammonia. Thus the soft handle of the textile material can be better preserved while the flameproof finish is made especially fast to washing. The well-known Proban (trade name) flameproofing process utilises this discovery and it gives a flameproof finish which is highly washfast.

All the newer processes for rendering cellulose fibre materials flameproof leave the fabric with fixed contents of phosphorus and nitrogen; a phosphorus content of 1 to 2 per cent is essential. Preferably the ratio of nitrogen to phosphorus should approach 3:1. Lightweight cotton fabrics require treatment with a higher amount of flame retardant agent and with some agents it is possible to reduce the amount of phosphorus compound applied as the amount of nitrogen compound is increased.

A very early used flame retardant process consisted of heating cotton fabric with a mixture of phosphoric acid and urea (or dicyandiamide) but it has now been found much better to use methyl phosphonic acid instead of the phosphoric acid. Incidentally this change confers crease resistance simultaneously with the flameproof effect.

A typical flameproof treatment consists of drying into cotton fabric a solution containing a zinc borofluoride catalyst and the flameproofing agent tris(1-azidinyl) phosphine and then heating (curing) the fabric at about 150°C for 3 min. Since some cellulose crosslinking occurs at the same time a moderate degree of crease resistance is also obtained. The crosslinking takes the following course:

$$
\begin{array}{cccccc}
 & & O & & & \\
CH_2 & & || & CH_2 & & \\
| & N\!-\!P\!-\!N & | & + & 2\ Cell\ OH \\
CH_2 & & | & CH_2 & & \\
 & & N & & cellulose\ molecules
\end{array}
$$

$$CH_2\!-\!CH_2$$
tris(1-azidinyl) phosphine

$\downarrow$curing stage

$$\text{Cell O CH}_2\text{CH}_2\text{—N—P—N—CH}_2\text{CH}_2\text{ O Cell}$$

with O double-bonded to P above, and N below P connected to

$$\text{CH}_2\text{—CH}_2$$

Crosslinked cellulose molecules

Whether or not crosslinking occurs it is found that always the application of a flameproofing agent to cotton or other cellulose fabric results in some loss of durability, softness and draping properties so that care is required in selecting the flame retardant and the conditions under which it is applied. From this viewpoint there is still a need for discovering new and improved flameproofing agents.

Progress has been made in the direction of providing a flame-proofing treatment which can be applied to garments in a dry cleaning machine. In one process cotton garments are treated with a liquor comprising perchloroethylene (this is a commonly used dry cleaning solvent) and about 2 per cent of the flameproofing agent tris(2:3-dibromopropyl) phosphate with $1\frac{1}{4}$ per cent of a polyvinyl acetate-acrylate resin as a binder. Following a drying-off of the solvent in the usual manner and a wash the garments not only have a washfast flameproof finish but they are left with an exceptionally high tear strength. The bromine content of the above agent contributes to the flameproof finish in addition to the usual effect of the phosphate.

Standard tests for measuring the flameproof properties of a fabric are available. One is based on igniting under specified conditions one end of a strip of fabric suspended vertically and noting the length of charred fabric which remains after flaming ceases; usually this length should not exceed about 6 in (150 mm) to pass the test. Another test is based on measuring the rate of burning.

A more recent test which is now favoured because of its simplicity but which gives useful results is to ignite with a match (hence the designation of the test as the 'match-test') the vertically suspended strip of fabric at its lower end and note where the flame becomes self-extinguished; the length of burnt fabric should not exceed 3 in (75 mm).

Quite a different test for measuring the flameproof properties of a fabric which has been flameproofed is to suspend the fabric within a glass cylinder and then pass a known mixture of oxygen and nitrogen through it. The oxygen: nitrogen ratio is varied until the minimum proportion of oxygen is present which will just sustain a steady burning of the fabric. The lower this proportion the less flameproof is the fabric.

The flameproofing of carpets has become important since the air they occlude in their pile surface can much assist their ignition and burning especially if made of cellulose fibres. A simple test used to determine their burning behaviour consists of applying a light to a pellet of guanidine placed in the carpet surface and noting the area of the burning so caused.

Flameproofing of wool

Until recently the flameproofing of wool has been considered as being less important than that for cotton, since wool has itself some useful resistance to burning. But it is now being given attention and it has been found that by application of titanium chloride from a boiling solution also containing citric acid to give a pH of about 3·5, the absorption by the wool of 0·4 per cent of titanium confers flameproofing without having any adverse effect on the durability and other properties of the wool.

In the above treatment practically all the titanium applied is absorbed into the wool with a liquor: wool ratio of 20:1. Oxalic and tartaric acids may be used instead of citric acid since they all form complex titanium compounds in the treating liquor as is necessary for their absorption into the wool.

Such flameproofing is very useful for wool used in the manufacture of carpets.

Instead of flameproofing wool by simple deposition of a flame-proofing agent in it such as THPC or a titanium or zirconium salt, it is now possible to confer a flameproof finish having the advantage of being washfast (and sometimes also fast to dry cleaning) by applying either simultaneously or in succession a flameproofing agent and a polymer. The function of the latter is to coat the wool fibres and make them non-felting and also bind the flameproofing agent to them. Only certain polymers are suitable and these include Synthappret LKF (a condensate of propylene oxide and a poly-functional isocyanate retaining free reactive isocyanate groups), Zesett T (a co-polymer of ethylene, vinyl acetate and methacrylyl chloride) and Primal HA–20 (a vinyl chloride co-polymer).

Soiling and soil removal from textiles

The much increased use of synthetic and other artificial fibres for carpet manufacture particularly those of the tufted type has shown that they usually pick up dirt more readily than the wool which has hitherto been almost exclusively used. Ordinary cotton and viscose rayon carpets are especially liable to become soiled in use. To a considerable degree this defect of wool substitute fibres is due to the presence in the fibres of oils from previous processing, but apart from this it is the nature of such fibres to become easily soiled. Recent research has therefore been directed towards discovering treatments for carpet fibres so that they less readily pick up dirt, and a fair degree of success has thus been obtained. Treatments which block the pores and superficial roughness of the fibres have been found to be most effective—with the spaces already occupied further occlusion of dirt particles is obviously much hindered. 'Elvan' (Courtaulds) is a special type of regenerated cellulose rayon now widely used in the manufacture of carpets. It is not only highly resilient but is resistant to soiling.

It has been observed that textile materials made of synthetic fibres can have a marked tendency to pick up soil from the surrounding dust-laden air. This has been correlated with the strong hydrophobicity of these fibres which causes them to accumulate static electricity so that the electrically charged dust particles are attracted to the fibres. This attraction can be at least partly neutralised by treating the synthetic fibre materials with anti-static agents, but these must be carefully chosen since some of them can cause any soil particles to be held more tenaciously — this is true for agents which are soft and/or greasy.

The soiling of textile materials must also be viewed from quite another direction. It has been found that in the laundering of soiled fabrics and garments in the usual type of detergent liquor it is possible for white or partly cleaned materials to pick up soil from the dirty liquor. This has been noticed with fabrics which have been finished so as to be crease resistant, or which contain impurities or finishing agents which are soft enough for the soil particles to become embedded therein and thus become more difficult to remove. To overcome such difficulties so-called 'soil-release' agents have become available. These can be applied to the textile materials at a convenient stage during their production, or immediately before washing, and they have the effect of loosening the soil in the fabric or garment and of hindering its reabsorption. Some soil-release agents such as Permalose (ICI Ltd) applied to polyester fibres are highly washfast. Other substances, such as sodium carboxymethyl cellulose, have soil-suspending properties and can be added to the detergent liquor so that there is less risk of soil becoming redeposited on the materials being washed.

It is interesting to recall that in the early days of weaving and knitting nylon fabrics, those left uncovered on the machines overnight or during the weekend were often found to have attracted dust (arising from an electrostatic charge left on the fabric) thus to be stained resistant to washing. Oil stains left on nylon and polyester fabrics were found after a heat treatment to be so firmly fixed as to make their removal very difficult. Dyers and finishers of synthetic fibre goods have by experience learned to treat the staining of these fibres with respect.

It has been observed that an oil stain in a cotton or other hydrophile fibre fabric can be pushed out by a detergent liquor soaking into the fibres. By contrast this does not happen with a hydrophile polyester fibre fabric since only a little of the liquor is absorbed. This observed fact has suggested how to facilitate oil stain removal. It is to make at least the surface of the fibres hydrophile and thus 'kill two birds with one stone'. By making the fibres hydrophile they are also made antistatic and thus less liable to attract soil while at the same time removal of the soil is made easier.

Obviously any such treatment must be washfast or at least not suffer appreciable change during say 30 to 40 successive washes in a domestic washing machine. An ingenious method of treatment has been devised to secure this washing fastness. It involves, as in the case of Permalose of ICI Ltd, a substance whose molecule contains one or more groups which are identical or similar to groups present in the molecules of the fibre to which the substance is to be applied. Under the conditions of application, which will normally involve a high temperature brief curing stage, the so-termed soil-release agent can coalesce or co-crystallise with the fibre and thus become very strongly bonded. Since the agent selected for this use is intentionally hydrophile this character will be transmitted to the surface of the hydrophobic fibres and permit easier removal of any oil stain.

At the present time various soil-resist and soil-release agents are available and they are widely used.

Wash-and-wear finishing of wool fabrics and garments

In dealing with the felting properties of wool and treatments for reducing or eliminating them, mainly by chemical modification of the epithelial scales (which as a continuous layer cover each fibre), or by rendering them much smoother by coating with a deposited, tenaciously adhering polymer film, it has been pointed out that such treatments by themselves could not make really washfast any creases or pleats subsequently hot pressed into the fabric. Although the fabric could be non-felting and thus shrink-resistant in washing, such creases and pleats would be removed by washing at a rate proportional to the severity of the washing treatment and in fact could be removed by simple repeated wetting and drying.

It was further noted that if the modified configuration of the fabric (creases, etc.) were hot pressed into the fabric with a reducing agent present, then the resulting creases could be fast to wetting and mild washing but not to domestic machine or other form of drastic washing; and there is a call for this increased washfastness.

This increase of crease stability is caused by a breaking of the lateral so-called disulphide bonds (—S—S—) which hold the wool molecules together side-by-side, so that in the subsequent essential hot pressing (with moisture present) they can reform in positions which accommodate and stabilise the distorted fibre molecular structure present at the creases. However, not all the broken bonds are reformed in this way; some are formed stabilising the creases but on balance crease stabilisation is not sufficiently complete to make the creases fast to machine washing. Nevertheless, the bonds are sufficiently modified to make them normal in the fabric structure so that if they are in any way flattened (by ironing, stretching, etc.) they will endeavour to reform whenever the existing conditions do not restrain this. The manner in which bond breakage and reforming take place is as follows:

$$—S—S— \xrightarrow[\text{treatment}]{\text{Reduction}} —SH+HS— \xrightarrow[\text{with steaming}]{\text{Hot pressing}} —S—S—$$

Two adjacent long wool molecules crosslinked by a disulphide bond	Showing break of crosslink by formation of two thiol groups	Showing crosslinks reformed in new positions during hot pressing (steaming) to fix the introduced creases

An early found method for counteracting the unwanted influence of the non-stabilising reformed bonds just mentioned consisted of an aftertreatment of the hot-pressed and reduced wool fabric with substances reactive to the free thiol groups.

Quite good crease resistance to machine washing can thus be obtained by first making the wool fabric or garment non-felting by chlorination for example, and then drying into the creased and/or pleated wool material an aqueous solution of sodium bisulphite.

Next it is hot pressed in a Hoffmann press (a 30-s steam and 30-s bake cycle) and finally washed with a liquor containing an oxidant (sodium perborate) or a crosslinking agent (formaldehyde or a polymerisable agent) to react with all the free thiol groups which remain and which could reduce the washing fastness of the creases.

Fortunately a simpler method has now been discovered for simultaneously making wool materials non-felting while fixing in them fully washfast creases or pleats and/or a smooth textured surface. It avoids the need for a separate preliminary non-felting treatment such as chlorination. This new method comprises three stages.

(1) The wool fabric is optionally temporarily set by dry heat or steam pressing to a given shape with creases and pleats.

(2) The thus prepared fabric is impregnated with a solution of a special polymer such as Synthappret LKF or Zeset TP in a volatile (dry cleaning) organic solvent such as tetrachloroethylene which is then removed by suitably heating to leave the fabric configuration undisturbed (the organic solvent does not swell the wool fibres as would an aqueous solvent).

(3) The wool fabric is then subjected (for Synthappret LKF) to pressing with steaming for 5 min or it is stored for 7 days at room temperature to allow (in either case) the polymer to acquire a fully insolubilised state or (for Zeset TP) simple heating in an oven at 100°C for 10 min.

In the last stage the polymers spread evenly over the wool fibre surface to form a tenaciously adhering film which confers non-felting properties and a satisfactory machine washfastness of the shape, creases and pleats.

In one way the use of an organic solvent may appear to be a disadvantage, but on the other hand if the necessary special machine is available the process is very effective and simple to carry out and indeed it is claimed to be possible to apply it in an ordinary dry-cleaning establishment. An important advantage is that it eliminates the necessity for a separate preliminary non-felting treatment, or for a treatment with sodium bisulphite or some other reducing agent.

The above method can be modified advantageously to use aqueous liquors and so avoid the use of organic solvents. It is based on an early observation of McPhee of the International Wool Secretariat that mildly chlorinating wool fabric without making it fully non-felting does modify the fibre surface to enable an applied polymeris-able Hercosett 57 resin to spread over it in a special manner such that the fabric can then be made completely non-felting by hot pressing as in a Hoffmann press. A further important advantage now recognised is that creases and pleats introduced into the resin-impregnated fabric on hot pressing become fast to repeated machine washing. The resin is therefore not effective without the pre-chlorination and it is not necessary or desirable to pre-chlorinate the wool strongly to make it fully non-felting. In the hot pressing the resin becomes insoluble and probably chemically combined with the wool fibre.

CHAPTER SIX

The care of clothes and simple identification tests

It has been mentioned previously that the textile fibres now in common use are those which over many years, some for many centuries, have been proved most durable and resistant to the deteriorating influences to which they are normally subject. This high degree of durability is evident in that many cotton, silk, linen and wool fabrics have been preserved to us from remote periods and from civilisations which have passed. But, in spite of this, everyone is familiar with the fact that clothes do wear out and become impoverished.

Most of the textile materials which we handle or wear in the form of fabrics and garments wear well and have a fairly long useful life, but we cannot expect them to last for ever. Each type of material has a limited life if it is used without being specially protected from deterioration. This is mainly because each type of fibre, cotton, linen, wool, or whatever it may be, has some weakness, and in everyday wear it gradually succumbs to this. However, if in use we avoid exposing textile fibres longer than is necessary to those influences which are most harmful, then their life can be prolonged. Thus the care of clothes, as we here understand it, is not to avoid their use but rather to avoid their ill-use longer than is necessary. All fabrics and garments have to suffer ill-use at some time or other.

Influences harmful to textiles

Before considering the weaknesses of the individual types of textile fibre it is useful to refer to their general weaknesses and the conditions which are harmful to most textile materials. These weaknesses have already been touched upon in Chapter 1 from another angle. All textile materials are adversely affected by sunlight, dampness, various organisms and wear and tear, and in ordinary use they will from time to time be exposed to such conditions.

Sunlight Sunlight falling on a textile material sets up chemical changes which always reduce the strength and durability of the fibres. Sunlight in fact causes gradual decomposition of a fibre into substances which are simpler in their molecular structure and in the final stage acquire solubility in water or dilute alkalis. This

Figure 6.1 Cotton fibres which have been attacked by mildew

latter point is interesting, for sometimes the weakness of a curtain material is not noticed until it is washed in a soap and soda liquor when it becomes much weaker for the simple reason that some of it dissolves in the washing liquor. In the case of a perfectly pure textile fibre the decomposing action of sunlight is generally slow. If, however, small amounts of certain substances (including certain dyes and some delustrants such as titanium dioxide) are present, either as impurity or as a substance necessarily present to give the textile material its colour, handle or other desired characteristic, then the harmful action of the light is accelerated. Clothes should always be stored in a subdued light or in the dark.

Synthetic fibres discolour to about the same extent as natural fibres.

The ultra-violet content of sunlight is most destructive to textile materials and it has been noticed that the synthetic fibres such as nylon, polyester (Terylene and Dacron) and polypropylene are especially liable to become weaker and more brittle during exposure to light. This deterioration is greater as the temperature during exposure is higher. Fortunately, much of this ultra-violet light is filtered off by window glass so that synthetic fibre curtains in normal use are much more durable than would otherwise be expected. However, it is now possible to give considerable protection against light deterioration by incorporating in the synthetic fibres a small proportion of so-called u.v. absorbers. These are colourless hydroxybenzophenone compounds which react with the ultra-violet light before the fibres and thus protect them. These u.v. absorbers can, at the same time, hinder the fading of any dyes present in the fibres.

Manganese, copper and phosphorus compounds have been found useful as additives to molten nylon immediately before its spinning into fibres for giving them protection against the harmful effects of exposure to sunlight and to high temperatures.

Dampness   It is harmful to clothes to keep them damp for prolonged periods for two reasons. Firstly, water itself can gradually assist the decomposition of textile fibres by swelling the individual fibres and assist any harmful substance present to attack them. But it is for a second reason that moisture is most particularly harmful. Within the clothes themselves and in the air surrounding them are to be found all kinds of bacteria, mildew and other organisms capable of decomposing textile fibres. These remain inactive if the textile material is sufficiently dry. If, however, the materials become damp, then these organisms at once become active and by living on the fibres and forming harmful substances within them produce deterioration and possibly serious loss of strength. Thus the synthetic fibres, because of their high hydrophobicity (this much restricts their water absorption), resist strongly any attack by micro-organisms.

An interesting example of the harmful effect of dampness is to be found in the behaviour of cotton goods. In the export of these it has been found most important to ensure that the packed fabrics have a lower moisture content than 8 to 9 per cent. If they are damper than this then it is almost certain that they will develop mildew during transport. For similar reasons fabrics buried in the moist ground soon perish by reason of the accelerated attack upon them of various organisms assisted and made possible by the moisture present.

High temperature   Exposure to high temperatures has a damaging influence on all kinds of textile materials. This damage may result either from decomposition of the fibre substance if the temperature is sufficiently high, or from the activating effect which a high temperature can have on any harmful substances present in the fabric. For example, the presence of a very weak acid substance such as vinegar in cotton material will not be harmful at ordinary temperatures, but if the fabric is kept at the boiling point of water or even at body temperature, this acid will weaken the fabric considerably. Of course, in the case of certain textile materials such as acetate, Orlon and nylon, a temperature high enough to melt or soften these fibres must be avoided. Similarly nylon and other synthetic fibres rapidly oxidise in air and become discoloured at high temperatures.

Perspiration　Perspiration can be harmful to most textile materials because it can be both acid and alkaline. Fresh perspiration is usually acid but on becoming stale it ferments and becomes alkaline owing to the formation of ammonia. Since most fibres are adversely affected either by acid or by alkali it is evident that sooner or later perspiration left in materials made of vegetable or animal fibres will become impoverished. This action of perspiration is the greater since it is usually present in a garment where this most frequently becomes warm, say, in the armpits. The ordinary harmful action of perspiration is increased by the fact that perspiration is an excellent nutrient medium for all kinds of bacteria which can flourish there and also attack the textile fibres.

Wear and tear　It is easy to understand how everyday wear and tear can weaken a textile material. The abrasion, rubbing and flexing to which it is subject have the effect of breaking the fibres or of displacing them in a thread so that this becomes weakened. When one thread in a woven or knitted fabric gives way, then extra strain is thrown on the adjacent threads and so the breakdown proceeds at an ever-increasing rate until a hole appears. There is very little that the average person can do about this except to act according to the old maxim, 'a stitch in time saves nine'.

Deterioration of various textile materials

Turning now to the different types of fibres, there are a few useful facts concerning these which are worth remembering. Thus cellulose fibres (cotton, linen, viscose and cuprammonium rayons) are easily deteriorated by acids but are fairly resistant to alkalis. In contrast, the animal or protein fibres (silk and wool) are easily attacked by alkalis and are fairly resistant to acids. None of these fibres is harmed by organic solvents such as carbon tetrachloride, trichlorethylene and white spirit or petrol.

Acetate fibres are changed by exposure to alkalis, for example a warm solution of washing soda, so that they become regenerated cellulose. In this change it usually becomes weakened in addition to having its textile properties completely changed. They are harmed by acids in much the same way as cotton or viscose rayon. Acetate materials soften when ironed at too high a temperature, so this should be avoided. Nylon, Vinyon, Courtelle, Terylene and other synthetic fibres are susceptible in the same way to high temperatures.

Acetate, nylon and Vinyon fibres are also swollen or dissolved by various organic solvents especially when these are hot, so care must be taken in treating these fibres until a preliminary test has shown that no damage is likely to ensue. On the other hand, all these fibres are highly resistant to attack by bacteria and other organisms even in the damp state.

Viscose, cuprammonium and acetate fibres temporarily lose about one-third of their strength when wet, but on drying they regain their original strength. In the wet state these fibres are very easily cut. Thus care should be exercised in washing such rayon materials. Nylon, Vinyon, Terylene and Saran materials are scarcely affected in this way by wetting. Their wet and dry strengths are about equal.

All the synthetic fibres are highly resistant to attack by bacteria and fungi; they are also repellent to insects and moths.

Defects of wool Alone among all the textile fibres, wool is subject to two serious defects. It is readily attacked by moths and it is liable to felt during ordinary washing.

Many people have a mistaken idea as to how moths attack wool but the matter is quite simple. Moths alight on a wool material and lay eggs. Later these eggs hatch out into grubs or worms and it is these and not the moths which do the damage. The grubs are ravenously hungry and they eat the wool fibres as wool is an excellent food for them. These grubs cannot live on other fibres such as cotton and linen and would starve on them. Naturally, holes in the wool material are soon formed as the grubs eat the fibres, and thus are produced the so-called 'moth-eaten' wool fabrics with which we are all unfortunately acquainted.

There is only one certain preventative against moth attack. It is to have present in the wool material substances which either render the wool fibres absolutely uneatable by the grubs or which actually poison them when eaten. Several moth-proofing substances are known and available. The most satisfactory solution of the moth damage problem is to have these applied during the dyeing and finishing of wool materials and this is now widely done. It may be remarked here that the placing of mothballs, printed newspapers or cigar ends among wool materials, which is sometimes recommended, is of very doubtful value and cannot be relied on. Apparently the idea behind the use of these is that the moths will dislike their odour and keep away and not lay their eggs in the wool. But we must remember that human beings have different reactions to smells and some have no sense of smell whatever. It is possible that moths too differ in the same way, and so the use of an objectionable odour to keep away moths from clothes is likely to be uncertain.

That interesting property of wool by which the fibres tend to close up into a more compact form during washing and which is known as felting has been described previously (p. 208). Felting can be utilised with advantage in the finishing of wool materials but generally a wool fabric or garment which felts in the hands of the public is something of a nuisance. This is because with each wash it shrinks and thickens so as to lose shape and size. It is much to be preferred that one or other of the different methods used for reducing the felting power of wool shall be applied at a suitable stage in the finishing of wool materials.

It is a fact, however, that at the present time a considerable proportion of wool materials in everyday use felt in washing. How can this be reduced to a minimum? There is one simple means for achieving this. It is that of squeezing or rubbing the wool material as little as possible during washing. Felting is a closing-up of the fibres and obviously this is much assisted by squeezing the wet garment or fabric. With every squeeze the material is made more compact and on release of the pressure the fibres never spring back completely to their original open form. One additional precaution may be taken to avoid felting, especially in those cases where it is impossible to avoid a certain amount of rubbing or squeezing to remove obstinately retained dirt. Felting takes place slowly in the cold but most rapidly in a soap liquor just about as hot as the hand can bear. If rubbing is used the soap liquor should be just warm, not hot.

Much has been written about the washing of wool materials in which emphasis is laid on the choice of a good soap, the avoiding of hard water, slow drying or quick drying and so on. Most of this can only be termed nonsense as most readers have probably found out. The recommendation which is far and away the most important is that squeezing and rubbing in a hot soap liquor should be avoided since this promotes and assists the natural tendency of the fibres to close up during such treatment and remain closed up.

It may be noted generally that the deterioration of all types of fibres results in a rupture of the long chain molecules of which fibres are composed so that these are shortened, with a consequent loss of strength.

Simple fibre identification tests

Not so long ago it was comparatively easy to identify the various textile fibres in use since there were so few of them, and they had easily distinguishable differences. Times have now changed a great deal. Today there are at least three times as many fibres as previously, and often it requires much technical skill to differentiate between them, especially in mixtures of three or more fibres. Thus the ordinary person can distinguish between the different fibres only in fairly simple cases. Nevertheless, the following suggestions will be found useful.

If a microscope is available then this is of great assistance for under a fairly high magnification it is seen that the various textile fibres differ considerably in shape and thickness, so that they can be recognised easily. The microscopical appearances of the different textile fibres have been described previously in Chapter 2, and reference should be made to this.

But, apart from the use of a microscope, much can be learnt from a few simple tests. Thus wool fibres burn readily with a very pungent odour like that of burning horn or feathers; a hard ash is left. Silk burns similarly but the residual ash is friable. This burning test is not really for the purpose of distinguishing between wool and silk (this can be done by simple comparison of the fabrics or garments) but to distinguish these fibres from cotton, linen, viscose, cuprammonium and acetate which burn with a pungent odour quite different from that of burning wool.

Another test to distinguish between the animal (wool and silk) and the vegetable (cotton and linen) fibres is that of heating them in a solution of caustic soda. The animal fibres dissolve whilst cotton, linen, viscose, cuprammonium and acetate remain undissolved. Silk dissolves but not so easily as wool under these conditions.

Acetate fibres tightly twisted together burn vigorously with spluttering whilst black molten globules are thrown off from time to time. It is easy to distinguish acetate fibres from all the other fibres in this way.

The introduction of a number of different synthetic fibres has now made the simple identification of different textile fibres even more difficult. However, a scheme is available whereby these fibres can be identified. It is based on the shape of the fibre cross-sections and their solubility in different organic solvents.

Acetate fibre differs from viscose and cuprammonium rayons by being soluble in acetone.

Most of the rayons can be distinguished from the natural and synthetic fibres by observing that they can be much more easily broken when wetted. The wool fibres have about the same strength wet or dry.

The identification of synthetic fibres now that there are so many different kinds has become complex. As a class they can usually be detected on account of their readiness to soften and melt when heated, but if it is desired to identify particular fibres, and say differentiate between nylon, Terylene, Orlon, etc. then special chemical and physical tests must be employed. Sometimes a very useful indication of the kind of synthetic fibre can be obtained from its density—by throwing a few of the fibres on to a liquid of known specific gravity and noting whether or not the fibres float, a useful inference as to the type of fibre can be drawn. That this is possible is evident from the considerable differences in density of the various fibres. For example, the densities of polypropylene, nylon and cotton are 0·92, 1·14 and 1·52, respectively.

The Shirley Kit of coloured staining solutions can be very useful for indentifying the various fibres, especially when not in the form of fibre mixtures.

BIBLIOGRAPHY

The following books will be found useful by those readers who wish to study more fully various aspects of textile production and processing which, necessarily, have been only lightly dealt with in this present book.

BEEVERS, H. *The Practice of Bradford Open Drawing* (National Trade Press, London, 1954).

BIRD, C.L. *Theory and Practice of Wool Dyeing,* 4th Edition (The Society of Dyers and Colourists, Bradford, 1972).

CARROL-PORCGYNSKI, C. Z. *Manual of Man-Made Fibres* (Astex Publishing Co, Guildford, 1960).

CHAMBERLAIN, J. *Principles of Machine Knitting* (Textile Institute, Manchester, 1951).

CLARKE, W. *Introduction to Textile Printing,* 3rd Edition (Butterworths, London, 1971).

COCKETT, S. R. and HILTON, K. A. *The Dyeing of Cellulosic Fibres and Related Processes* (Leonard Hill (Books), London, 1961).

COULSON, A. F. W. *Manual of Cotton Spinning* (Textile Institute, Manchester, 1955).

FORD, J. E., PEARSON, G. and SMITH, R. M. *Identification of Textile Materials* (Textile Institute, Manchester, 1970).

GARNER, W. *Textile Laboratory Manual* (National Trade Press, London, 1951).

GRIFFIN, T. F. *Practical Worsted Combing* (National Trade Press, London, 1953).

HAIGH, D. *Dyeing and Finishing of Knitted Goods* (Hosiery Trade Journal, Leicester, 1971).

HALL, A. J. *A Handbook of Textile Finishing,* 3rd Edition (Butterworths, London, 1966).

HALL, A. J. *Modern Textile Auxiliaries* (Skinner, London, 1952).

HALL, A. J. *A Handbook of Textile Dyeing and Printing,* 3rd Edition (National Trade Press, London, 1955).

HALL, A. J. *A Student's Textbook of Textile Science,* 3rd Edition (Mills and Boon, London, 1963).

HILL, R. (Editor). *Fibres from Synthetic Polymers* (Elsevier Publishing Co, Amsterdam).

HOWITT, F. O. *Bibliography of the Technical Literature of Silk* (Hutchinson, London, 1947).

IRWIN, J. and BRETT, K. B. *Origins of Chintz* (H.M.S.O., London, 1970).

KNECHT, E. and FOTHERGILL, J. B. *Principles and Practice of Textile Printing* (Griffin, London, 1952).

KORNREICH, E. *Introduction to Fibres and Fabrics* (National Trade Press, London, 1952).

KRCEMA, R. *Manuel of Non-wovens* (Textile Trade Press Ltd, Manchester, 1972).

LENNOX-KERR, P. *Carpet Substrates* (Textile Trade Press Ltd, Manchester, 1972).

LENNOX-KERR, P. *Needle-felted Fabrics* (Textile Trade Press Ltd, Manchester, 1972).

LOMAX, I. *Textile Testing* (Longmans Green, London, 1949).

MARSH, J. T. *Introduction to Textile Finishing* (Chapman and Hall, London, 1950).

MARSH, J. T. and WOOD, F. C. *Introduction to the Chemistry of Cellulose* (Chapman and Hall, London, 1946).

MONCRIEFF, R. W. *Man-made Fibres*, 5th Edition (Butterworths, London, 1970).

MONCRIEFF, R. W. *Mothproofing* (Leonard Hill, London, 1950).

MONCRIEFF, R. W. *Wool Shrinkage and Its Prevention* (National Trade Press, London, 1953).

MORTON, W. E. and WRAY, G. R. *Introduction to the Study of Spinning*, 3rd Edition (Longmans Green, London, 1962).

PALING, D. F. *Warp Knitting Technology* (Harlequin Press, Manchester, 1952).

PETERS, R. H. *Textile Chemistry*, Vol. 1 (Elsevier Publishing Co., London, 1963).

PRESTON, J. M., *et al*. *Fibre Science* (Textile Institute, Manchester, 1932).

RADCLIFFE, J. W. **Woollen and Worsted Yarn Manufacture** (Emmott & Co Ltd, Manchester, 1960).

ROBINSON, A. T. C. and MARKS, R. *Woven Cloth Construction* (Textile Institute, Manchester, 1967).

ROBINSON, R. S. *A History of Dyed Textiles* (Studio Vista, London, 1969).

SCHOFIELD, J. and J. C. *The Finishing of Wool Goods* (Schofield, Huddersfield, 1935).

SCHWARTZ, A. M. and PERRY, J. W. *Surface Active Agents: Their Chemistry and Technology* (Interscience, New York and London, 1948).

SCHWARTZ, A. M., PERRY, J. W. and BERCH, J. *Surface Active Agents and Detergents* (Interscience, New York and London, 1958).

SHERMAN, J. V. and S. L. *The New Fibres* (Macmillan, London, 1946).

SPIBEY, H. *The British Wool Manual* (Columbine Press, Buxton, 1969).

STRONG, J. H. *Foundations of Fabric Structure* (National Trade Press, London, 1953).

TAUSSIG, W. *Screen Printing* (Clayton Aniline Co, Manchester, 1950).

Textile Terms and Definitions (Textile Institute, Manchester, 1970).

TROTMAN, E. R. *Dyeing and Chemical Technology of Textile Fibres* (Longmans Green, London, 3rd Edition, 1970).

VICKERSTAFF, T. *Physical Chemistry of Dyeing* (Oliver and Boyd, London, 1950).

WALKER, H. *Worsted Drawing and Spinning*, Vol. 1 (Textile Institute, Manchester, 1954).

WARD, D. T. *Tufting of Textiles* (Textile Business Press Ltd, Manchester, 1969).

WHITTAKER, C. M. and WILCOX, C. C. *Dyeing and Coal Tar Dyestuffs* (Baillière, Tindall and Cox, London, 1964).

WILCOX, C. C. and ASHWORTH, J. L. *Whittaker's Dyeing with Coal Tar Dyestuffs* (Baillière, Tindall and Cox, London, 1964).

WRIGHT, R. H. *Modern Textile Design and Production* (National Trade Press, London, 1952).

Review of Textile Progress (Textile Institute and Society of Dyers and Colourists, Manchester, Vol 1, 1950 to Vol 18, 1968).

The Identification of Textile Materials (Textile Institute, Manchester, 1951).

Numerous books covering various aspects of textile manufacture and processing are available as published by the Textile Institute (Manchester) and The Society of Dyers and Colourists (Bradford).

Monthly journals that record current developments in the textile industry include *Textile Month* (Manchester), *The Textile Manufacturer* (Manchester), *British Knitting Industry* (Manchester), *The Hosiery Trade Journal* (Leicester), *The Dyer* (London) and *Textile World* (Atlanta, USA).

Index

433